INTRODUCTION TO
QUANTUM-STATE
ESTIMATION

INTRODUCTION TO
QUANTUM-STATE ESTIMATION

Yong Siah Teo

Palacký University, Czech Republic

World Scientific

NEW JERSEY • LONDON • SINGAPORE • BEIJING • SHANGHAI • HONG KONG • TAIPEI • CHENNAI

Published by

World Scientific Publishing Co. Pte. Ltd.
5 Toh Tuck Link, Singapore 596224
USA office: 27 Warren Street, Suite 401-402, Hackensack, NJ 07601
UK office: 57 Shelton Street, Covent Garden, London WC2H 9HE

Library of Congress Cataloging-in-Publication Data
Teo, Yong Siah, author.
  Introduction to quantum-state estimation / Yong Siah Teo, Palacký University, Czech Republic.
    pages cm
  Includes bibliographical references and index.
  ISBN 978-9814678834 (hardcover : alk. paper) -- ISBN 978-9814678841 (softcover : alk. paper)
  1. Quantum theory. 2. Estimation theory. I. Title.
  QC174.12.T44 2015
  530.12--dc23
                        2015005913

British Library Cataloguing-in-Publication Data
A catalogue record for this book is available from the British Library.

Copyright © 2016 by World Scientific Publishing Co. Pte. Ltd.

In-house Editor: Ng Kah Fee

Typeset by Stallion Press
Email: enquiries@stallionpress.com

To my parents, teachers, colleagues and friends.

Preface

Quantum-state estimation is an important subject in quantum information theory that focuses on the verification of the quality of a quantum source through statistical inference from data collected using a quantum measurement scheme. The body of literature in this field is still growing, with a substantial amount addressing practical issues that still lack complete solutions. Some novel ideas that have emerged include recent developments in scalable partial estimation schemes for many-body quantum states and quantum channels of large Hilbert-space dimensions, as well as the assignment of error bars for quantum-state estimators.

In order to understand the subject matter well, it is necessary to start from the basics. The book begins with a background formalism for quantum-state estimation to lay the foundation. Next, a substantial amount of time will be spent on exploring some specific estimation schemes, directing the reader's attention to a few common techniques such as the maximum-likelihood and maximum-entropy estimation schemes, two rather intuitive techniques suitable for an introductory survey on the subject. After which, the tomographic performance of state estimation schemes as well as other key experimental aspects of quantum-state tomography will be investigated, where the ideas of continuous-variable measurements are introduced. It is also during this time that preparations are made for the final discussion on quasi-probability distribution functions, which are natural representations for measurement data of continuous quantum degrees of freedom.

This book is intended to serve as an instructive and self-contained medium for advanced undergraduate and postgraduate students to grasp the basics of quantum-state estimation, so that future independent studies can be subsequently carried out with relative ease. Any reader with a solid foundation in quantum mechanics, linear algebra and calculus would be able to follow the book comfortably. The material presented are partly a reflection of my perspectives and experience gained from carrying out research in the field, some of which were presented in 2013 at a summer school hosted in Palacký University in Olomouc, Czech Republic. In writing this book, I did not follow any other reference material. However, as the book covers aspects of contemporary research, I have supplied a short reading list at the end of every chapter as a source for additional information.

Hints and sample solutions to all problems are included for reference, although they by no means represent the only approaches, only exemplifying ones, since there is typically more than one method to solve a given problem.

A goal of this book is to demonstrate some of the mathematical calculations that are typically performed in the field of quantum-state estimation. Mastering some elementary aspects of complex analysis and special mathematical functions would certainly expedite the learning process for later chapters, although the essential ideas from these topics are reiterated wherever necessary to make the discussions coherent. It is my hope that, after going through the details, the reader will find these mathematical techniques, which are commonly employed in other areas in physics, more easily accessible.

I wish to express my sincere gratitude to my former Honours-Year and Ph.D. supervisor Prof. Berthold-Georg Englert at the Centre for Quantum Technologies, National University of Singapore, for his ever insightful teachings, patience and friendship. His passion to eloquently convey physical concepts and mathematical techniques has greatly influenced my understanding in physics. Much more progress ensued through collaborations with my colleagues and supervisors Prof. Jaroslav Řeháček and Prof. Zdeněk Hradil at the Department of Optics in Palacký University, with whom I have eventually become good friends, and learnt plenty of technical skills that deepened my knowledge. They represent many experts whom I have worked with and gained numerous insights from.

The World Scientific Publishing Company has provided its utmost professional assistance and services for the completion of this book. I would like to especially thank the desk editor Kah Fee Ng for his patience, understanding and proactive contributions throughout the process of editing this book, the cover artist Lionel Seow for the captivating cover design, the copy editor Prathima from the subsidiary editorial services company ACES for meticulously transferring the contents of the book to the master copy, as well as the in-house typesetter Yolande Koh and those from Stallion Press for typesetting the contents and removing typographical artifacts. They have graciously accommodated my periodically demanding editorial requests.

Most certainly, I thank all my teachers, colleagues and friends who helped me in any way during my education, and my family members who have been providing encouragement and support all this time. In particular, I would like to acknowledge Jibo Dai, Wei Hou Tan, Yink Loong Len, Prof. Englert, and Prof. Jan Peřina for proofreading the book and giving valuable comments and suggestions to improve the material.

Olomouc, February 2015   Y. S. Teo

Acknowledgments

The initial preparation of the material for the book started in Singapore in the fall of 2012 while I was a research assistant in Prof. Berthold-Georg Englert's group at the Centre for Quantum Technologies, National University of Singapore, which was shortly after my Ph.D. graduation from the Centre. Research during this period was supported by the Ministry of Education and the National Research Foundation.

The typesetting and editing of the book were mostly carried out during the period when I was a postdoctoral fellow at Palacký University in Olomouc, Czech Republic, under the project POST-UP (Podpora vytváření excelentních výzkumných týmů a intersektorální mobility na Univerzitě Palackého v Olomouci) that was supported by the European Social Fund and the state budget of the Czech Republic, Project No. CZ.1.07/2.3.00/30.0004 (POST-UP I).

Contents

Acronyms

APD	avalanche photodiode
BCH	Baker–Campbell–Hausdorff
BM	Bayesian-mean
BS	beam splitter
HWP	half-wave plate
LIN	linear-inversion
LS	least-squares
MP	Moore–Penrose
ML	maximum-likelihood
MLME	maximum-likelihood-maximum-entropy
PBS	polarizing beam splitter
PNRD	photon-number resolving detector
POM	probability-operator measurement
PPBS	partially polarizing beam splitter
QWP	quarter-wave plate
SIC POM	symmetric informationally complete POM
SPAD	single-photon avalanche diode

Symbols and Notations

Elementary Review

i	imaginary unit.
π	≈ 3.14159265, Archimedes's constant.
e	≈ 2.71828182, Euler's number.
$k!$	factorial of k.
z^*, $\mathrm{Re}\,\{z\}$, $\mathrm{Im}\,\{z\}$	conjugate, real, imaginary parts of z.
$\log(\cdot)$	logarithm.
$\sin(\cdot)$, $\cos(\cdot)$, $\tan(\cdot)$, $\ldots$	trigonometric functions.
$\sinh(\cdot)$, $\cosh(\cdot)$, $\tanh(\cdot)$, $\ldots$	hyperbolic functions.
$\sum_j$, $\prod_j$	summation and product, with index j.
$\boldsymbol{v}$, $\boldsymbol{v}^{\mathsf{T}}$, $\boldsymbol{v}^{\dagger}$ (bold)	column $\boldsymbol{v}$, transpose, adjoint.
$\boldsymbol{v} \cdot \boldsymbol{w} \equiv \boldsymbol{v}^{\dagger}\boldsymbol{w}$, $\boldsymbol{v} \times \boldsymbol{w}$	dot and cross product of $\boldsymbol{v}$ and $\boldsymbol{w}$.
A, A^{T}, $\mathsf{A}^{\dagger}$ (bold, *sans*-serif)	dyadic (matrix) A, transpose, adjoint.
$\mathbf{1}$	unit dyadic (matrix).
$\boldsymbol{0}$, $\mathbf{0}$	zero column, zero dyadic (matrix).
$\mathsf{A} \cdot \boldsymbol{v} \equiv \mathsf{A}\boldsymbol{v}$, $\boldsymbol{v} \cdot \mathsf{A} \equiv \boldsymbol{v}^{\dagger}\mathsf{A}$	matrix-column multiplcation.
A, A^{T}, $A^{\dagger}$ (upper case)	operator A, transpose, adjoint.
1, 0	identity and zero operator (number).
$\dfrac{\mathrm{d}}{\mathrm{d}x}$, $\dfrac{\partial}{\partial x}$	(partial) derivative with respect to x.
∇	three-dimensional gradient vector.

Elementary Review

$$\int dx \qquad \text{integral with respect to } x.$$

$$\int (dr) \qquad \text{volume integral with respect to } r.$$

$$f(x)\Big|_{x=(\to)x_0} \equiv \lim_{x\to x_0} f(x) \qquad \text{limit of (evaluate) } f(x) \text{ at } x = x_0.$$

Metric Units

K   kelvin, a unit of temperature.
ns   nanosecond, a unit of time.
V   volt, a unit of voltage.

$\mathbb{H}$ positive operator describing bias of a quadratic form 7

$H_n(x)$ degree-n Hermite polynomial in x 150

$\mathcal{I}$ identity superoperator 18

I_{diff} $\propto$ photocurrent difference in perfect homodyne detection 140

I'_{diff} $\propto$ photocurrent difference in imperfect homodyne detection 147

$I_{\text{gauss}}(\,\cdot\,)$ two-dimensional Gaussian integral with complex parameters 186

$\ker(\,\cdot\,)$ kernel 17

κ_1, κ_2 real constants in a dual-operator calculation 20

κ dimension of $\boldsymbol{x}$ 108

K number of eigenstates of ρ of nonzero eigenvalues 45

$|k\rangle\langle k|$ D-dimensional von Neumann measurement outcomes 89

$\mathcal{K}$ number of output ports for a multi-port device 127

$\mathcal{L}(\mathbb{D};\rho)$ likelihood for $\mathbb{D}$ and ρ with Π_j 7

λ real number taking values between zero and one 43

$|\lambda_j\rangle, \lambda_j$ eigenkets and eigenvalues of ρ 45

$\mathcal{L}'_{N_t}(\mathbb{D};\rho)$ likelihood for $\mathbb{D}$ and ρ, with Π'_j and N_t 57

$\mathcal{L}'(\mathbb{D};\rho)$ generalized likelihood for $\mathbb{D}$ and ρ with Π'_j 63

$\overline{\mathcal{L}}(\mathbb{D};\rho)$ marginalized likelihood for $\mathbb{D}$ and ρ with Π'_j 68

L number of random states generated 72

$|L\rangle$ $= (-|H\rangle\,i + |V\rangle)\dfrac{1}{\sqrt{2}}$ 114

$L_n^{(\nu)}(x)$ degree-n associated Laguerre polynomial in x of index ν 170

$L_n(x)$ degree-n Laguerre polynomial in x 173

Squeezed Coherent States

α_{B} $= \alpha \cosh r + \alpha^* e^{i\theta} \sinh r$, the Bogolyubov transformed amplitude.

$\delta X, \delta P$ uncertainties of X and P quadrature operators.

γ complex parameter.

$\gamma(z)$ $= \cosh r - e^{i\theta} \sinh r$.

$L_0, L_\pm$ operators that obey a particular set of commutation relations.

r magnitude of the squeeze parameter z.

$S(z)$ squeeze operator defined by the complex parameter z.

θ phase of the squeeze parameter z.

w $= e^{i\theta} \tanh r$.

z $= r e^{i\theta}$, the complex squeeze parameter.

$\zeta(z)$ $= \dfrac{1 - e^{i\theta} \tanh r}{1 + e^{i\theta} \tanh r}$.

Chapter 1

Preliminaries of Quantum-State Estimation

1.1 Overview of quantum-state estimation

Quantum-state preparation is the first important step for any protocol that makes use of a source of quantum systems. For instance, a quantum-state teleportation protocol that is carried out with optical equipment requires a source that produces two photons in a maximally-entangled quantum state; that is, each photon in every photon pair possesses maximum quantum correlation with each other. In order to verify the integrity of the quantum state that appropriately describes the source prepared, one carries out *quantum-state tomography* on the source. Measurements are performed on a collection of identical copies of quantum systems (electrons, photons, *etc.*) that are prepared by the source. These measurements are generically described by a set of positive operators $\{\Pi_j \geq 0\}$ that compose a *probability-operator measurement* (POM), where $\sum_j \Pi_j = 1$. Here, we remind the reader that $\Pi_j \geq 0$ means that $\langle \, |\Pi_j| \, \rangle \geq 0$ for any $| \, \rangle$.

A familiar special case would be the mutually exclusive measurement outcomes in quantum mechanics. The corresponding $\Pi_j = |j\rangle \langle j|$ are prototypically orthonormal pure states that form a complete set — $\Pi_j \Pi_k = \delta_{j,k} \Pi_j$ and $\sum_j \Pi_j = \sum_j |j\rangle \langle j| = 1$ —, so that repeated measurements on the same quantum system, if physically possible,* always result in the same outcome for any Π_j. Such measurements are routinely implemented in practice. For the electron-spin measurement in a Stern–Gerlach[†] experiment, either the "up spin" or the "down spin" outcome is

---

*A photon, for instance, disappears after a photodetection.
[†]Otto Stern (1888–1969) and Walther Gerlach (1889–1979).

measured at any given instance through a path deflection by a spatially inhomogeneous magnet. For the photon-polarization measurement, either the "horizontal" or "vertical" polarization is detected with a partially polarizing beam splitter (see Chapter 4) and a photodetector.

The POM description therefore generalizes these mutually exclusive measurements to measurements with outcomes that are not necessarily mutually exclusive. We shall postpone the discussion on realizing such generalized measurements to Chapter 4. For now, we shall take this general description for granted. A reason for this generalization will become clear very soon.

The measurement data obtained are used to infer the identity of the source. Such a procedure of state inference is known as *quantum-state estimation*. In quantum mechanics, a quantum state encodes all information the observer has about a source *after* the measurement has been performed. This means that *all* quantum-mechanical probabilities for any future measurement can be predicted with this state. For a D-dimensional Hilbert* space, a quantum state is mathematically expressed as a *statistical operator* ρ with the properties

$$\rho \geq 0,$$

$$\mathrm{tr}\{\rho\} = 1. \tag{1.1.1}$$

It is easy to see why a POM is needed for state inference. Since any statistical operator is necessarily Hermitian,[†] we have altogether D^2 real parameters to characterize. Together with the unit-trace constraint, the number of independent parameters that specifies ρ is $D^2 - 1$. A mutually exclusive measurement alone is therefore not enough to fully characterize the source, since there can exist a set of at most D mutually exclusive outcomes. The probability p_j for each outcome Π_j is related to ρ by the Born[‡] rule, namely $p_j = \mathrm{tr}\{\rho\Pi_j\}$. The definitions for ρ and Π_j therefore ensure that $p_j \geq 0$ and $\sum_j p_j = 1$.

An observer, after measuring (a finite number) N copies of quantum systems, can obtain a *state estimator* (*aut* estimator[§]) $\widehat{\rho}$ from the measurement data, and this would typically be different from that obtained by another observer, after measuring his or her own copies with a different

---

*David Hilbert (1862–1943).

[†]Charles Hermite (1822–1901).

[‡]Max Born (1882–1970).

[§]All estimators are labeled with the " ⌢ " symbol.

measurement. This is not surprising since the quantum state directly reflects the knowledge about the source an observer gains, and this knowledge includes any *a priori* information available about the physical situation. Nevertheless, the different answers obtained by the two observers are respectively correct, because as far as each observer is concerned, all future predictions can be made accurately with his or her own statistical operator.

To be more concrete, assume that the maximally-entangled two-photon source generates photon pairs in a way that, for every pair, one photon traverses to the laboratory of an observer named Amy, and the other photon traverses to that of another observer named Bernadette. Equipment is available in both laboratories, which are extremely far away from the source, to carry out photon-polarization measurements. This maximally-entangled two-photon source is prepared in the pure state $\rho = |\ \rangle\langle\ |$ described by the ket

$$|\ \rangle = \left(|H\rangle_A |V\rangle_B + |V\rangle_A |H\rangle_B\right)\frac{1}{\sqrt{2}}, \qquad (1.1.2)$$

where $|H\rangle$ refers to the *horizontal-polarization* ket and $|V\rangle$ refers to the *vertical-polarization* ket. The mutual exclusiveness, or orthonormality, of the two normalized kets — $\langle H|V\rangle = 0$ — reflects the physical fact that only one of the two polarizations of a photon is measurable at any instant. The ket $|H\rangle_A |V\rangle_B$, sometimes written as $|H\rangle_A \otimes |V\rangle_B$ or simply $|HV\rangle$, refers to the tensor product of two polarization kets representing the respective measurement outcomes by Amy and Bernadette for the photon pairs. In other words, the POM for the photon-polarization measurement available to both Amy and Bernadette is the set $\{|H\rangle\langle H|, |V\rangle\langle V|\}$ with $|H\rangle\langle H| + |V\rangle\langle V| = 1$, and the probability for Amy and Bernadette to measure respectively the H and V polarization outcomes, for instance, is given by

$$\mathrm{tr}\{\rho |H\rangle_A |V\rangle_B\,_A\langle H|\,_B\langle V|\} = \frac{1}{2} \qquad (1.1.3)$$

according to the Born rule. The same goes for the V, H polarization outcomes, so that Amy and Bernadette can never measure the same polarization outcome simultaneously.

Amy, who is kept out of the loop by Bernadette, the one who bought the source, that her photons are part of maximally-entangled photon pairs, would naturally treat her photons as though they are emitted from a single-photon source that produces photons of random polarizations. To Amy, she would infer, based on her measurement data, that the quantum state of each of her photons is best described by the statistical operator $\rho_{\text{Amy}} = \dfrac{1}{2}$, the maximally-mixed state. This *reduced* statistical operator is a result of the ignorance about Bernadette's measurement data, which is equivalently expressed as the partial trace

$$
\begin{aligned}
\rho_{\text{Amy}} = \text{tr}_{\text{B}}\{\rho\} = \text{tr}_{\text{B}} & \left\{ (|\text{H}\rangle_{\text{A}}\,|\text{V}\rangle_{\text{B}} + |\text{V}\rangle_{\text{A}}\,|\text{H}\rangle_{\text{B}}) \frac{1}{2} \left({}_{\text{A}}\langle\text{H}|\,{}_{\text{B}}\langle\text{V}| + {}_{\text{A}}\langle\text{V}|\,{}_{\text{B}}\langle\text{H}| \right) \right\} \\
= & \frac{1}{2} \left(|\text{H}\rangle_{\text{A}}\,{}_{\text{B}}\langle\text{V}|\text{V}\rangle_{\text{B}}\,{}_{\text{A}}\langle\text{H}| + |\text{H}\rangle_{\text{A}}\,{}_{\text{B}}\langle\text{H}|\text{V}\rangle_{\text{B}}\,{}_{\text{A}}\langle\text{V}| \right. \\
& \left. + |\text{V}\rangle_{\text{A}}\,{}_{\text{B}}\langle\text{V}|\text{H}\rangle_{\text{B}}\,{}_{\text{A}}\langle\text{H}| + |\text{V}\rangle_{\text{A}}\,{}_{\text{B}}\langle\text{H}|\text{H}\rangle_{\text{B}}\,{}_{\text{A}}\langle\text{V}| \right) \\
= & |\text{H}\rangle_{\text{A}}\,\frac{1}{2}\,{}_{\text{A}}\langle\text{H}| + |\text{V}\rangle_{\text{A}}\,\frac{1}{2}\,{}_{\text{A}}\langle\text{V}| = \frac{1}{2}.
\end{aligned}
\tag{1.1.4}
$$

Bernadette, on the other hand, has more information about the actual physical situation, namely that her photons as well as Amy's are part of a maximally-entangled source. Therefore, her inference of Amy's quantum state would be different. To Bernadette, the correct descriptions for Amy's measured polarizations are simply those that are opposite of what she would obtain. Whenever Bernadette records her measured polarization datum, say H, she would infer that the corresponding quantum state for Amy's photon is $|\text{V}\rangle\langle\text{V}|$, and Bernadette would be right, of course.

Both observers are correct in their descriptions of each photon sent to Amy even though they are very different. This can be checked through numerous repetitions of photon-polarization measurements, where Amy would continue to find that her data are consistent with the prediction made by the single-photon statistical operator ρ_{Amy}, and Bernadette would always predict the correct polarization Amy would measure after inspecting her own recorded datum.

In real situations where the number of copies, N, is always finite, the choice of estimation schemes becomes important and will affect the tomographic accuracy of the state estimation. Ideally, as the number of copies approaches infinity, the different estimation procedures will yield estimators that tend to the same asymptotic statistical operator ρ as long as the measurements that are used completely characterize the source. This asymptotic operator is also known as the *true state* of the source, which we shall denote by $\rho_{\text{true}} \equiv \rho$ from now on, a terminology that is borrowed from statistics.

Although we acknowledge the existence of an asymptotic statistical operator ρ_{true} in quantum-state estimation, its subjective nature is implicit in making statistical predictions and interpretation. So, even if we shall be using the phrase "true state of a source" in many instances, the term "true state" is nothing more than a particular label that refers to the asymptotic statistical operator ρ_{true}, which is simply a mathematical tool for making statistical predictions. Attempts to frivolously endow *any* statistical operator with more attributes than necessary, as an effort to treat this operator like a property in classical physics say, can ultimately lead to unwarranted paradoxes and physical inconsistencies, a consequence that would be too difficult to bear for an honest physicist.

1.2 Basic aspects of estimation theory

1.2.1 *Cost functions*

A standard formalism of estimation theory is based on the mathematics of function optimization. Typically, an objective function involving the *cost function* $\mathbb{C}\,(\rho_{\text{true}}, \widehat{\rho})$ of an estimator $\widehat{\rho}$ for the unknown true state ρ_{true} that describes a source is minimized based on the measurement data $\mathbb{D}$.

To carry out the minimization procedure, the cost function may first be averaged over all measurement data $\mathbb{D}$ to obtain an estimator that represents an averaged solution over the statistical fluctuation. In addition, since the true state ρ_{true} ($=\rho$ here for notional simplicity) is always unknown, it is particularly useful to have an objective function which, upon its optimization, yields an "optimal" estimator regardless of the identity of the true state.

One way to achieve this property is to perform an average over all possible true states that plausibly describe the source, where a suitable *prior probability distribution* $p(\rho)$ is chosen to weight all the possible statistical operators. To this end, we introduce the *average cost function*

$$\overline{\mathbb{C}}(\widehat{\rho}) = \int (\mathrm{d}\tau_D)\, p(\rho) \sum_{\mathbb{D}} \mathbb{C}\,(\rho, \widehat{\rho})\,. \qquad (1.2.1)$$

Let us clarify the elements in Eq. (1.2.1). The space of all statistical operators ρ is known as the *state space*, which is a $(D^2 - 1)$-dimensional operator space that can be represented by a (D^2-1)-dimensional real vector space of columns $\boldsymbol{\omega}$, such that the real components ω_j of one such column are coefficients of *traceless* Hermitian operators Ω_j that define ρ as

$$\rho = \frac{1}{D} + \sum_{j=1}^{D^2-1} \omega_j \Omega_j\,. \qquad (1.2.2)$$

These traceless operators are mutually *trace-orthonormal* so that they normalize to the Kronecker* delta — $\mathrm{tr}\{\Omega_j\Omega_k\} = \delta_{j,k}$ — and $\mathrm{tr}\{\rho\,\Omega_j\} = \omega_j$. Thus, $(\mathrm{d}\tau_D)$ is a pre-chosen volume element for the $(D^2 - 1)$-dimensional state space, which consists of $D^2 - 1$ independent integration variables.

The $(D^2 - 1)$-dimensional vector space for ρ is not to be confused with the Hilbert space for ρ, which is a D-dimensional complex vector space. Although the dimension of the state space is $D^2 - 1$, we must always remember that the Hermitian-operator space is spanned by a set of D^2 Hermitian basis operators. A familiar example would be the *Bloch*[†] representation of a *single-qubit* $(D = 2)$ quantum state

$$0 \le \rho = \frac{1 + \boldsymbol{s} \cdot \boldsymbol{\sigma}}{2}, \tag{1.2.3}$$

where $\boldsymbol{s}$ is the three-dimensional real Bloch vector such that $|\boldsymbol{s}| \le 1$, and $\boldsymbol{\sigma}$ is a column of *Pauli*[‡] operators σ_x, σ_y and σ_z that are commonly represented by the three matrices

$$\sigma_x \mathrel{\widehat{=}} \begin{pmatrix} 0 & 1 \\ 1 & 0 \end{pmatrix}, \ \sigma_y \mathrel{\widehat{=}} \begin{pmatrix} 0 & -\mathrm{i} \\ \mathrm{i} & 0 \end{pmatrix}, \ \sigma_z \mathrel{\widehat{=}} \begin{pmatrix} 1 & 0 \\ 0 & -1 \end{pmatrix}. \tag{1.2.4}$$

The notation $\sum_{\mathbb{D}}$ should be understood as the *conditional average* over all possible measurement data $\mathbb{D}$ for a given statistical operator ρ. For a function $f(\mathbb{D})$ of the data $\mathbb{D}$, its conditional average $\overline{f(\mathbb{D})}$ is defined as

$$\overline{f(\mathbb{D})} = \sum_{\mathbb{D}} f(\mathbb{D}) \equiv \int (\mathrm{d}\mathbb{D})\, \mathcal{L}(\mathbb{D}; \rho)\, f(\mathbb{D}), \tag{1.2.5}$$

such that

$$\int (\mathrm{d}\mathbb{D})\, \mathcal{L}(\mathbb{D}; \rho) = 1. \tag{1.2.6}$$

The *likelihood* $\mathcal{L}(\mathbb{D}; \rho)$ that appears as a weight in the conditional average is the probability of acquiring the measurement data $\mathbb{D}$ *given* that the source is described by the statistical operator ρ. Here, the summation notation is an abbreviation for the integral average in Eq. (1.2.5).

One can also define the corresponding conditional probability distribution $p(\mathbb{D}|\rho)$, which involves the measurement data, in terms of the likelihood

---

*Leopold Kronecker (1823–1891).
[†]Felix Bloch (1905–1983).
[‡]Wolfgang Ernst Pauli (1900–1958).

$\mathcal{L}(\mathbb{D}; \rho)$ inasmuch as

$$p(\mathbb{D}|\rho) = \frac{\mathcal{L}(\mathbb{D}; \rho)}{\int (d\tau_D) p(\rho') \mathcal{L}(\mathbb{D}; \rho')}. \tag{1.2.7}$$

This conditional distribution is constructed such that it is normalized to unity over all quantum states. Naturally, the normalization constant of $p(\mathbb{D}|\rho)$ would include the prior distribution $p(\rho)$, which reflects the prior knowledge one has about the source. Accordingly, we have the *prior* $(d\rho) \equiv (d\tau_D) p(\rho)$ that consolidates both the geometrical aspect of and the prior information about the state space. After inserting all the elements into Eq. (1.2.1), the average cost function now reads

$$\overline{\mathbb{C}}(\hat{\rho}) = \int (d\mathbb{D}) \int (d\rho) \, p(\mathbb{D}|\rho) \, \mathbb{C}(\rho, \hat{\rho}) = \int (d\mathbb{D}) \frac{\int (d\rho) \, \mathcal{L}(\mathbb{D}; \rho) \, \mathbb{C}(\rho, \hat{\rho})}{\int (d\rho') \, \mathcal{L}(\mathbb{D}; \rho')}.$$
$$\tag{1.2.8}$$

1.2.1.1 *Cost for Bayesian-mean estimation*

To proceed, we need to decide on the form $\mathbb{C}(\rho, \hat{\rho})$ is going to take, for the estimator $\hat{\rho}$ strongly depends on the cost function. The typical quadratic function

$$\mathbb{C}_1(\rho, \hat{\rho}) = \frac{1}{2\|\mathbb{G}\|_2} \text{tr}\left\{ \left(\frac{\rho + \mathbb{H}}{1 + \text{tr}\{\mathbb{H}\}} - \hat{\rho} \right) \mathbb{G} \left(\frac{\rho + \mathbb{H}}{1 + \text{tr}\{\mathbb{H}\}} - \hat{\rho} \right) \right\} \tag{1.2.9}$$

in ρ and $\hat{\rho}$ defined by the positive operators $\mathbb{G}$ and $\mathbb{H}$ quantifies a biased distance between ρ and $\hat{\rho}$. Here, $\|\mathcal{A}\|_2$ refers to the operator two-norm

$$\|\mathcal{A}\|_2 = \max_{|y\rangle} \sqrt{\frac{\langle y| \mathcal{A}^\dagger \mathcal{A} |y\rangle}{\langle y|y\rangle}} \tag{1.2.10}$$

of the complex operator $\mathcal{A}$. This is equivalently the largest eigenvalue of the operator $\sqrt{\mathcal{A}^\dagger \mathcal{A}}$, or the largest *singular value* of $\mathcal{A}$. To show this, we note that the positive operator

$$\mathcal{A}^\dagger \mathcal{A} = \sum_j |q_j\rangle \, |a_j|^2 \, \langle q_j| \tag{1.2.11}$$

has positive eigenvalues $|a_j|^2$ and a complete set of orthonormal eigenkets $\{|q_j\rangle\}$. Hence, for any unnormalized ket $|y\rangle$,

$$\sqrt{\frac{\langle y|\,\mathcal{A}^\dagger\mathcal{A}\,|y\rangle}{\langle y|y\rangle}} = \sqrt{\text{tr}\left\{\mathcal{A}^\dagger\mathcal{A}\,\frac{|y\rangle\,\langle y|}{\langle y|y\rangle}\right\}} \quad \left(|\,\rangle\langle\,| = \frac{|y\rangle\,\langle y|}{\langle y|y\rangle}\right)$$

$$= \sqrt{\sum_j |a_j|^2\,|\langle q_j|\,\rangle|^2}$$

$$\leq \left(\max_{\{a_j\}}\{|a_j|\}\right)\underbrace{\sqrt{\sum_j |\langle q_j|\,\rangle|^2}}_{=\,1}$$

$$= \max_{\{a_j\}}\{|a_j|\}. \tag{1.2.12}$$

This cost function is also bounded by one, and this follows from

$$\text{tr}\left\{\left(\frac{\rho+\mathbb{H}}{1+\text{tr}\{\mathbb{H}\}} - \hat{\rho}\right)\mathbb{G}\left(\frac{\rho+\mathbb{H}}{1+\text{tr}\{\mathbb{H}\}} - \hat{\rho}\right)\right\}$$

$$= \text{tr}\left\{\left(\frac{\rho+\mathbb{H}}{1+\text{tr}\{\mathbb{H}\}} - \hat{\rho}\right)^2\right\}\text{tr}\left\{\frac{\left(\frac{\rho+\mathbb{H}}{1+\text{tr}\{\mathbb{H}\}} - \hat{\rho}\right)^2}{\text{tr}\left\{\left(\frac{\rho+\mathbb{H}}{1+\text{tr}\{\mathbb{H}\}} - \hat{\rho}\right)^2\right\}}\mathbb{G}\right\}$$

$$\leq \text{tr}\left\{\left(\frac{\rho+\mathbb{H}}{1+\text{tr}\{\mathbb{H}\}} - \hat{\rho}\right)^2\right\}\|\mathbb{G}\|_2$$

$$\leq \left[\text{tr}\left\{\left(\frac{\rho+\mathbb{H}}{1+\text{tr}\{\mathbb{H}\}}\right)^2\right\} + \text{tr}\{\hat{\rho}^2\}\right]\|\mathbb{G}\|_2 \leq 2\,\|\mathbb{G}\|_2. \tag{1.2.13}$$

In establishing the first inequality, we note that

$$\text{tr}\{\rho\,\mathbb{G}\} = \text{tr}\left\{\rho\sum_j |g_j\rangle\,\underbrace{g_j}_{\geq\,0}\,\langle g_j|\right\}$$

$$\leq \left(\max_{\{g_j\}}\{g_j\}\right)\underbrace{\sum_j \langle g_j|\,\rho\,|g_j\rangle}_{=\,1} = \|\mathbb{G}\|_2. \tag{1.2.14}$$

Such a function gives a nonzero "cost" whenever ρ is not equal to $\widehat{\rho}$ and the special case in which $\mathbb{G} = 1$ and $\mathbb{H} = 0$ yields the familiar *Hilbert–Schmidt** *distance.*

An extreme case for this cost function is given by

$$\mathbb{C}_2(\rho, \widehat{\rho}) = -\delta(\rho - \widehat{\rho}), \tag{1.2.15}$$

or the negative of the Dirac[†] delta function of $\rho - \widehat{\rho}$ that metes out a singularly large negative penalty when ρ equals $\widehat{\rho}$, and no penalty otherwise.

With $\mathbb{C}(\rho, \widehat{\rho}) = \mathbb{C}_1(\rho, \widehat{\rho})$, the variation $\delta\, \mathbb{C}_1(\rho, \widehat{\rho})$ can be calculated to be

$$\delta\, \mathbb{C}_1(\rho, \widehat{\rho}) = -\frac{1}{2\,\|\mathbb{G}\|_2}\text{tr}\left\{\delta\widehat{\rho}\left[\mathbb{G}\left(\frac{\rho+\mathbb{H}}{1+\text{tr}\{\mathbb{H}\}} - \widehat{\rho}\right) + \left(\frac{\rho+\mathbb{H}}{1+\text{tr}\{\mathbb{H}\}} - \widehat{\rho}\right)\mathbb{G}\right]\right\}. \tag{1.2.16}$$

The total variation of Eq. (1.2.8) for this cost function then works out to be

$$\delta\, \overline{\mathbb{C}}_1(\widehat{\rho})$$

$$= -\frac{1}{2\,\|\mathbb{G}\|_2}\int (\text{d}\mathbb{D})$$

$$\times\, \text{tr}\left\{\delta\widehat{\rho}\,\frac{\displaystyle\int (\text{d}\rho)\,\mathcal{L}(\mathbb{D};\rho)\left[\mathbb{G}\left(\frac{\rho+\mathbb{H}}{1+\text{tr}\{\mathbb{H}\}} - \widehat{\rho}\right) + \left(\frac{\rho+\mathbb{H}}{1+\text{tr}\{\mathbb{H}\}} - \widehat{\rho}\right)\mathbb{G}\right]}{\displaystyle\int (\text{d}\rho')\,\mathcal{L}(\mathbb{D};\rho')}\right\}$$

$$= -\frac{1}{2\,\|\mathbb{G}\|_2}\text{tr}\left\{\int (\text{d}\mathbb{D})\,\delta\widehat{\rho}\,\{\mathbb{G}\,[\widehat{\rho}_{\text{B}}\,(\mathbb{H},\mathbb{D}) - \widehat{\rho}] + [\widehat{\rho}_{\text{B}}\,(\mathbb{H},\mathbb{D}) - \widehat{\rho}]\,\mathbb{G}\}\right\}, \tag{1.2.17}$$

where

$$\widehat{\rho}_{\text{B}}\,(\mathbb{H},\mathbb{D}) = \frac{\displaystyle\int (\text{d}\rho)\,\mathcal{L}(\mathbb{D};\rho)\frac{\rho+\mathbb{H}}{1+\text{tr}\{\mathbb{H}\}}}{\displaystyle\int (\text{d}\rho')\,\mathcal{L}(\mathbb{D};\rho')}. \tag{1.2.18}$$

---

*David Hilbert (1862–1943) and Erhard Schmidt (1876–1959).
[†]Paul Adrien Maurice Dirac (1902–1984).

Since minimizing $\overline{\mathbb{C}}_1(\widehat{\rho})$ requires that $\delta\overline{\mathbb{C}}_1(\widehat{\rho}) = 0$, we thus have $\widehat{\rho} = \widehat{\rho}_{\text{B}}(\mathbb{H}, \mathbb{D})$. We shall drop the argument $\mathbb{D}$ for the sake of simplicity. The statistical operator $\widehat{\rho}_{\text{B}}(\mathbb{H})$ is known as the *Bayesian*[*]-*mean* (*BM*) *estimator* of ρ_{true} for the positive operator $\mathbb{H}$. A common variant of the BM estimator of interest is defined as $\widehat{\rho}_{\text{B}} = \widehat{\rho}_{\text{B}}(\mathbb{H} = 0)$.

In general, the integral average strongly depends on the definition of the prior $(\mathrm{d}\rho)$, which has no unique form whatsoever even when some constraints are reasonably imposed on the volume element $(\mathrm{d}\tau_D)$ for the true states. For example, when $D = 2$ and spherical coordinates (s, ϑ, φ) are used, a common choice of $(\mathrm{d}\rho)$ is given by

$$(\mathrm{d}\rho) = (\mathrm{d}\tau_2)\, p(s),$$
$$(\mathrm{d}\tau_2) = \mathrm{d}\varphi\, \mathrm{d}\vartheta\, \sin\vartheta\, \mathrm{d}s,$$

$$(1.2.19)$$

where the prior distribution $p(s)$ is now a function of s only. Here, the variable s describes the length of the Bloch vector s that characterizes ρ as in Eq. (1.2.3) (refer to Fig. 1.1).

The angular variables ϑ and φ, together, specify the orientation of s. The differential vectorial element $\mathrm{d}s$ is simply

$$\mathrm{d}s = e_s\,(\mathrm{d}s)_s + e_\vartheta\,(\mathrm{d}s)_\vartheta + e_\varphi\,(\mathrm{d}s)_\varphi$$
$$= e_s\,\mathrm{d}s + e_\vartheta s \sin\vartheta\, \mathrm{d}\vartheta + e_\varphi s\, \mathrm{d}\varphi,$$

$$(1.2.20)$$

where the orthogonal unit vectors are

$$e_s \stackrel{\frown}{=} \begin{pmatrix} \sin\vartheta\, \cos\varphi \\ \sin\vartheta\, \sin\varphi \\ \cos\vartheta \end{pmatrix}, \ e_\vartheta \stackrel{\frown}{=} \begin{pmatrix} \cos\vartheta\, \cos\varphi \\ \cos\vartheta\, \sin\varphi \\ -\sin\vartheta \end{pmatrix} \text{ and } e_\varphi \stackrel{\frown}{=} \begin{pmatrix} -\sin\varphi \\ \cos\varphi \\ 0 \end{pmatrix}.$$

$$(1.2.21)$$

The corresponding differential volume element (refer to Fig. 1.2) is defined by the volume of the differential cuboid formed by the unit vectors:

$$(\mathrm{d}s) = (\mathrm{d}s)_s\,(\mathrm{d}s)_\vartheta\,(\mathrm{d}s)_\varphi\, \underbrace{e_s \cdot e_\vartheta \times e_\varphi}_{= 1}$$
$$= \mathrm{d}\vartheta\, \sin\vartheta\, \mathrm{d}\varphi\, \mathrm{d}s\, s^2.$$

$$(1.2.22)$$

If ρ undergoes a unitary transformation, the orientation of the Bloch vector changes, while its magnitude remains a constant. The corresponding vector s is thus rotated to s', with the variable s unchanged as well as

---

[*]Thomas Bayes (1701–1761).

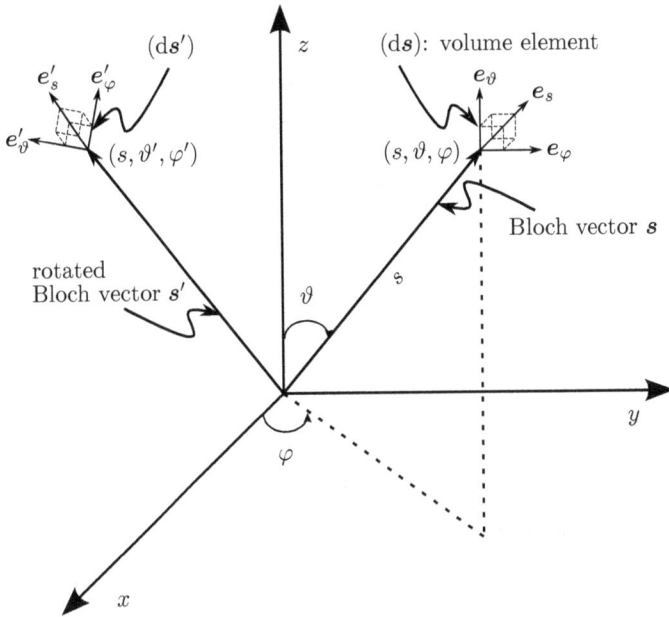

Fig. 1.1   Bloch-vector rotation illustrated with the spherical coordinate system.

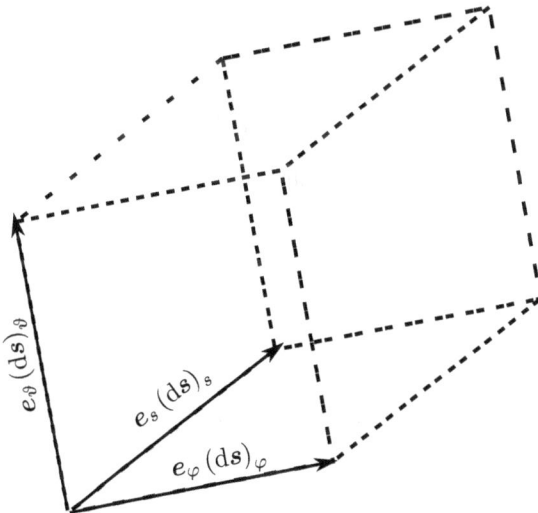

Fig. 1.2   Differential volume element for the spherical coordinate system.

the transformations $\vartheta \rightarrow \vartheta'$ and $\varphi \rightarrow \varphi'$ effected. Under any unitary transformation, the unit vectors $e_s \rightarrow e'_s$, $e_\vartheta \rightarrow e'_\vartheta$ and $e_\varphi \rightarrow e'_\varphi$ are always mutually orthogonal, which is a property of this coordinate system. Since a rotation of s does not change the differential volume element of the cuboid, we have $\mathrm{d}\vartheta \sin\vartheta \, \mathrm{d}\varphi = \mathrm{d}\vartheta' \sin\vartheta' \, \mathrm{d}\varphi'$. The prior $(\mathrm{d}\rho)$ defined in Eq. (1.2.19) is thus invariant under any unitary transformation.

Even for a fixed choice of $(\mathrm{d}\tau_2)$ with this coordinate system, the function $p(s)$ can still take arbitrary forms for different observers even if the prior information about the source known to each observer is the same. For instance, suppose that all of the observers agree that the source should produce quantum systems in a quantum state that is nearly pure, then any function that ultimately increases monotonically with the value of s would be a suitable candidate for the prior probability distribution $p(s)$. While this is in accordance with the Bayesian spirit, the subjective choice of the prior itself is still debatable. Furthermore, the operator integrals in Eq. (1.2.18) is usually difficult to compute.

1.2.1.2 *Cost for maximum-likelihood (ML) estimation*

One may also consider $\mathbb{C}(\rho,\widehat{\rho}) = \mathbb{C}_2(\rho,\widehat{\rho})$ in Eq. (1.2.15) that states the limiting form of the quadratic cost function in Eq. (1.2.9). The corresponding expression for $\overline{\mathbb{C}}_2(\widehat{\rho})$ then simplifies to

$$\overline{\mathbb{C}}_2(\widehat{\rho}) = -\int (\mathrm{d}\mathbb{D}) \, \frac{\mathcal{L}(\mathbb{D};\widehat{\rho})}{\displaystyle\int (\mathrm{d}\rho') \, \mathcal{L}(\mathbb{D};\rho')} . \tag{1.2.23}$$

Thus, minimizing Eq. (1.2.23) amounts to looking for the estimator $\widehat{\rho} = \widehat{\rho}_{\mathrm{ML}}$ that maximizes the likelihood $\mathcal{L}(\mathbb{D};\widehat{\rho})$. This estimator is the *maximum-likelihood (ML) estimator*. So, to minimize the objective function $\overline{\mathbb{C}}_2(\widehat{\rho})$, we need a scheme to search for a positive operator $\widehat{\rho}_{\mathrm{ML}}$, of unit trace, such that the likelihood takes the largest value within the admissible space of quantum states.

There is an asymptotic connection between $\widehat{\rho}_{\mathrm{ML}}$ and $\widehat{\rho}_{\mathrm{B}}$; When N is sufficiently large, the likelihood peaks very strongly at $\widehat{\rho} = \widehat{\rho}_{\mathrm{ML}}$ according to the law of large numbers, with

$$p\,(\mathbb{D}|\widehat{\rho}) = \frac{\mathcal{L}(\mathbb{D};\widehat{\rho})}{\displaystyle\int (\mathrm{d}\rho') \, \mathcal{L}(\mathbb{D};\rho')} \longrightarrow \delta(\widehat{\rho} - \widehat{\rho}_{\mathrm{ML}}), \tag{1.2.24}$$

and so $\widehat{\rho}_{\mathrm{B}}$ approaches $\widehat{\rho}_{\mathrm{ML}}$ in this limit.

Problem 1.1   Find the estimator $\hat{\rho}$ that minimizes the average cost function defined with

$$\mathbb{C}(\rho, \hat{\rho}) = \mathrm{tr}\{\hat{\rho}\,(\log\hat{\rho} - \log\rho)\}, \qquad (1.2.25)$$

subject to the constraint that $\mathrm{tr}\{\hat{\rho}\} = 1$. What can be said about this estimator?

In a typical quantum-state tomography experiment, one measures N copies of quantum systems with measurement outcomes that can be described by a POM consisting of M outcomes Π_j. For simplicity, we shall assume here that the detections are all perfect with no detection inefficiencies, that is $\sum_j \Pi_j = 1$. The measurement data $\mathbb{D}$ is a list of detection occurrences $\{n_j\}$ such that $\sum_j n_j = N$ (refer to Fig. 1.3). One may also define the corresponding set of measurement frequencies

$$\sum_{j=1}^{M} f_j = \sum_{j=1}^{M} \frac{n_j}{N} = 1. \qquad (1.2.26)$$

Let us consider the case in which the POM *completely* characterizes a given source. We say that this POM is *informationally complete*. This means that the POM contains $M \geq D^2$ outcomes that span the entire Hermitian-operator space of all possible statistical operators ρ for a D-dimensional Hilbert space.

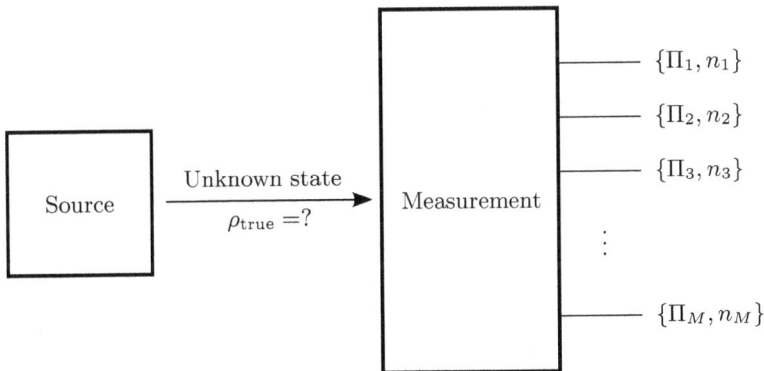

Fig. 1.3   Schematic diagram illustrating the principles of quantum-state tomography.

In this case, a *unique* estimator for ρ_{true} can be reconstructed since any D-dimensional ρ can be expressed in terms of a linear combination of these POM outcomes.

The simplistic ideal situation in the laboratory, which we shall take as the scenario of interest for more concrete discussions on state estimation, is when the source emits multiple copies of quantum systems such that each copy "is prepared in" ρ_{true}. By this, we mean that the observer describes each copy with the same $\widehat{\rho} \approx \rho_{\text{true}}$ after state estimation, and all statistical predictions can be made rather accurately with such a method of description.

As the detection of each copy is then independent of the rest, the number of occurrences n_j follow a multinomial distribution and so the corresponding likelihood $\mathcal{L}(\{n_j\}; \rho)$ for a *particular sequence* of detections observed in the experiment is

$$\mathcal{L}(\{n_j\}; \rho) = \prod_{j=1}^{M} p_j^{n_j},$$

$$p_j = \text{tr}\{\rho \Pi_j\} \geq 0. \qquad (1.2.27)$$

The concept of informational completeness of a POM is intimately related to the degree of linear independence of its outcomes, and this concept is best reviewed with an alternative formalism for quantum operators.

1.2.2   *Superoperators*

1.2.2.1   *Formalism*

If one temporarily ignores the positivity constraint that is always imposed on all quantum states, there exists a relatively simple construction for the ML estimator under certain situations. To do this, we introduce a transposition mapping on a given D-dimensional complex operator $\Psi = |a\rangle \gamma \langle b^*|$, defined by the complex numbers a, b and γ, to a ket of a higher dimension:

$$\Psi = |a\rangle \gamma \langle b^*| \longrightarrow |\Psi\rangle\rangle \equiv |a\rangle |b^*\rangle \gamma. \qquad (1.2.28)$$

The symbol "$|\Psi\rangle\rangle$" denotes the *superket* of the operator Ψ, which is a ket that resides in the larger D^2-dimensional Hilbert space and is derived from an operator in a D-dimensional Hilbert space. Just like the usual quantum-mechanical bras, we define the *superbra*

$$\langle\langle \Psi^\dagger| = |\Psi\rangle\rangle^\dagger = \gamma^* \langle a^*| \langle b|. \qquad (1.2.29)$$

The procedure of the adjoint operation on a ket, which consists of a transposition and a complex conjugation, has been explicitly illustrated with the notation $|a\rangle^\dagger = \langle a^*|$, and this notation shall henceforth be used to provide clarity in subsequent calculations.

Analogously to operators, one can define a D^2-dimensional *superoperator* $|\Phi\rangle\rangle\langle\langle\Psi^\dagger|$ from the superkets $|\Psi\rangle\rangle$ and $|\Phi\rangle\rangle$. The simple identity

$$\langle\langle\Psi^\dagger|\Phi\rangle\rangle = \text{tr}\{\Psi^\dagger\Phi\}, \qquad (1.2.30)$$

follows from these notations. To check Eq. (1.2.30) for the operators $\Psi = |a\rangle\,\gamma\,\langle b^*|$ and $\Phi = |a'\rangle\,\gamma'\,\langle b'^*|$ for instance, compare the left-hand side that proceeds as

$$\begin{aligned}
\langle\langle\Psi^\dagger|\Phi\rangle\rangle &= (\gamma^*\,\langle a^*|\,\langle b|)\,(|a'\rangle\,|b'^*\rangle\,\gamma') \\
&= \gamma^*\gamma'\,\langle a^*|a'\rangle\,\langle b|\,b'^*\rangle \\
&= \gamma^*\gamma'\,\langle a^*|a'\rangle\,\langle b'^*|b\rangle,
\end{aligned} \qquad (1.2.31)$$

with the right-hand side that swiftly gives

$$\begin{aligned}
\text{tr}\{\Psi^\dagger\Phi\} &= \text{tr}\{|b\rangle\,\gamma^*\,\langle a^*|a'\rangle\,\gamma'\,\langle b'^*|\} \\
&= \gamma^*\gamma'\,\langle a^*|a'\rangle\,\langle b'^*|b\rangle.
\end{aligned} \qquad (1.2.32)$$

Problem 1.2 There exists another representation for the superkets/ superbras. Given a $D \times D$ matrix

$$\Psi \,\hat{=}\, \begin{pmatrix} \boldsymbol{\psi}_1 & \boldsymbol{\psi}_2 & \cdots & \boldsymbol{\psi}_D \end{pmatrix}, \qquad (1.2.33)$$

that represents a D-dimensional complex operator, with $\boldsymbol{\psi}_j$ being $D \times 1$ complex matrix columns, the superket $|\Psi\rangle\rangle$ can be represented by the column

$$|\Psi\rangle\rangle \equiv \text{vec}(\Psi) \,\hat{=}\, \begin{pmatrix} \boldsymbol{\psi}_1 \\ \boldsymbol{\psi}_2 \\ \vdots \\ \boldsymbol{\psi}_D \end{pmatrix} \qquad (1.2.34)$$

that is formed by stacking the matrix columns $\boldsymbol{\psi}_j$. This procedure is also known as vectorization. Show that such a representation is also consistent with Eq. (1.2.30). How is this representation related to that given in Eq. (1.2.28)?

Problem 1.3 By using the definition in Eq. (1.2.28), show that for any operators A, B and C,

$$|ACB\rangle\rangle = A \otimes B^{\mathrm{T}} |C\rangle\rangle. \qquad (1.2.35)$$

Also, state a similar result for the representation in **Problem 1.2**.

Problem 1.4 Show, from Eq. (1.2.28), that for the two operators A and B that map ket spaces in the way $A : \mathcal{V}_1 \to \mathcal{V}$ and $B : \mathcal{V} \to \mathcal{V}_2$,

$$(\langle\langle 1_{\mathcal{V}} | \otimes 1_{\mathcal{V}_2} \otimes 1_{\mathcal{V}_1}) |B^{\mathrm{T}} \otimes A\rangle\rangle = |BA\rangle\rangle. \qquad (1.2.36)$$

1.2.2.2 *Dual operators and quantum-state estimation*

Under this formalism, we can invoke the theory of vector spaces to study the linear independence of the POM outcomes. The first step is to note that for a set consisting of D_{LI} outcomes that are *all* linearly independent, if the equation

$$\sum_{j=1}^{D_{\mathrm{LI}}} |\Pi_j\rangle\rangle \, c_j = 0 \qquad (1.2.37)$$

is to be satisfied for a given column of coefficients $c = (c_1, c_2, \ldots, c_{D_{\mathrm{LI}}})^{\mathrm{T}}$, then $c = 0$ since none of the outcomes can be expressed as a linear combination of the rest. In column and row notations, the above equation just amounts to

$$\underbrace{\left(|\Pi_1\rangle\rangle \quad |\Pi_2\rangle\rangle \quad \cdots \quad |\Pi_{D_{\mathrm{LI}}}\rangle\rangle \right)}_{\equiv \, V^\dagger} \begin{pmatrix} c_1 \\ c_2 \\ \vdots \\ c_{D_{\mathrm{LI}}} \end{pmatrix} = 0. \qquad (1.2.38)$$

Multiplying this equation by the adjoint column V on the left,

$$\underbrace{\begin{pmatrix} \langle\langle \Pi_1 | \\ \langle\langle \Pi_2 | \\ \vdots \\ \langle\langle \Pi_{D_{\mathrm{LI}}} | \end{pmatrix}}_{= \, V} \underbrace{\left(|\Pi_1\rangle\rangle \quad |\Pi_2\rangle\rangle \quad \cdots \quad |\Pi_{D_{\mathrm{LI}}}\rangle\rangle \right)}_{= \, V^\dagger} \begin{pmatrix} c_1 \\ c_2 \\ \vdots \\ c_{D_{\mathrm{LI}}} \end{pmatrix} = 0, \qquad (1.2.39)$$

with the understanding that $\mathbf{\Lambda}$ can be turned into $\mathbf{\Lambda}^\dagger$ by turning the column $\mathbf{\Lambda}$ into a row and taking the adjoint of all the objects in the row. Defining the D_{LI}-dimensional dyadic* $\mathbf{G} = \mathbf{V}\mathbf{V}^\dagger$, we note that $\mathbf{G}$ is a positive dyadic satisfying Eq. (1.2.37), with $\mathbf{c} = \mathbf{0}$ as the *only* possible solution.

In the language of linear algebra, we say that the *null space* of $\mathbf{G}$ (also known as the kernel), which is the space of all columns that give zero when multiplied by $\mathbf{G}$, has dimension zero as it contains only the zero column. It follows from the *rank-nullity theorem*, which is given as the statement

$$\text{rank}\{\mathbf{G}\} + \dim\{\ker(\mathbf{G})\} = \dim\{\mathbf{G}\}, \qquad (1.2.40)$$

that the rank of $\mathbf{G}$ must, in fact, be D_{LI}, which is the number of linearly independent outcomes in the set.

We have thus constructed a positive dyadic $\mathbf{G}$, with dyadic elements

$$\mathbf{G}_{jk} = \langle\!\langle \Pi_j | \Pi_k \rangle\!\rangle = \text{tr}\{\Pi_j \Pi_k\}, \qquad (1.2.41)$$

that has D_{LI} positive eigenvalues out of a set of D_{LI} linearly independent superkets $|\Pi_j\rangle\!\rangle$. This dyadic is known as the *Gram*[†] *dyadic*.

More generally, given an M-outcome POM, the quantity $D_{\text{LI}} \leq M$ represents the degree of linear independence of the POM. According to the general definition of the corresponding M-dimensional Gram dyadic $\mathbf{G} = \mathbf{V}\mathbf{V}^\dagger$ in Eq. (1.2.41) for any Π_j and **Problem 2.2**, the column $\mathbf{V}$ is therefore represented by a $M \times D^2$ matrix of rank D_{LI}. This means that $\{|\Pi_j\rangle\!\rangle\}$ spans a D_{LI}-dimensional row space of $\mathbf{V}$, or that this POM spans a D_{LI}-dimensional operator subspace.[‡] The largest value of D_{LI} is D^2 since this is clearly the highest rank for $\mathbf{V}$, which, incidently, is also the maximum number of linearly independent operators that can span the space of Hermitian operators as an operator basis. Therefore, an informationally complete POM that contains *at least* D^2 outcomes corresponds to a Gram dyadic $\mathbf{G}$ of maximal rank $D_{\text{LI}} = D^2$.

Going by a different route, one can define the *frame superoperator*

$$\mathcal{F} = \sum_{j=1}^{M} \frac{|\Pi_j\rangle\!\rangle \langle\!\langle \Pi_j|}{\text{tr}\{\Pi_j\}}. \qquad (1.2.42)$$

---

*As a brief recap, a general dyadic is an object formed from a linear combination of products of a column and a row. So $\mathbf{C}$ is a dyadic if $\mathbf{C} = \sum_j \mathbf{a}_j \mathbf{b}_j^\dagger$ for columns $\mathbf{a}_j$ and $\mathbf{b}_j$.

[†] Jørgen Pedersen Gram (1850–1916).

[‡] More about this subspace in Chapter 3.

With it, an equivalent criterion for a set of informationally complete POM outcomes $\{\Pi_j\}$ is that the frame superoperator $\mathcal{F}$ is invertible, since $\mathcal{F}$ can always be represented as a product of three matrices

$$\mathcal{F} \widehat{=} \underbrace{\left(|\Pi_1\rangle\rangle \; |\Pi_2\rangle\rangle \; \cdots \; |\Pi_M\rangle\rangle\right)}_{\text{rank } D^2 \leq M} \underbrace{\begin{pmatrix} \mathrm{tr}\{\Pi_1\} & 0 & \cdots & 0 \\ 0 & \mathrm{tr}\{\Pi_2\} & \cdots & 0 \\ \vdots & \vdots & \ddots & \vdots \\ 0 & 0 & \cdots & \mathrm{tr}\{\Pi_M\} \end{pmatrix}^{-1}}_{\equiv \, \boldsymbol{\Pi}^{-1} \text{ of rank } M} \begin{pmatrix} \langle\langle\Pi_1| \\ \langle\langle\Pi_2| \\ \vdots \\ \langle\langle\Pi_M| \end{pmatrix}$$

$$= \underbrace{\boldsymbol{V}^{\dagger}\boldsymbol{\Pi}^{-1}\boldsymbol{V}}_{\text{rank } D^2}, \tag{1.2.43}$$

which, in this case, is of rank D^2. This follows from the fact that given an $l \times m$ matrix $\boldsymbol{A}$ of rank m and an $m \times n$ matrix $\boldsymbol{B}$, then $\mathrm{rank}\{\boldsymbol{AB}\} = \mathrm{rank}\{\boldsymbol{B}\}$.

For any informationally complete measurement, there exist Hermitian *dual* operators Θ_j of the outcomes Π_j with the property that the corresponding superkets are related by

$$\sum_{j=1}^{M} |\Pi_j\rangle\rangle \, \langle\langle\Theta_j| = \mathcal{I} = \sum_{j=1}^{M} |\Theta_j\rangle\rangle \, \langle\langle\Pi_j|, \tag{1.2.44}$$

where $\mathcal{I}$ is the identity superoperator. This property can be written more compactly by introducing the column $\boldsymbol{V}$ of M superbras $\langle\langle\Pi_j|$ as defined in Eq. (1.2.39) and the column

$$\boldsymbol{W} = \begin{pmatrix} \langle\langle\Theta_1| \\ \langle\langle\Theta_2| \\ \vdots \\ \langle\langle\Theta_M| \end{pmatrix} \tag{1.2.45}$$

of superbras $\langle\langle\Theta_j|$, so that Eq. (1.2.44) is simply

$$\boldsymbol{W}^{\dagger}\boldsymbol{V} = \mathcal{I} = \boldsymbol{V}^{\dagger}\boldsymbol{W}. \tag{1.2.46}$$

Multiplying Eq. (1.2.46) by $\boldsymbol{V}$ on the left and $\boldsymbol{V}^{\dagger}$ on the right, one gets

$$\boldsymbol{V}\boldsymbol{W}^{\dagger}\boldsymbol{\mathsf{G}} = \boldsymbol{\mathsf{G}} = \boldsymbol{\mathsf{G}}\boldsymbol{W}\boldsymbol{V}^{\dagger}. \tag{1.2.47}$$

So far, we have established three equivalent relations stated respectively in Eq. (1.2.44), Eq. (1.2.46) and Eq. (1.2.47) for any informationally complete measurement. Since the maximum number of positive eigenvalues for the Gram dyadic $\mathbf{G}$ is D^2, the inverse of $\mathbf{G}$ exists when the measurement is *minimally complete*; That is, when the number of outcomes M *equals* D^2, all of which are linearly independent. For such kind of measurement, Eq. (1.2.47) implies that

$$\mathbf{V}\mathbf{W}^\dagger = \mathbf{1} = \mathbf{W}\mathbf{V}^\dagger, \tag{1.2.48}$$

or

$$\mathrm{tr}\{\Theta_j\Pi_k\} = \delta_{j,k}. \tag{1.2.49}$$

So, for minimally complete measurements, the dual operators are trace-orthonormal to the POM outcomes.

From Eq. (1.2.49), it immediately follows that any statistical operator ρ can be completely characterized by the dual operators Θ_j via the relation $\rho = \sum_j p_j \Theta_j$, where p_j are the true probabilities that is associated with the POM outcomes Π_j. For *overcomplete* measurements, where the number of outcomes M *exceeds* D^2, Eq. (1.2.49) does not hold in general.

From the definition of the frame superoperator stated in Eq. (1.2.42), one can define the dual superkets in a canonical way as

$$|\Theta_j\rangle\rangle = \frac{\mathcal{F}^{-1} |\Pi_j\rangle\rangle}{\mathrm{tr}\{\Pi_j\}}, \tag{1.2.50}$$

and it is straightforward to verify that this canonical definition is consistent with the dual property of $|\Theta_j\rangle\rangle$ stated in Eq. (1.2.44). Furthermore, it is a simple matter to verify that $\mathrm{tr}\{\Theta_j\} = 1$, since the superket $|1\rangle\rangle$ is an eigenvector of $\mathcal{F}$ with eigenvalue 1,

$$\mathcal{F} |1\rangle\rangle = \sum_{j=1}^{M} |\Pi_j\rangle\rangle \frac{1}{\mathrm{tr}\{\Pi_j\}} \langle\langle\Pi_j|1\rangle\rangle = \sum_{j=1}^{M} |\Pi_j\rangle\rangle = |1\rangle\rangle, \tag{1.2.51}$$

and this implies that

$$\mathrm{tr}\{\Theta_j\} = \langle\langle 1|\Theta_j\rangle\rangle = \langle\langle 1| \mathcal{F}^{-1}|\Pi_j\rangle\rangle \frac{1}{\mathrm{tr}\{\Pi_j\}} = \frac{\langle\langle 1|\Pi_j\rangle\rangle}{\mathrm{tr}\{\Pi_j\}} = 1. \tag{1.2.52}$$

If, in addition, the POM is *minimally complete*, the *canonical dual superkets* presented in Eq. (1.2.50) are the only unique dual superkets for $|\Pi_j\rangle\rangle$. This is implied by the simultaneous satisfaction of Eqs. (1.2.46) and (1.2.48), or the equivalent fact that $\boldsymbol{W}^\dagger = \boldsymbol{V}^{-1}$. For *overcomplete* measurements, there is more than one way of defining the set of dual superkets, all of which are plausible candidates for the purpose of state estimation depending on the situation at hand.

1.2.2.3 *Symmetric informationally complete measurements*

As an example, we consider a D-dimensional *symmetric informationally complete POM* (SIC POM) consisting of D^2 linearly independent rank-one outcomes Π_j such that $\mathrm{tr}\{\Pi_j\} = \dfrac{1}{D}$ and

$$\langle\!\langle \Pi_j | \Pi_k \rangle\!\rangle = \mathrm{tr}\{\Pi_j \Pi_k\} = \frac{D\delta_{j,k} + 1}{D^2(D+1)}. \tag{1.2.53}$$

To calculate $\mathcal{F}$ for such a POM, we first note that a convenient superket basis $\left\{ \left| \Pi_j^\perp \right\rangle\!\right\rangle \right\}$ can be constructed out of $|\Pi_j\rangle\!\rangle$ by defining

$$\left| \Pi_j^\perp \right\rangle\!\rangle = \kappa_1 \left(|\Pi_j\rangle\!\rangle - |1\rangle\!\rangle \kappa_2 \right). \tag{1.2.54}$$

The constants κ_1 and κ_2 can be found by enforcing the orthonormality condition $\left\langle\!\left\langle \Pi_j^\perp \middle| \Pi_k^\perp \right\rangle\!\right\rangle = \delta_{j,k}$, where

$$\begin{aligned}
\left\langle\!\left\langle \Pi_j^\perp \middle| \Pi_k^\perp \right\rangle\!\right\rangle &= \kappa_1^2 \left(\langle\!\langle \Pi_j | - \kappa_2 \langle\!\langle 1 | \right) \left(|\Pi_k\rangle\!\rangle - |1\rangle\!\rangle \kappa_2 \right) \\
&= \kappa_1^2 \left[\frac{D\delta_{j,k} + 1}{D^2(D+1)} - \frac{2\kappa_2}{D} + D\kappa_2^2 \right].
\end{aligned} \tag{1.2.55}$$

The respective values of these constants are thus given in terms of the equations

$$\kappa_1 = \sqrt{D(D+1)},$$
$$0 = D^3(D+1)\kappa_2^2 - 2D(D+1)\kappa_2 + 1. \tag{1.2.56}$$

The matrix elements of $\mathcal{F}$ in this basis are

$$\left\langle\!\left\langle \Pi_l^\perp \middle| \mathcal{F} \middle| \Pi_{l'}^\perp \right\rangle\!\right\rangle$$

$$= D(D+1) \sum_{j=1}^{D^2} D \left[\frac{D\delta_{j,l} + 1}{D^2(D+1)} \right] \left[\frac{D\delta_{j,l'} + 1}{D^2(D+1)} \right]$$

$$= D \left[\frac{\delta_{l,l'}}{D(D+1)} + \frac{D+2}{D^2(D+1)} + D(D+1)\kappa_2^2 - 2\left(\frac{D+1}{D} \right) \kappa_2 \right], \tag{1.2.57}$$

where at this point we may simplify the above expression using the second equation of (1.2.56) to obtain

$$\langle\!\langle \Pi_l^\perp | \mathcal{F} | \Pi_{l'}^\perp \rangle\!\rangle = \frac{\delta_{l,l'}}{D+1} + \frac{1}{D(D+1)}. \tag{1.2.58}$$

Since the solutions to the quadratic equation in Eq. (1.2.56) are

$$\kappa_2 = \frac{1}{D^2}\left(1 \pm \frac{1}{\sqrt{D+1}}\right), \tag{1.2.59}$$

so that the sum of all the orthonormal superkets yields

$$\sum_{l=1}^{D^2} |\Pi_l^\perp\rangle\!\rangle = \sqrt{D(D+1)} \sum_{l=1}^{D^2} \left(|\Pi_l\rangle\!\rangle - |1\rangle\!\rangle \kappa_2\right)$$

$$= |1\rangle\!\rangle \left(1 - D^2\kappa_2\right) \sqrt{D(D+1)} = \pm |1\rangle\!\rangle \sqrt{D}, \tag{1.2.60}$$

the frame superoperator is

$$\mathcal{F} = \sum_{l=1}^{D^2} \sum_{l'=1}^{D^2} |\Pi_l^\perp\rangle\!\rangle \langle\!\langle \Pi_l^\perp | \mathcal{F} | \Pi_{l'}^\perp \rangle\!\rangle \langle\!\langle \Pi_{l'}^\perp | = \frac{\mathcal{I} + |1\rangle\!\rangle \langle\!\langle 1|}{D+1}. \tag{1.2.61}$$

Here, we have made use of the relation

$$\sum_{l=1}^{D^2} |\Pi_l^\perp\rangle\!\rangle \langle\!\langle \Pi_l^\perp | = \mathcal{I} \tag{1.2.62}$$

for the set of orthonormal superkets that defines their completeness, a property that is to be directly verified in **Problem 1.5**.

To calculate the canonical dual operators, we need the inverse of the frame superoperator, and an *ansatz* for its structure is the form

$$\mathcal{F}^{-1} = a\mathcal{I} + |1\rangle\!\rangle\, b\, \langle\!\langle 1|, \tag{1.2.63}$$

with which we require

$$\mathcal{I} = \mathcal{F}\mathcal{F}^{-1}$$

$$= \frac{\mathcal{I} + |1\rangle\!\rangle \langle\!\langle 1|}{D+1} \left(a\mathcal{I} + |1\rangle\!\rangle\, b\, \langle\!\langle 1|\right)$$

$$= \frac{1}{D+1} \left[a\mathcal{I} + |1\rangle\!\rangle\, (a + b + Db)\, \langle\!\langle 1|\right]. \tag{1.2.64}$$

This requirement implies that $a = -(D+1)b$ and $a = D+1$, or

$$\mathcal{F}^{-1} = (D+1)\mathcal{I} - |1\rangle\!\rangle\,\langle\!\langle 1| \,, \tag{1.2.65}$$

which brings us to the answer

$$|\Theta_j\rangle\!\rangle = |\Pi_j\rangle\!\rangle\,D(D+1) - |1\rangle\!\rangle\,. \tag{1.2.66}$$

1.2.3   *A simple ML estimator for minimally complete measurements*

With all the necessary tools in place, we can define the estimator that maximizes the likelihood $\mathcal{L}(\{n_j\}; \rho)$. First, we note that the probabilities p_j that truly maximize $\mathcal{L}(\{n_j\}; \rho)$ are the measurement frequencies f_j. To show this easily, we vary the *log-likelihood* $\log \mathcal{L}(\{n_j\}; \rho)$ with respect to p_j, subject to the constraint that $\sum_j p_j = 1$. The log-likelihood possesses all stationary point(s) of the likelihood since the logarithm is just a monotonically increasing function that preserves all stationary points of its argument. With the help of a *Lagrange multiplier* μ, setting the variation of the *Lagrange* function* $\mathcal{D}_{\mathrm{L}}(\{n_j\}; \rho)$,

$$\delta\mathcal{D}_{\mathrm{L}}(\{n_j\}; \rho) = \delta\left[\log\mathcal{L}(\{n_j\}; \rho) - \mu\left(\sum_{j=1}^{M} p_j - 1\right)\right]$$

$$= \sum_{j=1}^{M} \delta p_j\left(N\frac{f_j}{p_j} - \mu\right), \tag{1.2.67}$$

to zero would determine the stationary point(s). Since the variations δp_j are arbitrary, setting the total variation to zero implies that

$$\frac{f_j}{\widehat{p}_j} = \frac{\mu}{N}, \tag{1.2.68}$$

where $\mu = N$ as the sum of all the estimated probabilities $\widehat{p}_j$ is one. This means that the extremal probabilities are $\widehat{p}_j = f_j$.

---

* Joseph-Louis de Lagrange (1736–1813).

For *minimally* complete measurements, it then follows from Eq. (1.2.49) that the most likely estimator is, in fact, equal to

$$\widehat{\rho}_{\text{LIN}} = \sum_{j=1}^{M} f_j \Theta_j, \tag{1.2.69}$$

since

$$\widehat{p}_j = \text{tr}\{\widehat{\rho}_{\text{LIN}}\, \Pi_j\} = \sum_{k=1}^{M} f_k \text{tr}\{\Theta_k \Pi_j\} = f_j. \tag{1.2.70}$$

For overcomplete measurements, Eq. (1.2.69) may still be used to define an estimator that is linear in f_j, although $\text{tr}\{\Pi_j \Theta_k\} \neq \delta_{j,k}$, and the dual operators defined according to Eq. (1.2.44) are in general not unique for a given set of POM outcomes. The reason for the subscript LIN will become clear in the next section.

1.2.4 *Linear inversion*

1.2.4.1 *Minimally complete measurements*

Equivalently, the estimator $\widehat{\rho}_{\text{LIN}}$ in Eq. (1.2.69) can be obtained by directly inverting the set of $M = D^2$ linear constraints $\text{tr}\{\widehat{\rho}_{\text{LIN}}\, \Pi_j\} = f_j$ for minimally complete measurement data. For this reason, $\widehat{\rho}_{\text{LIN}}$ is also known as the *linear-inversion (LIN) estimator*. An essential tool for linear inversion is a *complete* set of D^2 Hermitian, trace-orthonormal basis operators $\Gamma_j = \Gamma_j^\dagger$ that span the entire operator space, such that $\text{tr}\{\Gamma_j \Gamma_k\} = \delta_{j,k}$.

These trace-orthonormal operators are clearly linearly independent, for the corresponding Gram dyadic is simply the unit dyadic. The vector-space completeness of these operators, which is analogous to the completeness relation for a complete set of orthonormal basis kets, refers to the statement

$$\sum_{j=1}^{D^2} |\Gamma_j\rangle\!\rangle \langle\!\langle\Gamma_j| = \mathcal{I}. \tag{1.2.71}$$

The set of operators $\{\Pi_j^\perp\}$ constructed out of the SIC POM on page 20 is a perfect example of a complete set of trace-orthonormal basis operators for the D-dimensional Hilbert space.

Problem 1.5   Show that the mathematical property described by Eq. (1.2.71) is true for any set of D^2 trace-orthonormal Hermitian basis operators Γ_j. Verify this relation specifically for the set of single-qubit basis operators $\left\{ \dfrac{1}{\sqrt{2}}, \dfrac{\sigma_x}{\sqrt{2}}, \dfrac{\sigma_y}{\sqrt{2}}, \dfrac{\sigma_z}{\sqrt{2}} \right\}$ and the set $\left\{ |\Pi_{\tilde{j}}^{\perp}\rangle\!\rangle \right\}$ that is derived from a D-dimensional SIC POM.

Problem 1.6   Show that the set

$$\left\{ |j\rangle\langle j| \Big|_{j=1}^{D} , \; \frac{|j\rangle\langle k| + |k\rangle\langle j|}{\sqrt{2}} \Big|_{j=1}^{D} \Big|_{k>j} , \; \frac{|j\rangle\langle k| - |k\rangle\langle j|}{i\sqrt{2}} \Big|_{j=1}^{D} \Big|_{k>j} \right\}$$

$$(1.2.72)$$

that is formed from D orthonormal kets $|j\rangle$ is a complete set of trace-orthonormal Hermitian basis operators that span the D^2-dimensional space of Hermitian operators.

Since *any* D-dimensional Hermitian operator $\mathcal{A} = \mathcal{A}^{\dagger}$ is specified by exactly D^2 real parameters, it follows that $\mathcal{A}$ can be expressed as a linear combination of such a set of basis operators inasmuch as

$$\mathcal{A} = \sum_{j=1}^{D^2} \text{tr}\{\mathcal{A}\Gamma_j\}\Gamma_j,$$

$$(1.2.73)$$

where the coefficients $\text{tr}\{\mathcal{A}\Gamma_j\}$ are all the real D^2 parameters in this operator basis. We may therefore express the estimator $\widehat{\rho}_{\text{LIN}}$ and the POM outcomes Π_j in terms of an operator basis.

Before we do that, though, let us recall that one of the parameters for $\widehat{\rho}_{\text{LIN}}$ is always fixed relative to the others through the unit-trace constraint, so that the total number of independent parameters is in fact $D^2 - 1$. To incorporate this constraint, it is convenient to consider a set of basis operators such that one of them is $\dfrac{1}{\sqrt{D}}$, and the rest of the $D^2 - 1$ trace-orthonormal operators Ω_j are traceless operators, that is, a generalization of Eq. (1.2.2):

$$\widehat{\rho}_{\text{LIN}} = \frac{1}{D} + \sum_{k=1}^{D^2-1} t_k \Omega_k, \quad t_k = \text{tr}\{\widehat{\rho}_{\text{LIN}} \Omega_k\},$$

$$\Pi_j = \frac{\text{tr}\{\Pi_j\}}{D} + \sum_{k=1}^{D^2-1} q_{jk} \Omega_k, \quad q_{jk} = \text{tr}\{\Pi_j \Omega_k\}. \tag{1.2.74}$$

Clearly, we find that $\text{tr}\{\widehat{\rho}_{\text{LIN}}\} = 1$. The M linear constraints for the measurement data are then given by

$$f_j = \frac{\text{tr}\{\Pi_j\}}{D} + \sum_{k=1}^{D^2-1} q_{jk} t_k,$$

$$\text{or } f_j' \equiv \sum_{k=1}^{D^2-1} q_{jk} t_k, \tag{1.2.75}$$

where we have defined f_j' to be the new mock frequency data adjusted from the actual measured ones with the respective constant terms $\dfrac{\text{tr}\{\Pi_j\}}{D}$ that are irrelevant for the state estimation. For these mock data, $\sum_j f_j' = 0$.

Problem 1.7 Can one have a set of D^2 trace-orthonormal traceless Hermitian operators? Give reason(s) for your answer.

The set of linear constraints can hence be cast into the matrix equation

$$\underbrace{\begin{pmatrix} f_1' \\ f_2' \\ \vdots \\ f_M' \end{pmatrix}}_{\equiv \, \boldsymbol{f}'} = \underbrace{\begin{pmatrix} q_{11} & q_{12} & \cdots & q_{1D^2-1} \\ q_{21} & q_{22} & \cdots & q_{2D^2-1} \\ \vdots & \vdots & \ddots & \vdots \\ q_{M1} & q_{M2} & \cdots & q_{MD^2-1} \end{pmatrix}}_{\equiv \, \mathbf{Q}} \underbrace{\begin{pmatrix} t_1 \\ t_2 \\ \vdots \\ t_{D^2-1} \end{pmatrix}}_{\equiv \, \boldsymbol{t}} \tag{1.2.76}$$

that involves a real rectangular matrix $\mathbf{Q}$ of dimensions $M \times (D^2 - 1)$. Since all POM outcomes sum to the identity, the entries of every single column of $\mathbf{Q}$ sum to zero, which is anyway consistent with the sum of f_j'.

For a minimally complete POM, where there are $M = D^2$ outcomes that are all linearly independent, the first $D^2 - 1$ rows are all linearly independent, and the last row of the matrix $\mathbf{Q}$ supplies no useful information since it simply takes care of the unit-sum constraint for the POM. Correspondingly, the last measurement frequency f_{D^2} is always one minus the sum of the rest of the measurement data. As such, it is natural to write the rectangular matrix $\mathbf{Q}$ as an array of a square matrix $\widetilde{\mathbf{Q}}$ comprising the first $D^2 - 1$ linearly independent rows and the last row $\boldsymbol{q}^{\mathrm{T}} = (q_{D^2 1}\, q_{D^2 2}\, \cdots\, q_{D^2 D^2 - 1})$,

$$\mathbf{Q} = \begin{pmatrix} \widetilde{\mathbf{Q}} \\ \boldsymbol{q}^{\mathrm{T}} \end{pmatrix}. \tag{1.2.77}$$

The coefficients t_j can then be obtained by solving the system of linear equations. The matrix $\mathbf{Q}$, being rectangular, has no matrix inverse in the usual sense. Nevertheless, upon defining another rectangular matrix

$$\mathbf{Q}^- = \begin{pmatrix} \widetilde{\mathbf{Q}}^{-1} & \mathbf{0} \end{pmatrix}, \tag{1.2.78}$$

it is clear that $\mathbf{Q}^- \mathbf{Q} = \mathbf{1}_{D^2 - 1}$, where $\mathbf{1}_n$ is the $n \times n$ identity matrix. The matrix $\mathbf{Q}^-$ is known as a *generalized inverse* of $\mathbf{Q}$, a type of matrix that obeys the more general relation

$$\mathbf{Q}\mathbf{Q}^- \mathbf{Q} = \mathbf{Q}. \tag{1.2.79}$$

The coefficients are thus given by $\boldsymbol{t} = \mathbf{Q}^- \boldsymbol{f}'$. We may then write

$$\begin{aligned}
\widehat{\rho}_{\mathrm{LIN}} &= \frac{1}{D} + \sum_{k=1}^{D^2-1} t_k \Omega_k = \frac{1}{D} + \sum_{k=1}^{D^2-1} \left(\mathbf{Q}^- \boldsymbol{f}'\right)_k \Omega_k \\
&= \sum_{l=1}^{D^2} f_l \left(\frac{1}{D} + \sum_{k=1}^{D^2-1} \Omega_k \mathbf{Q}^-_{kl} \right) - \frac{1}{D} \sum_{l=1}^{D^2} \mathrm{tr}\{\Pi_l\} \sum_{k=1}^{D^2-1} \Omega_k \mathbf{Q}^-_{kl},
\end{aligned} \tag{1.2.80}$$

from which the dual operators for the minimally complete POM are defined as

$$\Theta_j = \frac{1}{D} + \frac{1}{D} \sum_{l=1}^{D^2} \mathrm{tr}\{\Pi_l\} \left[\sum_{k=1}^{D^2-1} \Omega_k \left(\mathbf{Q}^-_{kj} - \mathbf{Q}^-_{kl} \right) \right]. \tag{1.2.81}$$

As we have discussed before, for any informationally complete POM, the matrix $\mathbf{Q}$ always possesses both a row space and a column space that are of dimension $D^2 - 1$, as well as a cokernel of dimension $M - D^2 + 1$. Thus, for $M = D^2$, the cokernel of $\mathbf{Q}$ is spanned by a row of identical entries since $\sum_j q_{jk} = 0$. The matrix $\mathbf{Q}^-$ defined in Eq. (1.2.78) is but one of the many generalized inverses that has the property $\mathbf{Q}^-\mathbf{Q} = \mathbf{1}_{D^2-1}$. All these generalized inverses, which can always be decomposed into a generalized inverse $\mathbf{Q}^-_{\text{colQ}}$ in the column space of $\mathbf{Q}$ and a part $\mathbf{Q}^-_{\text{cokerQ}}$ in the cokernel of $\mathbf{Q}$ $(\mathbf{Q}^-_{\text{cokerQ}}\mathbf{Q} = \mathbf{0})$ inasmuch as

$$\mathbf{Q}^- = \mathbf{Q}^-_{\text{colQ}} + \mathbf{Q}^-_{\text{cokerQ}},$$

$$\mathbf{Q}^-_{\text{cokerQ}} = \mathbf{Q}_0 \left(\mathbf{1}_M - \mathbf{Q}\mathbf{Q}^-_{\text{colQ}}\right) \text{ for an arbitrary matrix } \mathbf{Q}_0, \qquad (1.2.82)$$

differ only in the part $\mathbf{Q}^-_{\text{cokerQ}}$. Consequently, for a minimally complete POM, every matrix row of $\mathbf{Q}^-_{\text{cokerQ}}$, in the computational basis, must consist of identical entries as $\mathbf{Q}^-_{\text{cokerQ}}$ is now a rank-one matrix that takes the form

$$\mathbf{Q}^-_{\text{cokerQ}} \propto \begin{pmatrix} * \\ * \\ \vdots \\ * \end{pmatrix} \begin{pmatrix} 1 & 1 & \cdots & 1 \end{pmatrix}. \qquad (1.2.83)$$

These statements imply that the difference $\mathbf{Q}^-_{kj} - \mathbf{Q}^-_{kl}$ is invariant under the variation of $\mathbf{Q}^-$ — the dual operators Θ_j of a minimally complete POM are unique, as they ought to be. So, the dual operators in Eq. (1.2.81) obtained from linear inversion are indeed the canonical ones in Eq. (1.2.50) for any minimally complete POM, with no other possibilities.

Problem 1.8   Verify that Eq. (1.2.49) holds for the dual operators in Eq. (1.2.81).

1.2.4.2   *Arbitrary measurements — with emphasis on overcomplete ones*

The linear inversion technique can be applied to a POM with any number of outcomes M, where $\mathbf{Q}$ is now a $M \times (D^2 - 1)$ matrix. Let us again define a $(D^2 - 1) \times M$ generalized inverse $\mathbf{Q}^-$ such that $t = \mathbf{Q}^- f'$. We may look for such a matrix with which the quantity

$$S_{\mathbf{Q}} = |\mathbf{Q}\,t - \boldsymbol{f}'|^2 \qquad (1.2.84)$$

is minimized. If all equations are consistent in Eq. (1.2.76), so that a solution for t that solves Eq. (1.2.76) does exist, the minimum of $S_{\mathbf{Q}}$ is zero. Otherwise, the matrix $\mathbf{Q}^-$ that minimizes $S_{\mathbf{Q}}$ defines a *least-squares* (LS) solution for Eq. (1.2.76). The procedure for linear inversion now turns into a problem of *LS estimation*.

After applying the principle of variation,

$$\delta S_{\mathbf{Q}} = \delta \left[\boldsymbol{f}'^{\mathrm{T}} \left(\mathbf{Q}\mathbf{Q}^- - \mathbf{1}_M \right)^{\mathrm{T}} \left(\mathbf{Q}\mathbf{Q}^- - \mathbf{1}_M \right) \boldsymbol{f}' \right]$$

$$= \boldsymbol{f}'^{\mathrm{T}} \delta \mathbf{Q}^{-\,\mathrm{T}} \mathbf{Q}^{\mathrm{T}} \left(\mathbf{Q}\mathbf{Q}^- - \mathbf{1}_M \right) \boldsymbol{f}' + \boldsymbol{f}'^{\mathrm{T}} \left(\mathbf{Q}\mathbf{Q}^- - \mathbf{1}_M \right)^{\mathrm{T}} \mathbf{Q}\,\delta \mathbf{Q}^-\,\boldsymbol{f}'. \qquad (1.2.85)$$

As $\delta S_{\mathbf{Q}} = 0$ at the minimum, we have the extremal equation

$$\mathbf{Q}^{\mathrm{T}} \left(\mathbf{Q}\mathbf{Q}^- - \mathbf{1}_M \right) \boldsymbol{f}' = \boldsymbol{0}. \qquad (1.2.86)$$

Certainly, in the situation where a solution does exist, then by virtue of the condition that the minimum of $S_{\mathbf{Q}}$ is zero, the extremal equation in Eq. (1.2.86) is trivially obeyed as long as Eq. (1.2.79) is satisfied, and there is a plethora of choices for $\mathbf{Q}^-$ that obeys this extremal equation.

However, such a situation never happens when $M > D^2$, for instance, for any given experiment where measurement data are collected, because statistical fluctuation is always present and inconsistencies in dependent equations are bound to happen. In such cases, we need to look for the solution to the equation

$$\mathbf{Q}^{\mathrm{T}} \mathbf{Q} \mathbf{Q}^- = \mathbf{Q}^{\mathrm{T}}. \qquad (1.2.87)$$

For any $M \times D^2$ real matrix $\mathbf{Q}$, there exists a *singular-value decomposition*

$$\mathbf{Q} = \mathbf{O}\mathbf{D}\mathbf{O}'^{\,\mathrm{T}} \qquad (1.2.88)$$

in terms of an $M \times (D^2 - 1)$ diagonal matrix $\mathbf{D}$ with elements $\mathbf{D}_{jk} = d_j \delta_{j,k}$ containing the ordered nonnegative singular values $d_1 \geq d_2 \geq d_3 \geq \cdots$ of $\mathbf{Q}$ — the eigenvalues of $\mathbf{Q}^{\mathrm{T}}\mathbf{Q}$ or $\mathbf{Q}\mathbf{Q}^{\mathrm{T}}$ —, and two orthogonal matrices $\mathbf{O}$ $(M \times M)$ and $\mathbf{O}'$ $[(D^2 - 1) \times (D^2 - 1)]$. This decomposition exists for any complex matrix $\mathcal{A} = \mathbf{UDU'}^\dagger$ in general, with orthogonal matrices replaced by unitary matrices $\mathbf{U}$ and $\mathbf{U'}$.

With this decomposition, Eq. (1.2.87) becomes

$$\mathbf{O'D}^\mathsf{T}\mathbf{DO'}^\mathsf{T}\mathbf{Q}^- = \mathbf{O'D}^\mathsf{T}\mathbf{O}^\mathsf{T}. \tag{1.2.89}$$

We find that in terms of the $(D^2 - 1) \times M$ diagonal matrix $\mathbf{D}^-$,

$$\mathbf{Q}^- = \mathbf{Q}^-_{\mathrm{MP}} \equiv \mathbf{O'D}^-\mathbf{O}^\mathsf{T},$$

$$\mathbf{D}^-_{jj} = d^-_j \equiv \begin{cases} \dfrac{1}{d_j} & \text{if } d_j > 0, \\ 0 & \text{if } d_j = 0. \end{cases} \tag{1.2.90}$$

It follows that $\mathbf{Q}^-_{\mathrm{MP}}$ does indeed possess the following properties:

$$\mathbf{QQ}^-_{\mathrm{MP}}\mathbf{Q} = \mathbf{Q},$$

$$\left(\mathbf{QQ}^-_{\mathrm{MP}}\right)^\mathsf{T} = \mathbf{QQ}^-_{\mathrm{MP}},$$

$$\mathbf{Q}^-_{\mathrm{MP}}\mathbf{QQ}^-_{\mathrm{MP}} = \mathbf{Q}^-_{\mathrm{MP}},$$

$$\left(\mathbf{Q}^-_{\mathrm{MP}}\mathbf{Q}\right)^\mathsf{T} = \mathbf{Q}^-_{\mathrm{MP}}\mathbf{Q}. \tag{1.2.91}$$

The first and third equations mathematize the fundamental properties of generalized inverses.

Such a matrix $\mathbf{Q}^-_{\mathrm{MP}}$ is called the *Moore–Penrose** (MP) pseudo-inverse* of $\mathbf{Q}$ and is unique for a given $\mathbf{Q}$.

Problem 1.9 Show this fact.

Problem 1.10 Write down the $\mathbf{Q}^-_{\mathrm{MP}}$ for $\mathbf{Q}$ that corresponds to an informationally complete POM.

---

*Eliakim Hastings Moore (1862–1932) and *Sir* Roger Penrose (1932–).

More generally, the pseudo-inverse $\mathcal{A}_{\mathrm{MP}}^-$ of $\mathcal{A}$ that satisfies the properties

$$\mathcal{A}\mathcal{A}_{\mathrm{MP}}^-\mathcal{A} = \mathcal{A},$$

$$\mathcal{A}_{\mathrm{MP}}^-\mathcal{A}\mathcal{A}_{\mathrm{MP}}^- = \mathcal{A}_{\mathrm{MP}}^-,$$

$$\left(\mathcal{A}\mathcal{A}_{\mathrm{MP}}^-\right)^\dagger = \mathcal{A}\mathcal{A}_{\mathrm{MP}}^-,$$

$$\left(\mathcal{A}_{\mathrm{MP}}^-\mathcal{A}\right)^\dagger = \mathcal{A}_{\mathrm{MP}}^-\mathcal{A}, \tag{1.2.92}$$

can be uniquely defined for any complex matrix $\mathcal{A}$. We have therefore established a general procedure for the linear inversion of Eq. (1.2.76): it amounts to finding the MP pseudo-inverse $\mathbf{Q}_{\mathrm{MP}}^-$ of $\mathbf{Q}$ that defines the linear constraints.

Problem 1.11 Consider an overcomplete POM with six outcomes defined by

$$\Pi_1 = \frac{1+\sigma_x}{6}, \qquad \Pi_3 = \frac{1+\sigma_y}{6}, \qquad \Pi_5 = \frac{1+\sigma_z}{6},$$

$$\Pi_2 = \frac{1-\sigma_x}{6}, \qquad \Pi_4 = \frac{1-\sigma_y}{6}, \qquad \Pi_6 = \frac{1-\sigma_z}{6}. \tag{1.2.93}$$

Using the single-qubit basis operators listed in **Problem 1.5**, write down the matrix $\mathbf{Q}$ for this POM and calculate the corresponding MP pseudo-inverse $\mathbf{Q}_{\mathrm{MP}}^-$.

This procedure would also give us a recipe for designing the set of LS dual operators Θ_j in terms of the MP pseudo-inverse of any $\mathbf{Q}$, with which the LS LIN estimator $\hat{\rho}_{\mathrm{LIN}}$ can be equivalently written as the linear combination given in Eq. (1.2.69). The corresponding LS Θ_js are given by

$$\Theta_j = \frac{1}{D} + \frac{1}{D}\sum_{l=1}^{M}\mathrm{tr}\{\Pi_l\}\sum_{k=1}^{D^2-1}\Omega_k\left[\left(\mathbf{Q}_{\mathrm{MP}}^-\right)_{kj} - \left(\mathbf{Q}_{\mathrm{MP}}^-\right)_{kl}\right]. \tag{1.2.94}$$

Problem 1.12 Calculate the LS dual operators for the six-outcome POM in **Problem 1.11**.

Problem 1.13 For overcomplete measurements, the canonical dual operators in Eq. (1.2.50) are generally not the LS dual operators. Show, however, that they are the same if $\mathrm{tr}\{\Pi_j\} = \dfrac{D}{M}$. This also means that the LS LIN estimator is also the canonical LIN estimator under this condition.

There is a serious drawback for linear inversion. Since physical resources are always finite, the measurement frequencies are no longer true probabilities. These measurement frequencies, in fact, vary over different experiments as a result of statistical fluctuation. Therefore, there is *no* guarantee that the resulting estimator $\widehat{\rho}_{\mathrm{LIN}}$ will be a positive operator.

Problem 1.14 As a sequel to **Problem 1.11**, suppose that the measurement data collected with this POM, after a measurement of $N = 5$ copies, are $\{n_1 = 1, n_2 = 0, n_3 = 1, n_4 = 1, n_5 = 1, n_6 = 1\}$. Compute $\widehat{\rho}_{\mathrm{LIN}}$ using the MP pseudo-inverse and state whether it is positive and consistent with the data. Suppose, after further data acquisition, the resulting measurement data for $N = 18$ copies are $\{n_1 = 6, n_2 = 0, n_3 = 3, n_4 = 3, n_5 = 2, n_6 = 4\}$. Compute the corresponding $\widehat{\rho}_{\mathrm{LIN}}$ using the MP pseudo-inverse and state, again, whether it is positive and consistent with the data.

Problem 1.15 For any M and D, verify that $\mathbf{\Pi} \geq \mathbf{G}$, where $\mathbf{G}$ is the Gram dyadic and $\mathbf{\Pi}$ is defined in Eq. (1.2.43), and give the POM for which the equality is satisfied. Next, express $\mathbf{G}$ in terms of $\mathbf{Q}$ and $\mathbf{\Pi}$. From this expression, or otherwise, state the obvious eigenvector of $\mathbf{\Pi} - \mathbf{G}$ that corresponds to the zero eigenvalue.

1.2.5 *Geometry of the quantum state space*

The LIN estimator $\widehat{\rho}_{\mathrm{LIN}}$ is an unbiased estimator for ρ_{true}, since

$$\overline{|\widehat{\rho}_{\mathrm{LIN}}\rangle\rangle} = \sum_{j=1}^{M} |\Theta_j\rangle\rangle \, \overline{f_j} = \sum_{j=1}^{M} |\Theta_j\rangle\rangle \, p_j^{\mathrm{true}} = \sum_{j=1}^{M} |\Theta_j\rangle\rangle \, \langle\langle\Pi_j|\rho_{\mathrm{true}}\rangle\rangle = |\rho_{\mathrm{true}}\rangle\rangle,$$

$$(1.2.95)$$

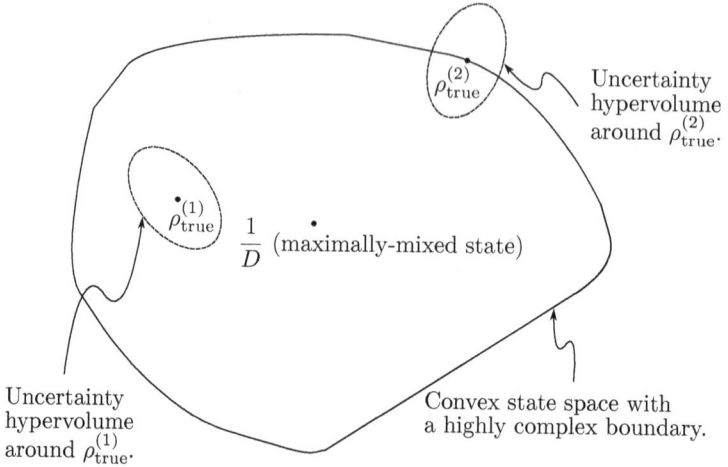

Fig. 1.4　Simplified pictorial description of the state space.

with $p_j^{\text{true}} = \text{tr}\{\rho_{\text{true}}\Pi_j\}$. This means that the set of all possible estimators $\widehat{\rho}_{\text{LIN}}$, for a given N, forming an *uncertainty hypervolume* is such that the centroid of this hypervolume is ρ_{true} (refer to Fig. 1.4). Statistically speaking, this uncertainty hypervolume marks the region containing the typical LIN estimators corresponding to the various possible sets of measurement data for a fixed value of N.

Because of this fact, $\widehat{\rho}_{\text{LIN}}$ is generally not a positive operator. Part of the uncertainty hypervolume around ρ_{true} that contains all estimators $\widehat{\rho}_{\text{LIN}}$ can lie outside the state space for a given finite N. As N increases, the hypervolume shrinks to a point in the state space when N approaches infinity. In other words, if ρ_{true} resides within the state space, the linear-inversion estimators $\widehat{\rho}_{\text{LIN}}$ will all be positive for sufficiently large N.*

If, on the other hand, ρ_{true} lies on the boundary of the state space, then as long as N is finite, no matter how small the uncertainty hypervolume is, there will always be LIN estimators that are not positive with significant probability. Such a state must therefore be *rank-deficient*, that is ρ_{true} possesses at least one zero eigenvalue or $\det\{\rho_{\text{true}}\} = 0$, since a tiny addition of negativity to its zero eigenvalues would result in a nonpositive operator, and is therefore consistent with a state on the state-space boundary.

---

*Under this condition, the probability that $\widehat{\rho}_{\text{LIN}}$ is nonpositive is essentially zero.

For an informationally complete POM with consistent measurement data such that $\text{tr}\{\hat{\rho}_{\text{LIN}}\Pi_j\} = f_j$ (typically only for minimally complete data), the LIN estimator $\hat{\rho}_{\text{LIN}}$ is the ML estimator we seek if $\hat{\rho}_{\text{LIN}} \geq 0$ for the data. If $\hat{\rho}_{\text{LIN}}$ lies outside the state space, it implies that the actual peak of $\mathcal{L}(\{n_j\}; \rho)$, which corresponds to the values $\hat{p}_j = f_j$ for the estimated probabilities, also lies outside the state space. In this case, the resulting positive ML estimator $\hat{\rho}_{\text{ML}}$ must be rank-deficient, and there is no closed-form expression for the positive estimator and numerical methods are needed to search for such an estimator. As a result, the probabilities from $\hat{\rho}_{\text{ML}}$ will not be the measurement frequencies f_j.

With typical overcomplete data, the measurement frequencies f_j are never consistent with any solution because of statistical fluctuation. As such, the peak of $\mathcal{L}(\{n_j\}; \rho)$ no longer corresponds to $\hat{p}_j = f_j$, and both $\hat{\rho}_{\text{LIN}}$ and $\hat{\rho}_{\text{ML}}$ are rather different in general. As with consistent data, this peak can lie outside the state space, in which case $\hat{\rho}_{\text{ML}}$ will also be rank-deficient corresponding to yet a different set of probabilities.

Regardless, the estimator $\hat{\rho}_{\text{ML}}$, like $\hat{\rho}_{\text{LIN}}$, is always a *consistent* estimator. This means that $\hat{\rho}_{\text{ML}}$ approaches ρ_{true} as N increases since, clearly, the measurement frequencies f_j approach the true probabilities p_j^{true}. In the limit of large N for overcomplete data, both $\hat{\rho}_{\text{ML}}$ and $\hat{\rho}_{\text{LIN}}$ approach each other owing to this statistical consistency. From hereon, it is understood that this estimator $\hat{\rho}_{\text{ML}}$ is the positive ML estimator we seek with the ML estimation scheme.

Problem 1.16 There is a distinction between an unbiased estimator and a consistent one. Can you give two examples of a biased estimator that is consistent?

1.3 Uncertainties and information

1.3.1 *Cramér–Rao limit for unbiased quantum-state estimation*

By denoting the columns

$$
\boldsymbol{r} = \begin{pmatrix} \text{tr}\{\rho_{\text{true}}\,\Gamma_1\} \\ \text{tr}\{\rho_{\text{true}}\,\Gamma_2\} \\ \vdots \\ \text{tr}\{\rho_{\text{true}}\,\Gamma_{D^2}\} \end{pmatrix} \quad \text{and} \quad \boldsymbol{t} = \begin{pmatrix} \text{tr}\{\hat{\rho}\,\Gamma_1\} \\ \text{tr}\{\hat{\rho}\,\Gamma_2\} \\ \vdots \\ \text{tr}\{\hat{\rho}\,\Gamma_{D^2}\} \end{pmatrix}, \tag{1.3.1}
$$

the covariance dyadic is defined as

$$\mathbf{C}\left(\widehat{\rho}, \rho_{\text{true}}\right) = \overline{\left(\boldsymbol{t} - \boldsymbol{r}\right)\left(\boldsymbol{t} - \boldsymbol{r}\right)^{\text{T}}} \geq \boldsymbol{0}, \qquad (1.3.2)$$

where the dyadic trace, denoted by Sp (*Spur*), of $\mathbf{C}\left(\widehat{\rho}, \rho_{\text{true}}\right)$ gives the familiar unnormalized Hilbert–Schmidt distance

$$\text{Sp}\left\{\mathbf{C}\left(\widehat{\rho}, \rho_{\text{true}}\right)\right\} = \overline{\left(\boldsymbol{t} - \boldsymbol{r}\right)^2} \equiv C_{\text{H-S}}\left(\widehat{\rho}, \rho_{\text{true}}\right), \qquad (1.3.3)$$

which quantifies the *mean squared error* between an estimator $\widehat{\rho}$ and ρ_{true}. Very generally, the structure of the covariance dyadic strongly depends on the type of estimators $\widehat{\rho}$ that are considered here, or the estimation strategy employed. In this section, we will analyze some properties of $\mathbf{C}\left(\widehat{\rho}, \rho_{\text{true}}\right)$ for estimators that are derived from an arbitrary unbiased estimation scheme, that is $\overline{\widehat{\rho}} = \rho_{\text{true}}$.

Using the definition in Eq. (1.2.5), we investigate an important property of $\mathbf{C}\left(\widehat{\rho}, \rho_{\text{true}}\right)$ by beginning with the defining property of all unbiased estimators:

$$\overline{\boldsymbol{t}} = \boldsymbol{r}. \qquad (1.3.4)$$

If we define the real differential operator $\dfrac{\partial}{\partial \boldsymbol{r}}$ with components $\dfrac{\partial}{\partial r_j}$, where r_j are the components of $\boldsymbol{r}$, such that the product of this differential operator and $\boldsymbol{r}$ gives*

$$\frac{\partial}{\partial \boldsymbol{r}} \boldsymbol{r} \equiv \begin{pmatrix} \dfrac{\partial}{\partial r_1} \\[2mm] \dfrac{\partial}{\partial r_2} \\[2mm] \vdots \\[2mm] \dfrac{\partial}{\partial r_{D^2}} \end{pmatrix} \begin{pmatrix} r_1 & r_2 & \cdots & r_{D^2} \end{pmatrix} = \mathbf{1}, \qquad (1.3.5)$$

---

*We have assumed the definition that the product notation $\boldsymbol{ab}$ for two real columnar objects $\boldsymbol{a}$ and $\boldsymbol{b}$ always refer to a multiplication of the **column** $\boldsymbol{a}$ to the left of the **row** of $\boldsymbol{b}$.

a differentiation of $\overline{t}$ with respect to r then results in

$$\frac{\partial}{\partial r}\,\overline{t} = \frac{\partial}{\partial r}\int (\mathrm{d}\mathbb{D})\,\mathcal{L}\,(\mathbb{D};\rho_{\mathrm{true}})\,t$$

$$= \int (\mathrm{d}\mathbb{D})\left\{\left[\frac{\partial}{\partial r}\,\mathcal{L}\,(\mathbb{D};\rho_{\mathrm{true}})\right]t\right\}$$

$$= \int (\mathrm{d}\mathbb{D})\left\{\mathcal{L}\,(\mathbb{D};\rho_{\mathrm{true}})\left[\frac{\partial}{\partial r}\,\log\mathcal{L}\,(\mathbb{D};\rho_{\mathrm{true}})\right]t\right\}. \qquad (1.3.6)$$

We thus have the first identity

$$\int (\mathrm{d}\mathbb{D})\left\{\mathcal{L}\,(\mathbb{D};\rho_{\mathrm{true}})\left[\frac{\partial}{\partial r}\,\log\mathcal{L}\,(\mathbb{D};\rho_{\mathrm{true}})\right]t\right\} = 1. \qquad (1.3.7)$$

Next, from the normalization condition for $\mathcal{L}\,(\mathbb{D};\rho_{\mathrm{true}})$ in Eq. (1.2.6), we can obtain the second identity

$$\int (\mathrm{d}\mathbb{D})\left\{\mathcal{L}\,(\mathbb{D};\rho_{\mathrm{true}})\left[\frac{\partial}{\partial r}\,\log\mathcal{L}\,(\mathbb{D};\rho_{\mathrm{true}})\right]r\right\} = 0 \qquad (1.3.8)$$

after a differentiation. Subtracting Eq. (1.3.8) from Eq. (1.3.7) gives

$$\int (\mathrm{d}\mathbb{D})\left\{\mathcal{L}\,(\mathbb{D};\rho_{\mathrm{true}})\left[\frac{\partial}{\partial r}\,\log\mathcal{L}\,(\mathbb{D};\rho_{\mathrm{true}})\right](t-r)\right\} = 1. \qquad (1.3.9)$$

To proceed, we invoke the *Cauchy–Schwarz** inequality for an integral of a pair of square-integrable functions $(f(x), g(x))$:

$$\left|\int \mathrm{d}x\, f(x)\, g(x)\right|^2 \le \int \mathrm{d}x\, |f(x)|^2 \int \mathrm{d}x\, |g(x)|^2. \qquad (1.3.10)$$

Baron Augustin-Louis Cauchy (1789–1857) and Karl Hermann Amandus Schwarz (1843–1921).

It then follows that, for a pair of numerical columns $\boldsymbol{x}$ and $\boldsymbol{y}$,

$$|\boldsymbol{x} \cdot \boldsymbol{y}|^2 = |\boldsymbol{x} \cdot \boldsymbol{1} \cdot \boldsymbol{y}|^2$$

$$= \left| \int (\mathrm{d}\mathbb{D}) \left\{ \mathcal{L}(\mathbb{D}; \rho_{\mathrm{true}}) \, \boldsymbol{x} \cdot \left[\frac{\partial}{\partial \boldsymbol{r}} \log \mathcal{L}(\mathbb{D}; \rho_{\mathrm{true}}) \right] (\boldsymbol{t} - \boldsymbol{r}) \cdot \boldsymbol{y} \right\} \right|^2$$

$$\leq \int (\mathrm{d}\mathbb{D}) \left\{ \mathcal{L}(\mathbb{D}; \rho_{\mathrm{true}}) \, \boldsymbol{x} \cdot \left[\frac{\partial}{\partial \boldsymbol{r}} \log \mathcal{L}(\mathbb{D}; \rho_{\mathrm{true}}) \right] \right.$$

$$\times \left. \left[\frac{\partial}{\partial \boldsymbol{r}} \log \mathcal{L}(\mathbb{D}; \rho_{\mathrm{true}}) \right] \cdot \boldsymbol{x} \right\}$$

$$\times \int (\mathrm{d}\mathbb{D}) \left\{ \mathcal{L}(\mathbb{D}; \rho_{\mathrm{true}}) \, \boldsymbol{y} \cdot (\boldsymbol{t} - \boldsymbol{r}) (\boldsymbol{t} - \boldsymbol{r}) \cdot \boldsymbol{y} \right\}$$

$$= \boldsymbol{x} \cdot \mathsf{F}(\rho_{\mathrm{true}}) \cdot \boldsymbol{x} \, \boldsymbol{y} \cdot \mathsf{C}(\widehat{\rho}, \rho_{\mathrm{true}}) \cdot \boldsymbol{y}, \tag{1.3.11}$$

where

$$\mathsf{F}(\rho_{\mathrm{true}}) = \overline{\left[\frac{\partial}{\partial \boldsymbol{r}} \log \mathcal{L}(\mathbb{D}; \rho_{\mathrm{true}}) \right] \left[\frac{\partial}{\partial \boldsymbol{r}} \log \mathcal{L}(\mathbb{D}; \rho_{\mathrm{true}}) \right]} \tag{1.3.12}$$

is the *Fisher* information dyadic* that is evaluated with the true state. A substitution of $\boldsymbol{x} = \mathsf{F}(\rho_{\mathrm{true}})^{-1} \cdot \boldsymbol{y}$ gives the inequality

$$\boldsymbol{y} \cdot \mathsf{F}(\rho_{\mathrm{true}})^{-1} \cdot \boldsymbol{y} \leq \boldsymbol{y} \cdot \mathsf{C}(\widehat{\rho}, \rho_{\mathrm{true}}) \cdot \boldsymbol{y} \tag{1.3.13}$$

that is satisfied for all $\boldsymbol{y}$.

This leads to the *Cramér–Rao[†] inequality*

$$\mathsf{F}(\rho_{\mathrm{true}})^{-1} \leq \mathsf{C}(\widehat{\rho}, \rho_{\mathrm{true}}), \tag{1.3.14}$$

which tells us that the mean squared error $C_{\text{H-S}}(\widehat{\rho}, \rho_{\mathrm{true}})$ is bounded from below by $\mathrm{Sp}\left\{ \mathsf{F}(\rho_{\mathrm{true}})^{-1} \right\}$ for unbiased estimators.

---

* *Sir* Ronald Aylmer Fisher (1890–1962).
[†] Harald Cramér (1893–1985) and Calyampudi Radhakrishna Rao (1920–).

Problem 1.17 Show that the Fisher information dyadic in Eq. (1.3.12) can, equivalently, be expressed as

$$\mathbf{F}(\rho_{\text{true}}) = -\overline{\frac{\partial}{\partial \boldsymbol{r}} \frac{\partial}{\partial \boldsymbol{r}} \log \mathcal{L}(\mathbb{D}; \rho_{\text{true}})}. \qquad (1.3.15)$$

Problem 1.18 Under multinomial statistics involving M outcomes and $N = \sum_j n_j$ measured copies, use the parametrization in Eq. (1.2.2) for a D-dimensional state ρ to establish that the Fisher information dyadic is

$$\mathbf{F}(\rho) = N \sum_{l=1}^{M} \frac{1}{p_l} \frac{\partial p_l}{\partial \boldsymbol{r}} \frac{\partial p_l}{\partial \boldsymbol{r}}. \qquad (1.3.16)$$

Thereafter, write $\mathbf{F}(\rho)$ as a function of $\mathbf{Q}$.

Problem 1.19 Show that for multinomial statistics, $\mathrm{Sp}\left\{\mathbf{F}(\rho)^{-1}\right\}$ is a concave function of ρ, that is

$$\mathrm{Sp}\left\{\mathbf{F}(\lambda\rho_1 + (1-\lambda)\rho_2)^{-1}\right\} \geq \lambda\,\mathrm{Sp}\left\{\mathbf{F}(\rho_1)^{-1}\right\} + (1-\lambda)\mathrm{Sp}\left\{\mathbf{F}(\rho_2)^{-1}\right\} \qquad (1.3.17)$$

for all states ρ_1 and ρ_2, with $0 \leq \lambda \leq 1$.

Problem 1.20 For $\rho = 1/D$ and $N = 1$, confirm that the Fisher information dyadic $\mathbf{F}_0 \equiv \mathbf{F}(1/D)$ obeys the inequalities

$$\mathrm{Sp}\{\mathbf{F}_0\} \leq D(D-1) \qquad (1.3.18)$$

$$\text{and} \quad \mathrm{Sp}\{\mathbf{F}_0^{-1}\} \geq \frac{(D+1)(D^2-1)}{D} \qquad (1.3.19)$$

for any informationally complete POM, where equalities are achieved when $\mathbf{F}_0 = \dfrac{D}{D+1}$.

1.3.2   *Asymptotic efficiency*

Being a consistent estimator, $\widehat{\rho}_{\mathrm{ML}}$ approaches the true state ρ_{true} as the number of copies N increases. Hence, the difference $\boldsymbol{t}_{\mathrm{ML}} - \boldsymbol{r}$ diminishes as N tends to infinity. This little observation allows us to study the behavior of the log-likelihood peak for various data about the true state in this limit. For this, we inspect the gradient of the log-likelihood. If ρ_{true} is within the state-space boundary, this inspection is facilitated by its Taylor* series

$$\frac{\partial}{\partial \boldsymbol{t}} \log \mathcal{L}\left(\mathbb{D}; \widehat{\rho}\right) \approx \frac{\partial}{\partial \boldsymbol{r}} \log \mathcal{L}\left(\mathbb{D}; \rho_{\mathrm{true}}\right) + (\boldsymbol{t} - \boldsymbol{r}) \cdot \frac{\partial}{\partial \boldsymbol{r}} \frac{\partial}{\partial \boldsymbol{r}} \log \mathcal{L}\left(\mathbb{D}; \rho_{\mathrm{true}}\right),$$

$$(1.3.20)$$

which, when evaluated at $\boldsymbol{t} = \boldsymbol{t}_{\mathrm{ML}}$, gives

$$(\boldsymbol{t}_{\mathrm{ML}} - \boldsymbol{r}) \approx \left[\frac{\partial}{\partial \boldsymbol{r}} \frac{\partial}{\partial \boldsymbol{r}} \log \mathcal{L}\left(\mathbb{D}; \rho_{\mathrm{true}}\right)\right]^{-1} \cdot \frac{\partial}{\partial \boldsymbol{r}} \log \mathcal{L}\left(\mathbb{D}; \rho_{\mathrm{true}}\right) \qquad (1.3.21)$$

since $\dfrac{\partial}{\partial \boldsymbol{t}_{\mathrm{ML}}} \log \mathcal{L}\left(\mathbb{D}; \widehat{\rho}_{\mathrm{ML}}\right) = \boldsymbol{0}$.

Under the assumption that each datum $\mathbb{D}_j$ obtained from measuring one copy, or sampling event, is *independent and identically distributed* (i.i.d.), the log-likelihood for N copies can always be written as a sum of the log-likelihoods for single copies, namely

$$\log \mathcal{L}\left(\mathbb{D}; \rho_{\mathrm{true}}\right) = \sum_{j=1}^{N} \log \mathcal{L}\left(\mathbb{D}_j; \rho_{\mathrm{true}}\right), \qquad (1.3.22)$$

since the joint probability of all detection events is a product of the probabilities of the individual detections. In this typical situation, we find that

$$\frac{\partial}{\partial \boldsymbol{r}} \frac{\partial}{\partial \boldsymbol{r}} \log \mathcal{L}\left(\mathbb{D}; \rho_{\mathrm{true}}\right) = N \left[\frac{1}{N} \frac{\partial}{\partial \boldsymbol{r}} \frac{\partial}{\partial \boldsymbol{r}} \sum_{j=1}^{N} \log \mathcal{L}\left(\mathbb{D}_j; \rho_{\mathrm{true}}\right)\right]$$

$$\approx N \overline{\frac{\partial}{\partial \boldsymbol{r}} \frac{\partial}{\partial \boldsymbol{r}} \log \mathcal{L}\left(\mathbb{D}_j; \rho_{\mathrm{true}}\right)}$$

$$= -N\mathbf{F}_1(\rho_{\mathrm{true}}) = -\mathbf{F}(\rho_{\mathrm{true}}), \qquad (1.3.23)$$

---

*Brook Taylor (1685–1731).

where we have used Eq. (1.3.15) in **Problem 1.17** to arrive at the expression involving the Fisher information dyadic $\mathbf{F}_1(\rho_{\text{true}})$ for a single copy. The last two equalities rely on the i.i.d. property of each measurement datum, which is another way of saying that the $\mathbf{F}_1(\rho_{\text{true}})$s are identical, so that the sum of all these dyadics gives the full information obtainable from the measurement data $\mathbb{D}$.

We then obtain

$$(\boldsymbol{t}_{\text{ML}} - \boldsymbol{r}) \approx -\mathbf{F}(\rho_{\text{true}})^{-1} \cdot \frac{\partial}{\partial \boldsymbol{r}} \log \mathcal{L}(\mathbb{D}; \rho_{\text{true}}), \qquad (1.3.24)$$

and it is now straightforward to compute the ML covariance dyadic $\mathbf{C}(\widehat{\rho}_{\text{ML}}, \rho_{\text{true}})$ from the right-hand side:

$$\mathbf{C}(\widehat{\rho}_{\text{ML}}, \rho_{\text{true}})$$

$$\approx \mathbf{F}(\rho_{\text{true}})^{-1} \cdot \underbrace{\overline{\left[\frac{\partial}{\partial \boldsymbol{r}} \log \mathcal{L}(\mathbb{D}; \rho_{\text{true}})\right]\left[\frac{\partial}{\partial \boldsymbol{r}} \log \mathcal{L}(\mathbb{D}; \rho_{\text{true}})\right]}}_{= \mathbf{F}(\rho_{\text{true}})} \cdot \mathbf{F}(\rho_{\text{true}})^{-1}$$

$$= \mathbf{F}(\rho_{\text{true}})^{-1}. \qquad (1.3.25)$$

This reveals that in the limit of large N, $\widehat{\rho}_{\text{ML}}$ achieves the Cramér–Rao bound asymptotically. This estimator is *asymptotically efficient*.

Returning very briefly to the discussion of uncertainty hypervolumes, one can deduce that if ρ_{true} is full-rank, then as N becomes sufficiently large, the shape of the hypervolume approaches a hyperellipsoid. The orientations of the main axes of such a hyperellipsoid are specified by the eigenvectors of the covariance dyadic of all the ML estimators that constitute the hyperellipsoid. The essential reason is that for sufficiently large N, the distribution of the ML estimators about ρ_{true} approaches a Gaussian* distribution according to the central limit theorem, which multi-dimensional level curves are characterized by the covariance dyadic.

Further reading

(1) B.-G. Englert, Remarks on some basic issues in quantum mechanics, *Z. Naturforsch.* **54a**, 11 (1999); On quantum theory, *Eur. Phys. J. D.* **67**, 238 (2013).

---

*Johann Carl Friedrich Gauss (1777–1855).

[Two review articles that serve as important reminders on the foundational understanding of quantum mechanics before moving on.]

(2) E. T. Jaynes, The well-posed problem, *Found. Phys.* **3**, 477 (1973).
[An article that stresses the inherent features of prior distributions and notions of invariance in statistical problems through the Bertrand paradox.]

(3) C. W. Helstrøm, *Quantum Detection and Estimation Theory*, Academic Press, New York (1976).
[A classic reference book on basic estimation theory and electromagnetic-field estimation.]

(4) M. Paris and J. Řeháček, *Quantum State Estimation*, Lecture Notes in Physics, Vol. 649, Springer, Berlin Heidelberg (2004).
[A coherent sequence of articles on quantum-state estimation.]

Chapter 2

Informationally Complete Estimation

2.1 ML scheme — perfect detections

2.1.1 *A brief recap on basic concepts*

The ML method, as we have discussed, gives a unique estimator for a set of informationally complete measurement data, as should any other estimation scheme that relies on such data. These data are obtained by using a measurement consisting of $M \geq D^2$ outcomes that span the operator space of positive operators.

A common example for single-qubits is the *six-outcome measurement* (introduced in **Problem 1.11**) that consists of $M = 6$ outcomes

$$\left\{ \frac{1 \pm \sigma_x}{6}, \frac{1 \pm \sigma_y}{6}, \frac{1 \pm \sigma_z}{6} \right\} \tag{2.1.1}$$

that span the $(2^2 = 4)$-dimensional space of positive operators. This six-outcome POM is the usual overcomplete measurement for single-qubits that is rather commonly used.

A good example of a minimally complete POM $(M = D^2)$ for the single-qubit would be the *tetrahedron measurement*, where the outcomes are described with the Bloch vectors

$$\boldsymbol{a}_1 \,\hat{=}\, \frac{1}{\sqrt{3}} \begin{pmatrix} 1 \\ 1 \\ 1 \end{pmatrix}, \quad \boldsymbol{a}_2 \,\hat{=}\, \frac{1}{\sqrt{3}} \begin{pmatrix} 1 \\ -1 \\ -1 \end{pmatrix}, \quad \boldsymbol{a}_3 \,\hat{=}\, \frac{1}{\sqrt{3}} \begin{pmatrix} -1 \\ 1 \\ -1 \end{pmatrix}, \quad \boldsymbol{a}_4 \,\hat{=}\, \frac{1}{\sqrt{3}} \begin{pmatrix} -1 \\ -1 \\ 1 \end{pmatrix},$$

$$\tag{2.1.2}$$

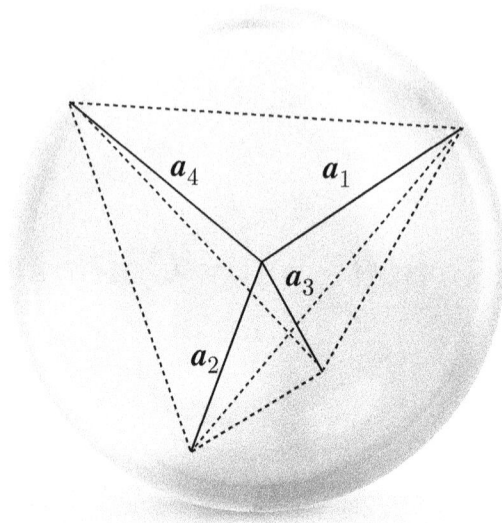

Fig. 2.1   Scaffold of a tetrahedron formed by the Bloch vectors $\boldsymbol{a}_1$, $\boldsymbol{a}_2$, $\boldsymbol{a}_3$ and $\boldsymbol{a}_4$.

that collectively form a regular tetrahedron (refer to Fig. 2.1), with

$$\boldsymbol{a}_j \cdot \boldsymbol{a}_k = \frac{4}{3}\delta_{j,k} - \frac{1}{3}. \tag{2.1.3}$$

Such a minimally complete POM that has the above symmetry falls under the class of SIC POMs for $D = 2$, a type of symmetric POM which we have looked at briefly in the previous chapter.

Problem 2.1   Show that the six-outcome measurement and the tetrahedron measurement outcomes are informationally complete.

As understood from Sec. 1.2.3, when minimally complete data are consistent with an estimator $\widehat{\rho}$ ($\mathrm{tr}\{\widehat{\rho}\,\Pi_j\} = f_j$), the likelihood $\mathcal{L}(\{n_j\}; \rho)$ always possesses a unique maximum at $\widehat{p}_j = f_j$ regardless of the measurement. This uniqueness extends also to overcomplete data where $\widehat{p}_j \neq f_j$, since $\widehat{p}_j$ is a linear function of the $D^2 - 1$ independent state parameters over which the likelihood is maximized. This confirms our understanding about informationally complete measurements. Although the likelihood has, at most, one peak, it is **not** a concave* function of the probabilities, for the function

$$\mathcal{L}(\{n_1 = 2, n_2 = 2\}; \rho) = p_1^2 \, p_2^2 = p_1^2 \, (1 - p_1)^2, \qquad (2.1.4)$$

for instance, has a bell-like structure. The likelihood is, however, *log-concave* in ρ, as

$$\log \mathcal{L}(\{n_j\}; \rho) = \sum_{j=1}^{M} n_j \log p_j \qquad (2.1.5)$$

is just a sum of the concave functions $\log p_j$.

We note, from an earlier discussion, that the actual likelihood peak can lie outside the state space (refer to Fig. 2.2). Hence, maximizing $\mathcal{L}(\{n_j\}; \rho)$ over positive ρs will lead to a rank-deficient estimator $\widehat{\rho}_{\mathrm{ML}}$ in this case.

2.1.2   The extremal equations

To maximize the likelihood over all possible states ρ, let us first vary the corresponding *log-likelihood* with respect to ρ:

$$\delta \log \mathcal{L}(\{n_j\}; \rho) = \sum_{j=1}^{M} n_j \frac{\delta p_j}{p_j} = N \sum_{j=1}^{M} f_j \frac{\mathrm{tr}\{\delta \rho \, \Pi_j\}}{p_j}$$

$$= N\mathrm{tr} \left\{ \delta\rho \sum_{j=1}^{M} \frac{f_j}{p_j} \Pi_j \right\} = N\mathrm{tr}\{R(\rho) \, \delta\rho\}, \qquad (2.1.6)$$

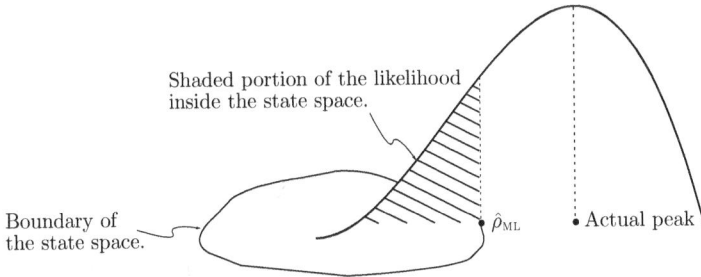

Fig. 2.2   Likelihood function and the state space.

---

*Recall that a function $g(x)$ is concave if $g(\lambda x_1 + (1 - \lambda)x_2) \geq \lambda \, g(x_1) + (1 - \lambda) \, g(x_2)$ for **all possible** $x = x_1$ and $x = x_2$, with $0 \leq \lambda \leq 1$. This also implies that $g(x)$ has only one unique maximum, with the exception of $g(x)$ being a constant value.

with

$$R(\rho) = \sum_{j=1}^{M} \frac{f_j}{p_j} \Pi_j. \qquad (2.1.7)$$

Next, to incorporate the trace and positivity constraints into the variable ρ, we use the parametrization

$$\rho = \frac{\mathcal{A}^\dagger \mathcal{A}}{\mathrm{tr}\{\mathcal{A}^\dagger \mathcal{A}\}}, \qquad (2.1.8)$$

where $\mathcal{A}$ is an auxiliary complex operator that can be represented by a general $n \times D$ matrix for $1 \le n \le D$.

Problem 2.2 Confirm that

$$\mathrm{rank}\{\mathcal{A}^\dagger \mathcal{A}\} = \mathrm{rank}\{\mathcal{A} \mathcal{A}^\dagger\} = \mathrm{rank}\{\mathcal{A}\} = \mathrm{rank}\{\mathcal{A}^\dagger\}. \qquad (2.1.9)$$

Therefore,

$$
\begin{aligned}
\delta\rho &= \frac{\delta\mathcal{A}^\dagger \mathcal{A} + \mathcal{A}^\dagger \delta\mathcal{A}}{\mathrm{tr}\{\mathcal{A}^\dagger \mathcal{A}\}} - \frac{\mathcal{A}^\dagger \mathcal{A}}{\mathrm{tr}\{\mathcal{A}^\dagger \mathcal{A}\}^2} \mathrm{tr}\{\delta\mathcal{A}^\dagger \mathcal{A} + \mathcal{A}^\dagger \delta\mathcal{A}\} \\
&= \frac{\delta\mathcal{A}^\dagger \mathcal{A} + \mathcal{A}^\dagger \delta\mathcal{A}}{\mathrm{tr}\{\mathcal{A}^\dagger \mathcal{A}\}} - \rho\, \mathrm{tr}\left\{ \frac{\delta\mathcal{A}^\dagger \mathcal{A} + \mathcal{A}^\dagger \delta\mathcal{A}}{\mathrm{tr}\{\mathcal{A}^\dagger \mathcal{A}\}} \right\}.
\end{aligned}
\qquad (2.1.10)
$$

Using this parametrization, we can look for the equation that is satisfied by $\widehat{\rho}_{\mathrm{ML}}$ such that

$$\delta \log \mathcal{L}\left(\{n_j\}; \rho\right)\big|_{\rho=\widehat{\rho}_{\mathrm{ML}}} = 0, \qquad (2.1.11)$$

since $\widehat{\rho}_{\mathrm{ML}}$ maximizes $\log \mathcal{L}\left(\{n_j\}; \rho\right)$. This is done by inserting Eq. (2.1.10) into Eq. (2.1.11). The left-hand side works out to be

$$\delta \log \mathcal{L}\left(\{n_j\}; \rho\right)\big|_{\rho=\widehat{\rho}_{\mathrm{ML}}}$$

$$= \mathrm{tr}\left\{ R(\rho) \left(\frac{\delta\mathcal{A}^\dagger \mathcal{A} + \mathcal{A}^\dagger \delta\mathcal{A}}{\mathrm{tr}\{\mathcal{A}^\dagger \mathcal{A}\}} - \rho\, \mathrm{tr}\left\{ \frac{\delta\mathcal{A}^\dagger \mathcal{A} + \mathcal{A}^\dagger \delta\mathcal{A}}{\mathrm{tr}\{\mathcal{A}^\dagger \mathcal{A}\}} \right\} \right) \right\}\Bigg|_{\rho=\widehat{\rho}_{\mathrm{ML}}}$$

$$= \mathrm{tr}\left\{\frac{R(\rho)\mathcal{A}^{\dagger} - \mathrm{tr}\{R(\rho)\rho\}\mathcal{A}^{\dagger}}{\mathrm{tr}\{\mathcal{A}^{\dagger}\mathcal{A}\}}\delta\mathcal{A} + \delta\mathcal{A}^{\dagger}\frac{\mathcal{A}R(\rho) - \mathrm{tr}\{R(\rho)\rho\}\mathcal{A}}{\mathrm{tr}\{\mathcal{A}^{\dagger}\mathcal{A}\}}\right\}\Bigg|_{\rho=\widehat{\rho}_{\mathrm{ML}}}$$

$$= \frac{1}{\mathrm{tr}\{\mathcal{A}^{\dagger}\mathcal{A}\}}\mathrm{tr}\big\{[R(\rho)-1]\,\mathcal{A}^{\dagger}\delta\mathcal{A} + \delta\mathcal{A}^{\dagger}\mathcal{A}\,[R(\rho)-1]\big\}\Big|_{\rho=\widehat{\rho}_{\mathrm{ML}}}, \quad (2.1.12)$$

where in the last equality, the value of

$$\mathrm{tr}\{R(\rho)\rho\} = \mathrm{tr}\left\{\sum_{j=1}^{M}\frac{f_j}{p_j}\Pi_j\rho\right\} = \sum_{j=1}^{M}\frac{f_j}{p_j}\underbrace{\mathrm{tr}\{\Pi_j\rho\}}_{= \, p_j} = \sum_{j=1}^{M}f_j = 1 \quad (2.1.13)$$

is used. Since $\delta\mathcal{A}$ and, correspondingly, $\delta\mathcal{A}^{\dagger} = (\delta\mathcal{A})^{\dagger}$ are arbitrary in the final variation of Eq. (2.1.12), it follows from Eq. (2.1.11) that

$$[R(\rho)-1]\,\mathcal{A}^{\dagger}\big|_{\rho=\widehat{\rho}_{\mathrm{ML}}} = \mathcal{A}\,[R(\rho)-1]\big|_{\rho=\widehat{\rho}_{\mathrm{ML}}} = 0 \quad (2.1.14)$$

or

$$R(\widehat{\rho}_{\mathrm{ML}})\,\widehat{\rho}_{\mathrm{ML}} = \widehat{\rho}_{\mathrm{ML}}R(\widehat{\rho}_{\mathrm{ML}}) = \widehat{\rho}_{\mathrm{ML}}, \quad (2.1.15)$$

with

$$R(\widehat{\rho}_{\mathrm{ML}}) = \sum_{j=1}^{M}\frac{f_j}{\widehat{p}_j^{\,\mathrm{ML}}}\Pi_j. \quad (2.1.16)$$

There is an interesting feature in Eq. (2.1.15). It tells us that $R(\widehat{\rho}_{\mathrm{ML}})$ commutes with $\widehat{\rho}_{\mathrm{ML}}$ and the products $R(\widehat{\rho}_{\mathrm{ML}})\,\widehat{\rho}_{\mathrm{ML}}$ and $\widehat{\rho}_{\mathrm{ML}}R(\widehat{\rho}_{\mathrm{ML}})$ are both equal to $\widehat{\rho}_{\mathrm{ML}}$. When $\widehat{\rho}_{\mathrm{ML}}$ is full-rank, $R(\widehat{\rho}_{\mathrm{ML}})$ is clearly the identity operator, since $\widehat{\rho}_{\mathrm{ML}}$ is invertible. If $\widehat{\rho}_{\mathrm{ML}}$ is rank-deficient, $R(\widehat{\rho}_{\mathrm{ML}})$ also behaves like the identity operator in the row space of $\widehat{\rho}_{\mathrm{ML}}$. Given that $\widehat{\rho}_{\mathrm{ML}}$ has $K < D$ nonzero eigenvalues, so that in its diagonal basis,

$$\widehat{\rho}_{\mathrm{ML}} = \sum_{j=1}^{K}|\lambda_j\rangle\,\lambda_j\,\langle\lambda_j|, \quad \sum_{j=1}^{K}\lambda_j = 1, \quad (2.1.17)$$

the fact that $R(\widehat{\rho}_{\mathrm{ML}})$ commutes with $\widehat{\rho}_{\mathrm{ML}}$ implies that they must share a common set of eigenstates. The spectral decomposition of $R(\widehat{\rho}_{\mathrm{ML}})$ would then take the form

$$R(\widehat{\rho}_{\mathrm{ML}}) = \sum_{j=1}^{K}|\lambda_j\rangle\,\langle\lambda_j| + R_0(\widehat{\rho}_{\mathrm{ML}}), \quad (2.1.18)$$

where $R_0(\widehat{\rho}_{\mathrm{ML}})$ is defined on the kernel of $\widehat{\rho}_{\mathrm{ML}}$, which is orthogonal to the row space of $\widehat{\rho}_{\mathrm{ML}}$ by definition. That is, $R_0(\widehat{\rho}_{\mathrm{ML}})\widehat{\rho}_{\mathrm{ML}} = 0$.

Problem 2.3   Write down $R_0(\widehat{\rho}_{\mathrm{ML}})$ and $R(\widehat{\rho}_{\mathrm{ML}})$ for any finite N when $\widehat{\rho}_{\mathrm{ML}}$ corresponds to the LIN estimator.

2.1.3   *A simple example with the von Neumann measurement — first digression*

So far, we have found the extremal equation for $\widehat{\rho}_{\mathrm{ML}}$. This equation is a nonlinear equation of $\widehat{\rho}_{\mathrm{ML}}$ since $R(\widehat{\rho}_{\mathrm{ML}})$ is itself related to $\widehat{\rho}_{\mathrm{ML}}$ in a nonlinear fashion. There is, however, a simple POM for which Eq. (2.1.15) can be solved directly. This POM is the *von Neumann*[*] measurement and, for simplicity, let us examine such a POM for single qubits, which consists of the two orthonormal outcomes

$$\Pi_0 = |0\rangle\langle 0| = \frac{1 + \sigma_z}{2},$$

$$\Pi_1 = |1\rangle\langle 1| = \frac{1 - \sigma_z}{2}, \tag{2.1.19}$$

where $|0\rangle$ and $|1\rangle$ are the eigenkets of the Pauli operator σ_z. One should remember this POM in the context of photon-polarization measurement $\{\Pi_0 = |\mathrm{H}\rangle\langle \mathrm{H}|, \Pi_1 = |\mathrm{V}\rangle\langle \mathrm{V}|\}$ introduced in Chapter 1. Their respective measurement frequencies are denoted by f_0 and f_1 such that $f_0 + f_1 = 1$. This POM is clearly not informationally complete because the number of measurement outcomes is not enough to span the four-dimensional space of Hermitian operators. We will return to the topic on *informationally incomplete* estimation in the next chapter. For now, we may just take this POM as an example and understand some properties of the solution to Eq. (2.1.15).

We shall first obtain the ML estimator. To do this, we parameterize the estimator in terms of the Bloch vector,

$$\widehat{\rho}_{\mathrm{ML}} = \frac{1 + \mathbf{s}_{\mathrm{ML}} \cdot \boldsymbol{\sigma}}{2}, \quad \mathbf{s}_{\mathrm{ML}} \mathrel{\widehat{=}} \begin{pmatrix} s_1 \\ s_2 \\ s_3 \end{pmatrix} \quad \text{and} \quad \boldsymbol{\sigma} = \begin{pmatrix} \sigma_x \\ \sigma_y \\ \sigma_z \end{pmatrix}, \tag{2.1.20}$$

---

[*]John von Neumann (1903–1957).

with the additional constraint $|s_{\mathrm{ML}}|^2 \leq 1$ to ensure its positivity. It turns out that, for this POM, the ML estimator $\widehat{\rho}_{\mathrm{ML}} \geq 0$ always supplies probabilities that are equal to the measurement frequencies, that is

$$\mathrm{tr}\left\{\frac{1 + s_{\mathrm{ML}} \cdot \boldsymbol{\sigma}}{2}\frac{1 + \sigma_z}{2}\right\} = f_0,$$

$$\mathrm{tr}\left\{\frac{1 + s_{\mathrm{ML}} \cdot \boldsymbol{\sigma}}{2}\frac{1 - \sigma_z}{2}\right\} = f_1. \qquad (2.1.21)$$

In terms of the third component of s_{ML}, we have

$$1 + s_3 = 2\,f_0,$$
$$1 - s_3 = 2\,f_1 = 2\,(1 - f_0), \qquad (2.1.22)$$

and solving for s_3 in Eq. (2.1.22) trivially gives the solution

$$s_{\mathrm{ML}} \,\widehat{=}\, \begin{pmatrix} s_1 \\ s_2 \\ 2\,f_0 - 1 \end{pmatrix}, \qquad (2.1.23)$$

such that

$$s_1^2 + s_2^2 \leq 4\,f_0\,(1 - f_0). \qquad (2.1.24)$$

It is clear that this inequality can always be satisfied, since $-1 \leq s_3 \leq 1$.

This is the first instance when we see the effects of an informationally incomplete POM in action; Two of the three components of the Bloch vector are arbitrary up to the positivity constraint. Since there exist infinitely many positive $\widehat{\rho}_{\mathrm{LIN}}$s that obey the inequality in Eq. (2.1.24), as long as $f_0 \neq 0, 1$, these estimators indeed solve the extremal equation for $\widehat{\rho}_{\mathrm{ML}}$. We thus have a *family* of $\widehat{\rho}_{\mathrm{ML}}$s for a fixed set of measurement data that maximize the likelihood. Such a family possesses a nice property. If we select two ML estimators $\widehat{\rho}_{\mathrm{ML}}$ and $\widehat{\rho}'_{\mathrm{ML}}$ that are represented by the respective Bloch vectors

$$s_{\mathrm{ML}} \,\widehat{=}\, \begin{pmatrix} s_1 \\ s_2 \\ 2\,f_0 - 1 \end{pmatrix} \quad \text{and} \quad s'_{\mathrm{ML}} \,\widehat{=}\, \begin{pmatrix} s'_1 \\ s'_2 \\ 2\,f_0 - 1 \end{pmatrix}, \qquad (2.1.25)$$

and take the resulting *convex sum* $\widehat{\rho}''_{\mathrm{ML}} = \lambda\widehat{\rho}_{\mathrm{ML}} + (1 - \lambda)\widehat{\rho}'_{\mathrm{ML}}$ with $0 \leq \lambda \leq 1$, we find that

$$\widehat{\rho}''_{\mathrm{ML}} = \frac{1 + s''_{\mathrm{ML}} \cdot \boldsymbol{\sigma}}{2}, \quad s''_{\mathrm{ML}} \,\widehat{=}\, \begin{pmatrix} \lambda s_1 + (1 - \lambda)s'_1 \\ \lambda s_2 + (1 - \lambda)s'_2 \\ 2\,f_0 - 1 \end{pmatrix}, \qquad (2.1.26)$$

which means that $\widehat{\rho}_{\mathrm{ML}}''$ is yet another positive ML estimator for the same data. These ML estimators thus constitute a *convex set*. Hence, in general, an informationally incomplete set of data gives rise to a convex set of ML estimators, since there is insufficient data to completely characterize ρ_{true}.

While we are at it, we shall, very briefly, explore an interesting relationship between the convex set and the positivity constraint. Suppose, for very small number of copies N, the measurement data obtained is such that $f_0 = 0$ and $f_1 = 1$, or *vice versa*. Then, according to the positivity constraint, $s_1^2 + s_2^2 = 0$. This implies that $s_1 = s_2 = 0$. Hence, we have the simple solutions

$$\widehat{\rho}_{\mathrm{ML}} = \frac{1 \mp \sigma_z}{2} \qquad (2.1.27)$$

for the respective measurement data $\{f_0 = 0, f_1 = 1\}$ and $\{f_0 = 1, f_1 = 0\}$. The brief lesson to be learned here is that, depending on the measurement data obtained, the positivity constraint can limit the size of the convex set to a single ML estimator.

For larger dimensions $D > 2$, the boundary of the convex set and that of the full state space becomes extremely complicated and much more work is required to understand this complex relationship.

Problem 2.4 A general single-qubit unitary operator U can be written as

$$U = \mathrm{e}^{\mathrm{i}(\varphi + \vartheta \boldsymbol{n} \cdot \boldsymbol{\sigma})} = \mathrm{e}^{\mathrm{i}\varphi} \left(\cos\vartheta + \mathrm{i}\boldsymbol{n} \cdot \boldsymbol{\sigma} \sin\vartheta \right), \qquad (2.1.28)$$

where $\boldsymbol{n}$ is a normalized real vector. Show that

$$U\boldsymbol{v} \cdot \boldsymbol{\sigma} U^{\dagger} = (\mathbf{O} \cdot \boldsymbol{v}) \cdot \boldsymbol{\sigma}, \qquad (2.1.29)$$

with the dyadic

$$\mathbf{O} = \cos(2\vartheta)\left(\mathbf{1} - \boldsymbol{n}\boldsymbol{n}^{\mathsf{T}}\right) + \sin(2\vartheta)\boldsymbol{n} \times \mathbf{1} + \boldsymbol{n}\boldsymbol{n}^{\mathsf{T}} \qquad (2.1.30)$$

and confirm that $\mathbf{O}$ is indeed orthogonal. The dyadic $\boldsymbol{n} \times \mathbf{1}$ is antisymmetric,

$$(\boldsymbol{n} \times \mathbf{1})^{\mathsf{T}} = -\boldsymbol{n} \times \mathbf{1} = \mathbf{1} \times \boldsymbol{n}^{\mathsf{T}}, \qquad (2.1.31)$$

and is represented by a matrix of columns obtained from cross products between $\boldsymbol{n}$ and each column of $\mathbf{1}$.

Problem 2.5 An informationally incomplete single-qubit POM consists of the two outcomes

$$\Pi_{\pm} = \frac{1 \pm (\sigma_x + \sigma_z)/\sqrt{2}}{2}, \tag{2.1.32}$$

with the respective measurement frequencies f_+ and $f_- = 1 - f_+$. Give the inequality that is satisfied by the components of the Bloch vectors for the ML estimators. Which orthogonal transformation on the Bloch vectors will reveal the actual geometry of the ML convex set?

2.1.4 *Solution to the extremal equations*

2.1.4.1 *Operator calculus*

In general, closed-form analytical expressions for $\widehat{\rho}_{\mathrm{ML}}$ do not exist, for the positivity constraint that is imposed on all quantum states makes analytical parametrization of $\widehat{\rho}_{\mathrm{ML}}$ extremely difficult. We shall resort to solving Eq. (2.1.15) numerically. To carry out this procedure, we return to Eq. (2.1.12) and note that for a given scalar function

$$f(\mathcal{A}, \mathcal{A}^{\dagger}) = \mathrm{tr}\{F(\mathcal{A}, \mathcal{A}^{\dagger})\}, \tag{2.1.33}$$

where $F(\mathcal{A}, \mathcal{A}^{\dagger})$ is an operator function of $\mathcal{A}$ and $\mathcal{A}^{\dagger}$,

$$\delta f(\mathcal{A}, \mathcal{A}^{\dagger}) = \mathrm{tr}\left\{\delta \mathcal{A} \frac{\partial f}{\partial \mathcal{A}} + \frac{\partial f}{\partial \mathcal{A}^{\dagger}} \delta \mathcal{A}^{\dagger}\right\}. \tag{2.1.34}$$

This follows from the definition of the total variation of f that is given by

$$\delta f(\mathcal{A}, \mathcal{A}^{\dagger}) = \left\langle \delta \mathcal{A}^{\dagger}, \frac{\partial f}{\partial \mathcal{A}} \right\rangle + \left\langle \frac{\partial f}{\partial \mathcal{A}}, \delta \mathcal{A}^{\dagger} \right\rangle, \tag{2.1.35}$$

with the inner product $\langle \mathcal{A}, \mathcal{B} \rangle$ of two complex operators $\mathcal{A}$ and $\mathcal{B}$ defined as

$$\langle \mathcal{A}, \mathcal{B} \rangle = \mathrm{tr}\{\mathcal{A}^{\dagger}\mathcal{B}\}. \tag{2.1.36}$$

Note the temporary notational interchange between $\mathcal{A}$ and $\mathcal{A}^{\dagger}$ here in Sec. 2.1.4.1 for the sake of argument.

Although the operator differentiation variables $\mathcal{A}$ and $\mathcal{A}^\dagger$ are in fact related, they behave like independent variables in Eq. (2.1.34). To understand this, it is convenient to think of these variables as a result of a linear transformation on a pair of *real* and *independent* operators $\mathcal{X}$ and $\mathcal{Y}$ that form the complex operators $\mathcal{A}$ and $\mathcal{A}^\dagger$, *viz.*

$$\mathcal{A} = \frac{\mathcal{X} + i\mathcal{Y}}{\sqrt{2}},$$

$$\mathcal{A}^\dagger = \frac{\mathcal{X}^{\mathrm{T}} - i\mathcal{Y}^{\mathrm{T}}}{\sqrt{2}}. \tag{2.1.37}$$

It is, in fact, possible to cast Eq. (2.1.37) into a compact matrix equation that describes a change of operator variables from the independent pair $(\mathcal{X}, \mathcal{Y})$ to another independent pair $(\mathcal{A}, \mathcal{A}^\dagger)$.

For this, we need to express the operator transpositions as linear maps on $\mathcal{X}$ and $\mathcal{Y}$. This can be done in the language of the superoperators, with which we introduce the *transposition superoperator* T for the operators $\mathcal{X}$ and $\mathcal{Y}$ respectively. By expressing the D-dimensional real operators

$$\mathcal{X} = \sum_{j=1}^{D} \sum_{k=1}^{D} |v_j\rangle \, x_{jk} \, \langle v_k| \quad \text{and} \quad \mathcal{Y} = \sum_{j=1}^{D} \sum_{k=1}^{D} |v_j\rangle \, y_{jk} \, \langle v_k| \tag{2.1.38}$$

in terms of some complete set of orthonormal $\{|v_j\rangle\}$, the corresponding transposition superoperator is defined as

$$T = \sum_{j=1}^{D} \sum_{k=1}^{D} |v_k\rangle \, |v_j\rangle \, \langle v_j| \, \langle v_k|, \tag{2.1.39}$$

so that $T\,|\mathcal{X}\rangle\rangle = |\mathcal{X}^{\mathrm{T}}\rangle\rangle$ and $T\,|\mathcal{Y}\rangle\rangle = |\mathcal{Y}^{\mathrm{T}}\rangle\rangle$.

In this way, we arrive at the matrix equation

$$\begin{pmatrix} |\mathcal{A}\rangle\rangle \\ |\mathcal{A}^\dagger\rangle\rangle \end{pmatrix} = \frac{1}{\sqrt{2}} \begin{pmatrix} I & iI \\ T & -iT \end{pmatrix} \begin{pmatrix} |\mathcal{X}\rangle\rangle \\ |\mathcal{Y}\rangle\rangle \end{pmatrix} \tag{2.1.40}$$

that turns the independent operator variables $\mathcal{X}$ and $\mathcal{Y}$ into $\mathcal{A}$ and $\mathcal{A}^\dagger$.

Problem 2.6   Show that the matrix

$$\frac{1}{\sqrt{2}} \begin{pmatrix} I & iI \\ T & -iT \end{pmatrix} \tag{2.1.41}$$

is unitary. Find its eigenvectors and eigenvalues.

We can define the partial derivative

$$\frac{\partial}{\partial \mathcal{A}} = \frac{1}{\sqrt{2}} \left(\frac{\partial}{\partial \mathcal{X}} - \mathrm{i} \frac{\partial}{\partial \mathcal{Y}} \right) \qquad (2.1.42)$$

with respect to the operator $\mathcal{A}$ in terms of the independent operators $\mathcal{X}$ and $\mathcal{Y}$, following the analogous definition for numerical variables.

2.1.4.2 *Complex vector calculus — second digression*

Equation (2.1.34) can be treated as a generalization of the usual vector calculus, for which we shall make a brief instructive detour. Let us define $\boldsymbol{v}$ as a D-dimensional column vector. The gradient vectors with respect to $\boldsymbol{v}$ and $\boldsymbol{v}^\dagger$ are now defined in such a way that

$$\frac{\partial}{\partial \boldsymbol{v}} \boldsymbol{v}^\dagger = \boldsymbol{0}^\mathrm{T}, \qquad \frac{\partial}{\partial \boldsymbol{v}^\dagger} \boldsymbol{v} = \boldsymbol{0}. \qquad (2.1.43)$$

Before we attempt to understand these complex gradient vectors, we need to recall some basic properties of, say, the usual three-dimensional position vector $\boldsymbol{r} = (x\,y\,z)^\mathrm{T}$.

In the theory of vectors, the vector $\boldsymbol{r}$ is typically defined to be a *contravariant vector*, which basically means that it obeys the transformation rule of a *column* vector. The corresponding gradient vector $\frac{\partial}{\partial \boldsymbol{r}}$, in a similar sense, obeys the transformation rule of a *row* vector and is known as a *covariant vector*. In simple terms, both $\boldsymbol{r}$ and $\frac{\partial}{\partial \boldsymbol{r}}$ are hereby regarded as column and row vectors respectively by convention. With this understanding, we can write down the well-known identity $\boldsymbol{\nabla} \boldsymbol{r} = \mathbf{1}$ in these notations, that is

$$\frac{\partial}{\partial \boldsymbol{r}^\mathrm{T}} \boldsymbol{r}^\mathrm{T} = \mathbf{1} = \boldsymbol{r} \frac{\partial}{\partial \boldsymbol{r}}, \qquad (2.1.44)$$

with the bottom arrow for the differential operator on the rightmost side of the equation indicating the differentiation direction.

For the complex version, we can write $v = (v_x + iv_y)\dfrac{1}{\sqrt{2}}$ in terms of the independent real column vectors v_x and v_y and simply replace the transposition by the adjoint:

$$\frac{\partial}{\partial v^\dagger} v^\dagger = 1 = v \frac{\partial}{\overleftarrow{\partial v}}. \qquad (2.1.45)$$

The relations in Eq. (2.1.45) allows us to parameterize the complex gradient vector* in terms of the real column vector v_x and v_y, that is

$$(\text{row vector}) \qquad \frac{\partial}{\partial v} = \frac{1}{\sqrt{2}}\left(\frac{\partial}{\partial v_x} - i\frac{\partial}{\partial v_y}\right),$$

$$(\text{column vector}) \qquad \frac{\partial}{\partial v^\dagger} = \frac{1}{\sqrt{2}}\left(\frac{\partial}{\partial v_x^{\mathrm{T}}} + i\frac{\partial}{\partial v_y^{\mathrm{T}}}\right), \qquad (2.1.46)$$

which is consistent with the identity

$$\begin{aligned}
\frac{\partial}{\partial v^\dagger} v^\dagger &= \frac{1}{\sqrt{2}}\left(\frac{\partial}{\partial v_x^{\mathrm{T}}} + i\frac{\partial}{\partial v_y^{\mathrm{T}}}\right)\frac{v_x^{\mathrm{T}} - iv_y^{\mathrm{T}}}{\sqrt{2}}\\
&= \frac{1}{2}\left(\frac{\partial}{\partial v_x^{\mathrm{T}}}v_x^{\mathrm{T}} + \frac{\partial}{\partial v_y^{\mathrm{T}}}v_y^{\mathrm{T}}\right)\\
&= \frac{1}{2}(1+1)\\
&= 1 \qquad\qquad\qquad\qquad (2.1.47)
\end{aligned}$$

as in Eq. (2.1.45). The quantity

$$\begin{aligned}
\frac{\partial}{\partial v} v &= \frac{1}{\sqrt{2}}\left(\frac{\partial}{\partial v_x} - i\frac{\partial}{\partial v_y}\right)\frac{v_x + iv_y}{\sqrt{2}}\\
&= \frac{1}{2}\Big(\underbrace{\frac{\partial}{\partial v_x}v_x}_{=\,D} + \underbrace{\frac{\partial}{\partial v_y}v_y}_{=\,D}\Big)\\
&= D \qquad\qquad\qquad\qquad (2.1.48)
\end{aligned}$$

counts the dimension of the complex vector v as in the three-dimensional divergence $\nabla \cdot r = 3$. Naturally, we can show that Eq. (2.1.43) is obeyed.

---

*Wilhelm Wirtinger (1865–1945).

We shall only demonstrate this for the first term and this does give

$$\frac{\partial}{\partial \boldsymbol{v}} \boldsymbol{v}^\dagger = \frac{1}{\sqrt{2}} \left(\frac{\partial}{\partial \boldsymbol{v}_x} - \mathrm{i} \frac{\partial}{\partial \boldsymbol{v}_y} \right) \frac{\boldsymbol{v}_x^\mathrm{T} - \mathrm{i} \boldsymbol{v}_y^\mathrm{T}}{\sqrt{2}}$$

$$= \frac{1}{2} \left(\frac{\partial}{\partial \boldsymbol{v}_x} \boldsymbol{v}_x^\mathrm{T} - \frac{\partial}{\partial \boldsymbol{v}_y} \boldsymbol{v}_y^\mathrm{T} \right)$$

$$= \boldsymbol{0}^\mathrm{T}. \tag{2.1.49}$$

This is in accordance with the fact that

$$\frac{\partial}{\partial \boldsymbol{v}_x} \boldsymbol{v}_x^\mathrm{T} = \frac{\partial}{\partial \boldsymbol{v}_y} \boldsymbol{v}_y^\mathrm{T} = \begin{pmatrix} \boldsymbol{e}_1^\mathrm{T} & \boldsymbol{e}_2^\mathrm{T} & \cdots & \boldsymbol{e}_D^\mathrm{T} \end{pmatrix}, \tag{2.1.50}$$

which is a "row-like" *tensor* formed from the D basis unit vectors $\boldsymbol{e}_j$ or a "flattened" identity matrix.

Problem 2.7   What is the action of

$$\frac{\boldsymbol{v}^\dagger}{|\boldsymbol{v}|^2} \left(-\frac{\mathrm{d}}{\mathrm{d}\lambda} \right)^{\boldsymbol{v} \frac{\partial}{\partial \boldsymbol{v}}} \frac{1}{\lambda} \Bigg|_{\lambda=1} \tag{2.1.51}$$

on the D-dimensional complex vector $\boldsymbol{v}$?

Consider the scalar function $f(\boldsymbol{v}, \boldsymbol{v}^\dagger) = |\boldsymbol{v}|^2 = \boldsymbol{v}^\dagger \boldsymbol{v}$ of $\boldsymbol{v}$ and $\boldsymbol{v}^\dagger$. With the above introduction, we can then evaluate the gradients of the function $f(\boldsymbol{v}, \boldsymbol{v}^\dagger)$ directly:

$$\frac{\partial f}{\partial \boldsymbol{v}^\dagger} = \frac{\partial}{\partial \boldsymbol{v}^\dagger} (\boldsymbol{v}^\dagger \boldsymbol{v}) = \mathbf{1}\boldsymbol{v} = \boldsymbol{v},$$

$$\frac{\partial f}{\partial \boldsymbol{v}} = (\boldsymbol{v}^\dagger \boldsymbol{v}) \frac{\partial}{\partial \boldsymbol{v}} = \boldsymbol{v}^\dagger \mathbf{1} = \boldsymbol{v}^\dagger. \tag{2.1.52}$$

Alternatively, we can also write $f(v, v^\dagger) = \mathrm{Sp}\{vv^\dagger\}$. Then the variation

$$\delta f(v, v^\dagger) = \mathrm{Sp}\{\delta v v^\dagger + v\,\delta v^\dagger\}. \tag{2.1.53}$$

To be consistent with the results in Eq. (2.1.52), the total differential for f must be defined as

$$\delta f = \left\langle \delta v^\dagger, \frac{\partial f}{\partial v} \right\rangle + \left\langle \frac{\partial f}{\partial v}, \delta v^\dagger \right\rangle$$

$$= \mathrm{Sp}\left\{ \delta v \frac{\partial f}{\partial v} + \frac{\partial f}{\partial v^\dagger} \delta v^\dagger \right\}, \tag{2.1.54}$$

so that $\dfrac{\partial f}{\partial v} = v^\dagger$ and $\dfrac{\partial f}{\partial v^\dagger} = v$. Hence, just as differentiating a scalar with respect to a vector gives another vector, differentiating a scalar with respect to an operator must yet be another operator.

2.1.4.3 *Operator calculus revisited — a steepest-ascent algorithm*

Therefore, with the definition of the operator derivatives from Eq. (2.1.34), we have

$$\frac{\partial \log \mathcal{L}(\{n_j\}; \rho)}{\partial \mathcal{A}} = \frac{[R(\rho) - 1]\,\mathcal{A}^\dagger}{\mathrm{tr}\{\mathcal{A}^\dagger \mathcal{A}\}}. \tag{2.1.55}$$

By drawing analogy from the three-dimensional gradient vector

$$\nabla \,\hat{=}\, \begin{pmatrix} \dfrac{\partial}{\partial x} \\[2mm] \dfrac{\partial}{\partial y} \\[2mm] \dfrac{\partial}{\partial z} \end{pmatrix}, \tag{2.1.56}$$

we can define the two-component complex column of operator derivatives

$$\partial = \begin{pmatrix} \dfrac{\partial}{\partial \mathcal{A}} \\[2mm] \dfrac{\partial}{\partial \mathcal{A}^\dagger} \end{pmatrix}, \quad \partial^\dagger = \begin{pmatrix} \dfrac{\partial}{\partial \mathcal{A}^\dagger} & \dfrac{\partial}{\partial \mathcal{A}} \end{pmatrix}, \tag{2.1.57}$$

along with the column

$$Z = \begin{pmatrix} \mathcal{A} \\ \mathcal{A}^\dagger \end{pmatrix}. \tag{2.1.58}$$

The variation of $\log \mathcal{L}(\{n_j\}; \rho)$ then reads

$$\delta \log \mathcal{L}(\{n_j\}; \rho) = \mathrm{tr}\{\delta \boldsymbol{Z}^{\mathrm{T}} \boldsymbol{\partial} \log \mathcal{L}(\{n_j\}; \rho)\}. \qquad (2.1.59)$$

In order to maximize $\log \mathcal{L}(\{n_j\}; \rho)$, we note that the variation $\delta \log \mathcal{L}(\{n_j\}; \rho)$ needs to be positive as long as the maximum of $\log \mathcal{L}(\{n_j\}; \rho)$ is not reached.

Hence, we need to enforce this positivity constraint in the numerical scheme to maximize the (log-)likelihood. The simplest way to accomplish this task is to define

$$\delta \boldsymbol{Z} = \begin{pmatrix} \delta \mathcal{A} \\ \delta \mathcal{A}^\dagger \end{pmatrix} = \frac{\epsilon}{2\,\mathrm{tr}\{\mathcal{A}^\dagger \mathcal{A}\}} \begin{pmatrix} \mathcal{A}\,[R(\rho) - 1] \\ [R(\rho) - 1]\,\mathcal{A}^\dagger \end{pmatrix} \propto \boldsymbol{\partial}^\dagger \log \mathcal{L}(\{n_j\}; \rho)^{\mathrm{T}}, \qquad (2.1.60)$$

where $\epsilon > 0$ is a small parameter. Correspondingly,

$$\delta \log \mathcal{L}(\{n_j\}; \rho) = \epsilon \frac{\mathrm{tr}\{[R(\rho) - 1]\,\rho\,[R(\rho) - 1]\}}{\mathrm{tr}\{\mathcal{A}^\dagger \mathcal{A}\}} \geq 0. \qquad (2.1.61)$$

In defining $\delta \boldsymbol{Z}$ to be a column that is parallel to the gradient operator $\boldsymbol{\partial}^\dagger \log \mathcal{L}(\{n_j\}; \rho)^{\mathrm{T}}$, we have adopted the *steepest-ascent method* in the complex operator space.

Since

$$\rho' = \frac{(\mathcal{A}^\dagger + \delta \mathcal{A}^\dagger)(\mathcal{A} + \delta \mathcal{A})}{\mathrm{tr}\{(\mathcal{A}^\dagger + \delta \mathcal{A}^\dagger)(\mathcal{A} + \delta \mathcal{A})\}}, \qquad (2.1.62)$$

where

$$\delta \mathcal{A} = (\delta \mathcal{A}^\dagger)^\dagger = \frac{\epsilon}{2} \frac{\mathcal{A}\,[R(\rho) - 1]}{\mathrm{tr}\{\mathcal{A}^\dagger \mathcal{A}\}}, \qquad (2.1.63)$$

we now have an equation that can be solved iteratively to obtain the ML estimator $\widehat{\rho}_{\mathrm{ML}}$. In each iterative step, the parameter ϵ should be optimized so that the increment $\delta \log \mathcal{L}(\{n_j\}; \rho)$ is maximized. In practice, however, it is much more convenient to fix $\dfrac{\epsilon}{\mathrm{tr}\{\mathcal{A}^\dagger \mathcal{A}\}} = \epsilon'$ to take a certain value. The final iterative equation for ML is then given by

$$\rho_{k+1} = \frac{\left\{1 + \frac{\epsilon'}{2}\,[R(\rho_k) - 1]\right\} \rho_k \left\{1 + \frac{\epsilon'}{2}\,[R(\rho_k) - 1]\right\}}{\mathrm{tr}\left\{\left\{1 + \frac{\epsilon'}{2}\,[R(\rho_k) - 1]\right\} \rho_k \left\{1 + \frac{\epsilon'}{2}\,[R(\rho_k) - 1]\right\}\right\}}, \qquad (2.1.64)$$

where ϵ' is an appropriately chosen parameter that defines the step size. This defines the steepest-ascent ML algorithm.

If one starts from a rank-one operator, $\rho_{k=0} = |\ \rangle\langle\ |$, then the subsequent ρ_ks will always be rank-one according to the structure of Eq. (2.1.64). In this way, the likelihood is maximized over all pure states and this has applications in some optimization problems.

In general, to obtain the correct ML estimator, one must start with a full-rank operator, usually the *maximally-mixed state* $\rho_{k=0} = \dfrac{1}{D}$, and the operator sequence of ρ_ks will converge to $\widehat{\rho}_{\mathrm{ML}}$ as long as ϵ' is small enough. Numerically, it is not possible to obtain the exact estimator. In view of this, the estimator $\widehat{\rho}_{\mathrm{ML}} = \rho_\kappa$ may be defined such that the operator two-norm $\|R(\rho_\kappa)\rho_\kappa - \rho_\kappa\|_2$ is smaller than some pre-chosen value that quantifies the precision of the estimator.

2.2   Generalized ML scheme — imperfect detections

2.2.1   *Imperfect measurements*

So far, we have looked at informationally complete POMs with outcomes having the property $\sum_j \Pi_j = 1$ — the POM outcomes represent perfect detections.

For an experiment, the POM should describe a set of outcomes that are typically *imperfect*. One type of imperfection is that each of the measurement outcomes detects a reduced number of copies of quantum systems as compared to the set of outcomes in the ideal situation. In this case, if the ideal set of outcomes are represented by $\{\Pi_j\}$, the actual set of outcomes may be defined as $\{\Pi'_j \equiv \eta_j \Pi_j\}$, where η_j is the *detection efficiency* of the jth outcome that is always nonnegative and less than unity: $0 \leq \eta_j \leq 1$.

So, while the ideal outcomes sum to the identity, the actual ones do not;

$$G \equiv \sum_{j=1}^{M} \Pi'_j < 1. \qquad (2.2.1)$$

A slightly more general case for imperfections of this type gives outcomes of the form

$$\Pi'_j = \sum_{k=1}^{M} \eta_{jk} \Pi_k, \qquad (2.2.2)$$

where $\sum_j \eta_{jk} < 1$, such that Eq. (2.2.1) is again obeyed. There exist, also, other types of imperfections, such as systematic errors (say a systematic deviation in the readings provided by a certain device), *etc.* In principle, there are some imperfections that, after an extensive calibration procedure, can always be incorporated so that one can identify a new set of outcomes $\{\Pi'_j\}$ that describes the actual measurement apparatus, with $G < 1$.

If $G < 1$, then $\sum_j p_j = \sum_j \mathrm{tr}\{\rho\,\Pi'_j\} < 1$ and consequently, whenever we use a set of imperfect detection outcomes, there always exists a loss of measurement probabilities. In other words, not all copies are detected. We now define N to be the number of *detected* copies and $N_t \geq N$ to be the *true* number of copies. An experienced reader would have noticed that there is a kind of imperfection, known as the *dark-count* imperfection, that can add fictitious counts to the actual number of copies N_t generated from the source. Here, we have implicitly assumed that the effect of dark counts has been heavily suppressed.

2.2.2 *ML estimation of the unknown number of copies*

Therefore, the actual situation is like this: Unlike the ideal case $G = 1$ where the particular detection sequence $\{\Pi_1, \Pi_2, \Pi_1, \Pi_5, \Pi_6, \Pi_3, \Pi_1, \ldots\}$, say, that is observed in an experiment accounts for all the copies, one will not know exactly which outcomes have missed detection in the case where $G < 1$. The likelihood for the latter situation is therefore given by

$$\mathcal{L}'_{N_t}(\{n_j\}; \rho) = \frac{N_t!}{N!\,(N_t - N)!} \left(\prod_{j=1}^{M} p_j^{n_j} \right) \left(1 - \sum_{j'=1}^{M} p_{j'} \right)^{N_t - N}, \quad (2.2.3)$$

where, still, $N = \sum_j n_j$. The combinatorial weight signifies the lack of information about the missing copies. Note that $\mathcal{L}'_{N_t = N}(\{n_j\}; \rho) = \mathcal{L}(\{n_j\}; \rho)$ when $N_t = N$, so that the new likelihood now describes the ideal situation where *all* copies are accounted for in the experiment. We define the quantity

$$\eta \equiv \sum_{j=1}^{M} p_j \leq 1 \qquad (2.2.4)$$

to represent the *overall efficiency* for the experiment. To find $\widehat{\rho}_{\mathrm{ML}}$, we vary the corresponding log-likelihood

$$\log \mathcal{L}'_{N_t}(\{n_j\}; \rho) = \sum_{j=1}^{M} n_j \log p_j + (N_t - N) \log(1 - \eta)$$

$$+ \log\left(\frac{N_t!}{N!\,(N_t - N)!}\right). \qquad (2.2.5)$$

There are now two variables appearing in the log-likelihood, namely $\widehat{\rho}_{\mathrm{ML}}$ and N_t. Since we have no information pertaining to N_t, a strategy for moving on is to treat N_t as an auxiliary parameter for $\widehat{\rho}_{\mathrm{ML}}$. Following the spirit of the ML method, it is reasonable to look for the estimate for N_t that maximizes $\log \mathcal{L}'_{N_t}(\{n_j\}; \rho)$. For that, let us first consider an approximation of Eq. (2.2.5) as it shall lead to a simple and intuitive solution for the ML estimate of the unknown true number of copies N_t. For this approximation, the *Stirling* formula*

$$\log y! \approx y \log y - y \qquad (2.2.6)$$

helps in the simplification of the term

$$\log\left(\frac{N_t!}{N!\,(N_t - N)!}\right) \approx \log\left(\frac{N_t^{N_t}}{N^N\,(N_t - N)^{N_t - N}}\right). \qquad (2.2.7)$$

Carrying out the variation on the approximated version of Eq. (2.2.5),

$$\delta \log \mathcal{L}'_{N_t}(\{n_j\}; \rho) = \sum_{j=1}^{M}\left(N \frac{f_j}{p_j} - \frac{N_t - N}{1 - \eta}\right)\delta p_j$$

$$+ \delta N_t \log\left(\frac{(1 - \eta)N_t}{N_t - N}\right), \qquad (2.2.8)$$

where the variables p_j and N_t are treated independently. Since at the maximum of $\log \mathcal{L}'_{N_t}(\{n_j\}; \rho)$, the variation must vanish, we obtain a very simple result,

$$\widehat{N}_t = \frac{N}{\eta}, \qquad (2.2.9)$$

---

*James Stirling (1692–1770).

from the second variation term in Eq. (2.2.8). Thus, the ML estimate for N_t is just the measured number of copies N divided by the overall efficiency η. This is not a surprising result since all we really know are the occurrences n_js and the most straightforward estimate for N_t that we can figure out given such limited information is given in Eq. (2.2.9).

Now that we have a simple and physically intuitive answer, let us compare this result to the more accurate estimate by varying the right-hand side of Eq. (2.2.5). For that we recall that $N_t!$ can be defined as the *Euler* factorial function*,

$$N_t! = \int_0^\infty ds\, e^{-s} s^{N_t}, \qquad (2.2.10)$$

implying that

$$\delta \log(N_t!) = \frac{\displaystyle\int_0^\infty ds\, e^{-s} s^{N_t} \log s}{N_t!} \, \delta N_t$$

$$= F(N_t)\, \delta N_t, \qquad (2.2.11)$$

where we define the *Digamma function*

$$F(x) = \frac{1}{x!} \int_0^\infty ds\, e^{-s} s^x \log s, \qquad (2.2.12)$$

where the archaic Greek letter "F" stands for "Digamma". Hence, maximizing the original log-likelihood in Eq. (2.2.8) directly gives

$$F(N_t - N) - F(N_t) = \log(1 - \eta) \qquad (2.2.13)$$

as the extremal equation for the ML estimate $N_t = \widehat{N}_t$. The overall function $F(N_t - N) - F(N_t)$ on the left-hand side of Eq. (2.2.13) can be further simplified by noting that

$$\frac{d}{ds}\left(e^{-s} \log s\right) = -e^{-s} \log s + s^{-1} e^{-s}, \qquad (2.2.14)$$

---

*Leonhard Euler (1707–1783).

from which

$$\int\limits_0^\infty ds\, e^{-s}\, s^{x-1}\, \log s = \frac{s^x\, e^{-s}\, \log s}{x}\bigg|_0^\infty \underbrace{}_{=\,0} - \int\limits_0^\infty ds\, \frac{s^x}{x}\left(-e^{-s}\, \log s + s^{-1}\, e^{-s}\right)$$

$$= \frac{1}{x}\int\limits_0^\infty ds\, e^{-s}\, s^x\, \log s - \frac{1}{x}\underbrace{\int\limits_0^\infty ds\, e^{-s}\, s^{x-1}}_{=\,(x-1)!}, \qquad (2.2.15)$$

or

$$\int\limits_0^\infty ds\, e^{-s}\, s^x\, \log s = x\int\limits_0^\infty ds\, e^{-s}\, s^{x-1}\, \log s + (x-1)!. \qquad (2.2.16)$$

This gives the recurrence relation

$$F(x) = F(x-1) + \frac{1}{x}, \qquad (2.2.17)$$

and we may proceed to reduce the left-hand side of Eq. (2.2.13) as follows:

$$F\left(\widehat{N}_t - N\right) - F\left(\widehat{N}_t\right) = -\left[F\left(\widehat{N}_t\right) - F\left(\widehat{N}_t - N\right)\right]$$

$$= -\frac{1}{\widehat{N}_t} - \left[F\left(\widehat{N}_t - 1\right) - F\left(\widehat{N}_t - N\right)\right]$$

$$= -\frac{1}{\widehat{N}_t} - \frac{1}{\widehat{N}_t - 1} - \left[F\left(\widehat{N}_t - 2\right) - F\left(\widehat{N}_t - N\right)\right]$$

$$\vdots$$

$$= -\sum_{k=0}^{N-1}\frac{1}{\widehat{N}_t - k}. \qquad (2.2.18)$$

Finally, the extremal equation is simplified to

$$\sum_{k=0}^{N-1}\frac{1}{\widehat{N}_t - k} = -\log(1-\eta). \qquad (2.2.19)$$

To solve this general equation for $\widehat{N}_t$, we can consider the function $\Phi\left(\widehat{N}_t\right)$ defined as

$$\Phi\left(\widehat{N}_t\right) = \left[\sum_{k=0}^{N-1} \frac{1}{\widehat{N}_t - k} + \log\left(1 - \eta\right)\right]^2 \tag{2.2.20}$$

and look for its minimum using either a steepest-ascent algorithm or other efficient optimization routines. Numerical results show that the approximate estimate $\widehat{N}_t = \dfrac{N}{\eta}$ is extremely accurate even for small N (refer to Fig. 2.3). We shall therefore take this intuitive answer as the ML estimate for N_t.

We acknowledge that $\widehat{N}_t$ is in general not an integer, a property that is certainly important if one is interested in estimating the true number of copies. For our purpose, $\widehat{N}_t$ is auxiliary in estimating ρ_{true}.

Problem 2.8 If one enforces $\widehat{N}_t$ to be an integer, derive the corresponding extremal equation for $\widehat{N}_t$.

Problem 2.9 Derive the identity

$$F(n) = -\gamma + H_n \tag{2.2.21}$$

for integer n, where γ is the Euler–Mascheroni* constant

$$\gamma = \left(H_m - \log m\right)\big|_{m \to \infty} \approx 0.577216 \tag{2.2.22}$$

and H_n is the order-n harmonic number

$$H_n = \sum_{k=1}^{n} \frac{1}{k}. \tag{2.2.23}$$

---

*Leonhard Euler (1707–1783) and Lorenzo Mascheroni (1750–1800).

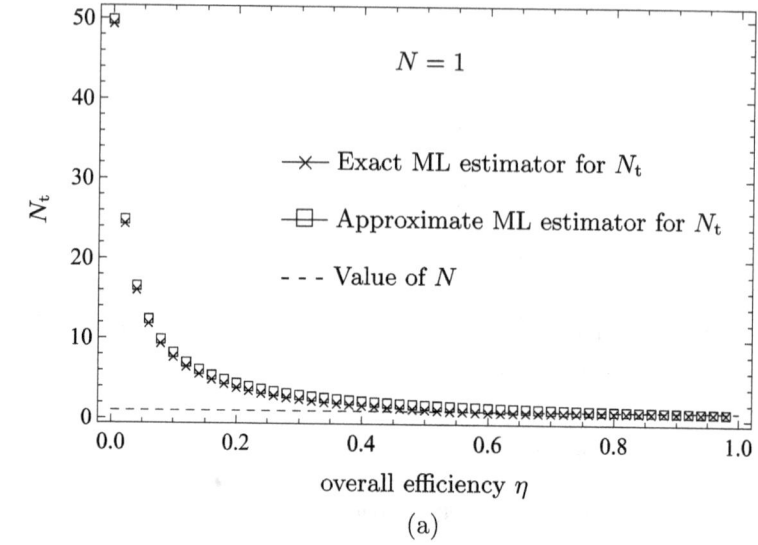

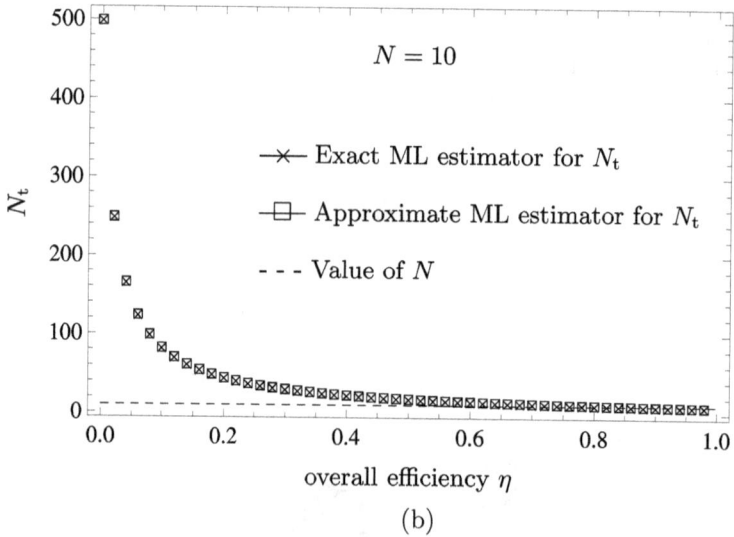

Fig. 2.3   Plots of different estimates for N_t against η. (a) $N = 1$; (b) $N = 10$.

Then, the log-likelihood is simplified to

$$\log \mathcal{L}'(\{n_j\}; \rho) \equiv \log \mathcal{L}'_{N_t=N/\eta}(\{n_j\}; \rho)$$

$$= \sum_{j=1}^{M} n_j \log p_j + N \frac{1-\eta}{\eta} \log(1-\eta)$$

$$+ \frac{N}{\eta} \log \left(\frac{N}{\eta}\right) - N \log N - N \frac{1-\eta}{\eta} \log \left(N \frac{1-\eta}{\eta}\right)$$

$$= \sum_{j=1}^{M} n_j \log p_j - N \log \eta$$

$$= \sum_{j=1}^{M} n_j \log \left(\frac{p_j}{\eta}\right). \qquad (2.2.24)$$

The *generalized log-likelihood* $\log \mathcal{L}'(\{n_j\}; \rho)$ for imperfect measurements is, therefore, a log-likelihood function of the *relative probabilities* $\frac{p_j}{\eta}$. Based on the principle of maximum likelihood, the subnormalized probabilities are to be replaced by their respective relative probabilities to account for missing copies.

2.2.3 *The extremal probabilities*

The condition that maximizes $\log \mathcal{L}'(\{n_j\}; \rho)$ is readily found to be

$$\frac{\widehat{p}_j^{\text{ML}}}{f_j} = \sum_{k=1}^{M} \widehat{p}_k^{\text{ML}} \qquad (2.2.25)$$

for the probabilities $\widehat{p}_j^{\text{ML}}$. These equations may easily be translated into the matrix equation

$$\begin{pmatrix} 1-f_1 & -f_1 & \cdots & -f_1 \\ -f_2 & 1-f_2 & \cdots & -f_2 \\ \vdots & \vdots & \ddots & \vdots \\ -f_M & -f_M & \cdots & 1-f_M \end{pmatrix} \begin{pmatrix} \widehat{p}_1^{\text{ML}} \\ \widehat{p}_2^{\text{ML}} \\ \vdots \\ \widehat{p}_M^{\text{ML}} \end{pmatrix} = 0, \quad \sum_{j=1}^{M} f_j = 1. \qquad (2.2.26)$$

Owing to the constraint on the sum of all the measurement frequencies, any row of the matrix in Eq. (2.2.26) is dependent on the rest of the $M-1$ rows. As a result, the matrix is rank-deficient. This means that the solution for the extremal probabilities, as far as Eq. (2.2.26) is concerned, is not unique as the matrix of measurement frequencies is not invertible, which is reassuring since the probabilities cannot be zero all at once.

As a matter of fact, given two sets of probabilities $\left\{\widehat{p}_j^{\,\mathrm{ML}(1)}\right\}$ and $\left\{\widehat{p}_j^{\,\mathrm{ML}(2)}\right\}$ that satisfy Eq. (2.2.25), these sets must be related by

$$\widehat{p}_j^{\,\mathrm{ML}(1)} = \underbrace{\frac{\sum_{k=1}^M \widehat{p}_k^{\,\mathrm{ML}(1)}}{\sum_{k'=1}^M \widehat{p}_{k'}^{\,\mathrm{ML}(2)}}}_{=\ \text{positive constant}} \widehat{p}_j^{\,\mathrm{ML}(2)}, \qquad (2.2.27)$$

that is, they must be a constant multiple of each other.

We thus have an infinite number of solutions for $\widehat{p}_j^{\,\mathrm{ML}}$ *if no other restrictions are imposed on these probabilities*, that is. However, for quantum systems, we require the probabilities to originate from quantum states described by statistical operators. This means that physical probabilities are functions of the *state parameters* that specify a quantum state.

Since the measurement is informationally complete, there are as many independent state parameters as there are independent linear equations in Eq. (2.2.26). Hence, the ML probabilities $\widehat{p}_j^{\,\mathrm{ML}}$ are *unique*, as expected for informationally complete POMs, and, as a result, only one ML estimator is obtained for any given measurement data that are collected with such a POM. An example would be the tetrahedron measurement, with each outcome having an detection efficiency η_j. Since there are $4-1=3$ independent extremal equations, this is exactly the number of state parameters needed to uniquely specify an ML estimator.

2.2.4 *Statistical consistency of the generalized ML estimator*

It is important to check that the ML estimator obtained using this generalized ML method is a consistent one. For this, we first look at the measurement frequencies f_j. Since the detections are imperfect, the parameter N_t describing the actual copy count is unknown. In typical experiments, the source emits copies of quantum systems in a statistical fashion, so that N_t is to be treated as a random variable that may take any integer value greater than or equal to N according to an *a priori* probability distribution $p(N_t)$ describing the source. As all data are distributed according to multinomial statistics for any given N_t, for a given number of measured copies N, the averages $\overline{f_j}$ are given by

$$\overline{f_j} = \overline{\left(\frac{n_j}{N}\right)} = \frac{N}{N} \sum_{N_t=N}^{\infty} p(N_t) \sum_{\left\{\begin{smallmatrix} n_1+\cdots+n_M=N, \\ n_0+N=N_t \end{smallmatrix}\right\}} n_j \frac{N_t!}{n_0!\,n_1!\cdots n_M!} p_0^{n_0} p_1^{n_1} \cdots p_M^{n_M},$$

(2.2.28)

where p_j is the true probability of occurrence for the outcome Π_j, n_0 is the unknown number of missing copies for the unmeasured outcome Π_0 with probability p_0.* The normalization constant can be evaluated straightforwardly by noting that n_0 takes only one value for a fixed N and any given N_t, that is

$$\frac{1}{N} = \sum_{N_t=N}^{\infty} p(N_t) \sum_{\left\{\begin{smallmatrix} n_1+\cdots+n_M=N, \\ n_0+N=N_t \end{smallmatrix}\right\}} \frac{N_t!}{n_0!\,n_1!\cdots n_M!} p_0^{n_0} p_1^{n_1} \cdots p_M^{n_M}$$

$$= \sum_{N_t=N}^{\infty} p(N_t) \binom{N_t}{N} p_0^{N_t-N} \sum_{\{n_1+\cdots+n_M=N\}} \frac{N!}{n_1!\cdots n_M!} p_1^{n_1} \cdots p_M^{n_M}$$

$$= (1-p_0)^N \sum_{N_t=N}^{\infty} p(N_t) \binom{N_t}{N} p_0^{N_t-N}.$$

(2.2.29)

Similarly, the multinomial sum in Eq. (2.2.28) amounts to

$$\sum_{\left\{\begin{smallmatrix} n_1+\cdots+n_M=N, \\ n_0+N=N_t \end{smallmatrix}\right\}} n_j \frac{N_t!}{n_0!\,n_1!\cdots n_M!} p_0^{n_0} p_1^{n_1} \cdots p_M^{n_M}$$

$$= \binom{N_t}{N} p_0^{N_t-N} \sum_{\{n_1+\cdots+n_M=N\}} n_j \frac{N!}{n_1!\cdots n_M!} p_1^{n_1} \cdots p_M^{n_M}$$

---

*Note that $p_0 + p_1 + \cdots + p_M = 1$.

$$= \binom{N_t}{N} p_0^{N_t-N} p_j \frac{\partial}{\partial p_j} \left(\underbrace{\sum_{\{n_1+\cdots+n_M=N\}} \frac{N!}{n_1!\cdots n_M!} p_1^{n_1}\cdots p_M^{n_M}}_{=(p_1+\cdots+p_M)^N} \right)$$

$$= N \binom{N_t}{N} p_0^{N_t-N} p_j (1-p_0)^{N-1}. \tag{2.2.30}$$

This means that the f_js are unbiased estimates for the true relative probabilities of the outcomes,

$$\overline{f_j} = \frac{p_j}{1-p_0}, \tag{2.2.31}$$

for any prior distribution $p(N_t)$.

Hence, just like the measurement frequencies for a perfect POM, the frequencies for an imperfect POM approach their average values — the true relative probabilities — as N tends to infinity according to the law of large numbers. It follows that $\widehat{\rho}_{\mathrm{ML}}$ also approaches the true state ρ_{true} in the large N limit.

2.2.5 *A steepest-ascent algorithm*

Therefore, the generalized log-likelihood in Eq. (2.2.24) has a unique and consistent maximum for informationally complete POMs, just like the log-likelihood in Eq. (2.1.5) for perfect detections. To locate this maximum, we can again derive an iterative equation using the steepest-ascent method. Substituting the value of N_t in Eq. (2.2.9) into the first variation term of Eq. (2.2.8), the variation of $\log \mathcal{L}'(\{n_j\}; \rho)$ in terms of ρ is

$$\delta \log \mathcal{L}'(\{n_j\}; \rho) = N \mathrm{tr} \left\{ \left[R'(\rho) - \frac{1}{\eta}G \right] \delta\rho \right\}, \tag{2.2.32}$$

where we note that the only difference between $\delta \log \mathcal{L}'(\{n_j\}; \rho)$ and $\delta \log \mathcal{L}(\{n_j\}; \rho)$ is the $\dfrac{G}{\eta}$ term that accounts for missing copies, which equals the identity for perfect detections. Here, $R'(\rho)$ takes the same expression as in Eq. (2.1.7), with the Π_js replaced by Π'_js.

By going through the same calculations for an ideal POM, we arrive at the extremal equation

$$\left[R'(\widehat{\rho}_{\mathrm{ML}}) - \frac{G}{\eta_{\mathrm{ML}}} \right] \widehat{\rho}_{\mathrm{ML}} = \widehat{\rho}_{\mathrm{ML}} \left[R'(\widehat{\rho}_{\mathrm{ML}}) - \frac{G}{\eta_{\mathrm{ML}}} \right] = 0, \quad \eta_{\mathrm{ML}} = \sum_{k=1}^{M} \widehat{p}_k^{\mathrm{ML}}. \tag{2.2.33}$$

Problem 2.10   Show these calculations explicitly.

The new iterative equation is then given by

$$\rho_{k+1} = \frac{\left\{1 + \frac{\epsilon'}{2}\left[R'(\rho_k) - \frac{1}{\eta_k}G\right]\right\}\rho_k\left\{1 + \frac{\epsilon'}{2}\left[R'(\rho_k) - \frac{1}{\eta_k}G\right]\right\}}{\mathrm{tr}\left\{\left\{1 + \frac{\epsilon'}{2}\left[R'(\rho_k) - \frac{1}{\eta_k}G\right]\right\}\rho_k\left\{1 + \frac{\epsilon'}{2}\left[R'(\rho_k) - \frac{1}{\eta_k}G\right]\right\}\right\}},$$

$$(2.2.34)$$

which is obtained by simply replacing $R(\rho_k) - 1$ in Eq. (2.1.64) by $R'(\rho_k) - \frac{G}{\eta_k}$.

Problem 2.11   For an imperfect POM consisting of outcomes $\Pi'_j = \eta_j \Pi_j$, where all η_js are equal and $\sum_j \Pi_j = 1$, compare the corresponding ML algorithm with that for the POM that is made up of the outcomes Π_j. What can we learn from this?

2.3   Another way of accounting for imperfections

The generalized ML method is a systematic way of obtaining a unique and consistent ML estimator by accounting for imperfect detections. This method does not rely on any assumption about the physical source at hand. The auxiliary parameter N_t, also known as the *nuisance parameter* in statistics, is chosen to be the ML estimate irrespective of the type of source used in the experiment. This approach is justified by a consistent application of the maximum-likelihood principle throughout the estimation procedure, which is reasonable if one has no additional knowledge about the source.

If an observer knows some properties about the source to a fair amount of accuracy, he or she can incorporate these properties into state estimation. Instead of singling out one particular value of N_t, the observer can choose to average the likelihood over this nuisance parameter using this prior knowledge. In the form of a prior distribution $p(n)$ of the nuisance parameter $N_t = n$, which was introduced in Sec. 2.2, the *marginal likelihood* is defined as

$$\overline{\mathcal{L}}\left(\{n_j\}; \rho\right) = \sum_{n=N}^{\infty} p(n)\, \mathcal{L}'_n\left(\{n_j\}; \rho\right). \qquad (2.3.1)$$

This way of accounting for imperfections goes with the spirit of Bayesian estimation, where all plausible values of the nuisance parameter are taken into account with the appropriate weight function that is determined from the prior distribution.

For instance, if one has reasons to believe that the source produces copies distributed according to Poissonian* statistics, so that the prior distribution takes the specific form

$$p(n) = e^{-\nu} \frac{\nu^n}{n!}, \tag{2.3.2}$$

then one can treat the nuisance parameter n as a Poissonian random variable of a properly calibrated mean number ν. The Poissonian distribution for n can now act as a prior distribution that represents the observer's prior knowledge about the source. The corresponding log-marginal likelihood is computed to be

$$
\begin{aligned}
\overline{\mathcal{L}}\left(\{n_j\}; \rho\right) &= e^{-\nu} \sum_{n=N}^{\infty} \frac{\nu^n}{n!} \mathcal{L}'_n\left(\{n_j\}; \rho\right) \\
&= e^{-\nu} \sum_{n=N}^{\infty} \frac{\nu^n}{n!} \frac{n!}{(n-N)!\, N!} \left(\prod_{j=1}^{M} p_j{}^{n_j}\right) (1-\eta)^{n-N} \\
&= \frac{e^{-\nu}}{N!} \left(\prod_{j=1}^{M} p_j{}^{n_j}\right) \sum_{n=N}^{\infty} \frac{[\nu(1-\eta)]^{n-N}}{(n-N)!} \\
&= \frac{e^{-\nu\eta}}{N!} \left(\prod_{j=1}^{M} p_j{}^{n_j}\right).
\end{aligned}
\tag{2.3.3}
$$

For the reasoning given in the previous section, one can now maximize this log-marginal likelihood to obtain a unique estimator. After varying the log-marginal likelihood,

$$\delta \log \overline{\mathcal{L}}\left(\{n_j\}; \rho\right) = -\nu \sum_{j=1}^{M} \delta p_j + \sum_{j=1}^{M} \frac{n_j}{p_j} \delta p_j, \tag{2.3.4}$$

---

*Siméon Denis Poisson (1781–1840).

the extremal equations for p_j read

$$\widehat{p}_j^{\mathrm{ML}} = \frac{n_j}{\nu}. \tag{2.3.5}$$

That the measurement frequencies $\frac{n_j}{N}$ converge to the true relative probabilities for large N is a result of the analysis in the previous subsection.

Of course, for finite N, the extremal equations in Eq. (2.3.5) may not be satisfied because the probabilities must originate from a quantum state. When this is the case, one needs to maximize the marginal likelihood over the legitimate state space. Using the parametrization in Eq. (2.1.8), the variation in Eq. (2.3.4) becomes

$$\delta \log \overline{\mathcal{L}}\left(\{n_j\}; \rho\right)$$

$$= \sum_{j=1}^{M} \left(-\nu + \frac{n_j}{p_j}\right) \mathrm{tr}\{\delta\rho\, \Pi'_j\}$$

$$= \mathrm{tr}\left\{\overline{R}(\rho,\nu)\frac{\delta A^\dagger A + A^\dagger \delta A}{\mathrm{tr}\{A^\dagger A\}} - \mathrm{tr}\{\overline{R}(\rho,\nu)\rho\}\frac{\delta A^\dagger A + A^\dagger \delta A}{\mathrm{tr}\{A^\dagger A\}}\right\},$$
$$\tag{2.3.6}$$

where

$$\overline{R}(\rho,\nu) = \sum_{j=1}^{M}\left(-\nu + \frac{n_j}{p_j}\right)\Pi'_j = -\nu\, G + N R'(\rho) \tag{2.3.7}$$

and

$$\mathrm{tr}\{\overline{R}(\rho,\nu)\rho\} = -\nu\eta + N. \tag{2.3.8}$$

The extremal equation for $\widehat{\rho}_{\mathrm{ML}}$ then reads

$$\overline{R}(\widehat{\rho}_{\mathrm{ML}},\nu)\widehat{\rho}_{\mathrm{ML}} = (N - \nu\eta_{\mathrm{ML}})\,\widehat{\rho}_{\mathrm{ML}} = \widehat{\rho}_{\mathrm{ML}}\overline{R}(\widehat{\rho}_{\mathrm{ML}},\nu). \tag{2.3.9}$$

The iterative algorithm is then defined by

$$\rho_{k+1}$$

$$= \frac{\left\{1 + \frac{\epsilon'}{2}\left[\overline{R}(\rho_k,\nu) + \nu\eta_k - N\right]\right\}\rho_k\left\{1 + \frac{\epsilon'}{2}\left[\overline{R}(\rho_k,\nu) + \nu\eta_k - N\right]\right\}}{\mathrm{tr}\left\{\left\{1 + \frac{\epsilon'}{2}\left[\overline{R}(\rho_k,\nu) + \nu\eta_k - N\right]\right\}\rho_k\left\{1 + \frac{\epsilon'}{2}\left[\overline{R}(\rho_k,\nu) + \nu\eta_k - N\right]\right\}\right\}}, \tag{2.3.10}$$

where the parameter ϵ' should now be scaled with either N or ν for better numerical stability.

How statistically biased this estimator is to ρ_{true} would depend on how accurate the value of ν is. If ν is not well-calibrated, then it too has to be estimated, and another prior distribution for this parameter may be introduced, and so forth.

2.4 Bayesian-mean estimation — some brief statements

In a typical experiment in tomography, it is desirable to measure a large number of copies N, so that any appropriate estimator obtained from informationally complete measurement data is as close to the true state as possible in the asymptotic limit. There are situations where the value of N is far from this limit.

For example, to perform tomography on quantum systems of large Hilbert-space dimensions, typically correlated many-body systems, the observer is required to measure all D^2 POM outcomes. Each outcome has to be sampled with a sufficiently large number of copies. Therefore in general, to obtain a good tomographic accuracy, N would have to increase at least polynomially with D. In addition, the generation of these correlated systems, out of a multi-photon source say, generally becomes rarer as D increases, and so the time needed to measure a fixed number of copies of these systems increases. Thus, probing highly complex systems would involve a relatively fewer number of copies.

In the regime of small N, the likelihood has a broad peak, and neighboring quantum states around the maximum give approximately the same likelihood as the maximal value. As such, it is statistically reasonable to consider all these neighboring states also as plausible states of the physical source. The idea of marginalization can thus be employed when the observer incorporates his or her prior distribution $p(\rho)$ about the true state of the source into state estimation.

The resulting estimator constructed from this procedure is the BM estimator introduced in Chapter 1, that is

$$\widehat{\rho}_{\text{B}} = \frac{\displaystyle\int (\mathrm{d}\rho)\,\mathcal{L}\left(\{n_j\};\rho\right)\rho}{\displaystyle\int (\mathrm{d}\rho)\,\mathcal{L}\left(\{n_j\};\rho\right)}. \tag{2.4.1}$$

In this perspective, let us discuss some aspects of this estimation scheme which were briefly touched on in the previous chapter.

As discussed before, the prior $(\mathrm{d}\rho)$ depends not only on the prior distribution, but also on the choice of the volume element $(\mathrm{d}\tau_D)$. If one argues that apart from $p(\rho)$, no other factors should distinguish one state from the rest, then one may insist on using a flat distribution for $(\mathrm{d}\tau_D)$. Or, if one enforces invariance under rotations, then a rotationally-invariant measure similar to that for single-qubits in Eq. (1.2.19) will fit the requirement.

There is another kind of prior that is invariant in form under a change of variables for the integration parameters. For a given function $\mathcal{G}$ that is characterized by $\boldsymbol{\theta}$, a column of parameters θ_j that parameterize the state space in the integral, this *Jeffreys* prior* is defined as

$$(\mathrm{d}\rho) \equiv \left(\prod_j \mathrm{d}\theta_j \right) \sqrt{\mathrm{Det}\left\{ \frac{\partial}{\partial\boldsymbol{\theta}} \frac{\partial}{\partial\boldsymbol{\theta}} \mathcal{G} \right\}}, \qquad (2.4.2)$$

where Det denotes the dyadic determinant in contrast with the operator determinant det.

To see that the Jeffreys prior remains form-invariant under a parameter change $\boldsymbol{\theta} \to \boldsymbol{\theta}'$, we remember that this change

$$\left(\prod_j \mathrm{d}\theta_j \right) = \left(\prod_j \mathrm{d}\theta'_j \right) \left| \mathrm{Det}\left\{ \frac{\partial\boldsymbol{\theta}}{\partial\boldsymbol{\theta}'} \right\} \right| \qquad (2.4.3)$$

is accompanied by the Jacobian[†] $\left| \mathrm{Det}\left\{ \dfrac{\partial\boldsymbol{\theta}}{\partial\boldsymbol{\theta}'} \right\} \right|$ for this change. It then follows that

$$(\mathrm{d}\rho) = \left(\prod_j \mathrm{d}\theta'_j \right) \left| \mathrm{Det}\left\{ \frac{\partial\boldsymbol{\theta}}{\partial\boldsymbol{\theta}'} \right\} \right| \sqrt{\mathrm{Det}\left\{ \frac{\partial}{\partial\boldsymbol{\theta}} \frac{\partial}{\partial\boldsymbol{\theta}} \mathcal{G} \right\}}$$

$$= \left(\prod_j \mathrm{d}\theta'_j \right) \sqrt{\mathrm{Det}\left\{ \frac{\partial\boldsymbol{\theta}}{\partial\boldsymbol{\theta}'} \right\} \mathrm{Det}\left\{ \frac{\partial}{\partial\boldsymbol{\theta}} \frac{\partial}{\partial\boldsymbol{\theta}} \mathcal{G} \right\} \mathrm{Det}\left\{ \frac{\partial\boldsymbol{\theta}}{\partial\boldsymbol{\theta}'} \right\}}$$

---

[*]*Sir* Harold Jeffreys (1891–1989).
[†]Carl Gustav Jacob Jacobi (1804–1851).

$$= \left(\prod_j d\theta'_j \right) \sqrt{\mathrm{Det} \left\{ \frac{\partial \boldsymbol{\theta}}{\partial \boldsymbol{\theta}'} \cdot \frac{\partial}{\partial \boldsymbol{\theta}} \frac{\partial}{\partial \boldsymbol{\theta}} \mathcal{G} \cdot \frac{\partial \boldsymbol{\theta}}{\partial \boldsymbol{\theta}'} \right\}}$$

$$= \left(\prod_j d\theta'_j \right) \sqrt{\mathrm{Det} \left\{ \frac{\partial}{\partial \boldsymbol{\theta}'} \frac{\partial}{\partial \boldsymbol{\theta}'} \mathcal{G} \right\}} = (d\rho') . \tag{2.4.4}$$

Although a common function for $\mathcal{G}$ is the negative of the log-likelihood, with which the Jeffreys prior is simply the square-root of the determinant of the Fisher information dyadic, other functions can also be exploited.

The other aspect that is rather difficult to handle, as pointed out in Sec. 1.2.1, is the operator integral. This integral is to be performed over the admissible state space, which means that the parameters can vary only up to the positivity constraint for ρ. As the boundary of the state space is highly complicated, it is impossible to do the integrations by hand. One is forced to do them numerically.

A simple-minded way of doing the integral is to generate very many random positive operators of unit trace — a Monte Carlo method —, compute the integrands for each of these randomly generated operators, and approximate the integrals by average sums using the relations

$$\int (d\rho) \mathcal{L}(\{n_j\}; \rho) \rho \approx \frac{1}{L} \sum_{l=1}^{L} \tau(\rho_l) p(\rho_l) \mathcal{L}(\{n_j\}; \rho_l) \rho_l,$$

$$\int (d\rho) \mathcal{L}(\{n_j\}; \rho) \approx \frac{1}{L} \sum_{l=1}^{L} \tau(\rho_l) p(\rho_l) \mathcal{L}(\{n_j\}; \rho_l), \tag{2.4.5}$$

where L is the number of random statistical operators and $\tau(\rho_l)$ is the distribution function for the volume element.

Unfortunately, this naive procedure fails for $D > 2$ and moderately large N. First, as dimension increases, it gets harder to sample the full state space homogeneously. Depending on the method of random sampling, the distribution of the random states is typically highly inhomogeneous for large D. Second, as N increases, the likelihood quickly becomes very sharply peaked, so the chances of random states corresponding to significant values of the likelihood are slim. More sophisticated Monte Carlo methods must be employed to sample the state space according to the distribution dictated by $(d\rho) \mathcal{L}(\{n_j\}; \rho)$.

Despite the difficulties in handling the BM scheme, the Bayesian method does provide a reasonable statistical description about the source. However, this does not mean that the ML scheme is "bad" in any sense. We end this discussion by pointing out that, in talking about various methods of state estimation, the likelihood always plays a central role in these techniques. After all, the different types of estimators constructed out of informationally complete measurement data must eventually converge to ρ_{true} in the large N limit. Only when N is finite does one need to take other factors into consideration when deciding which estimator to use for achieving a particular objective.

For instructive purposes, we shall continue to discuss quantum-state estimation in the next chapter with the ML scheme as the central theme to reveal some more important features.

Further reading

(1) J. Orear, *Notes on Statistics for Physicists*, University of California rep., UCRL-8417 (1958).
[An early treatise on the subject of the generalized ML method.]

(2) S. Kotz and N. L. Johnson, *Breakthroughs in Statistics: Foundations and Basic Theory*, Springer Series in Statistics, Springer, New York (1992).
[An invaluable collection of original articles written by influential pioneers in statistics, with external commentaries. Read, for instance, the 1922 article by R. A. Fisher that introduces the ML method.]

(3) R. D. Cousins, Why isn't every physicist a Bayesian?, *Am. J. Phys.* **63**, 398 (1995).
[An instructive review on frequentist and Bayesian statistical methods in the context of particle physics.]

Chapter 3

Informationally Incomplete Estimation

3.1 Likelihood and entropy — perfect detections

We have learned that the minimum number of POM outcomes needed to form an informationally complete measurement is the square of the Hilbert-space dimension D for the quantum systems, since the statistical operators that describe quantum states are positive. As the dimension increases, this minimum also increases rapidly. For quantum systems of large Hilbert spaces, it is typically not feasible to construct informationally complete measurements. In these cases, only measurements that are *informationally incomplete* are available.

3.1.1 *Informationally incomplete measurements*

Let us suppose, for the moment, that the measurements of interest are perfect $(G = 1)$. For simplicity, we start by analyzing single-qubit POMs. The first measurement is the familiar von Neumann measurement that is defined in Eq. (2.1.19). We have briefly looked at one such POM in Sec. 2.1 when we were discussing an example for which an exact solution to $\widehat{\rho}_{\mathrm{ML}}$ can be calculated. We recall that for this POM, the Bloch vector s_{ML} of the ML estimators is such that $s_1^2 + s_2^2 + (2f_0 - 1)^2 \leq 1$, where s_1 and s_2 are the first two components of s_{ML} and f_0 is the frequency of $|0\rangle\langle 0|$. Thus, the informational incompleteness renders s_1 and s_2 arbitrary in such a way that all ML estimators lying on a disc, of radius $\sqrt{1 - (2f_0 - 1)^2}$, are consistent with the ML probabilities, which are $\widehat{p}_j^{\mathrm{ML}} = f_j$ in this case. Since a convex sum of *any* two ML estimators also lies within this disc, we say that this disc of ML estimators form a convex set.

The second measurement that we shall now consider is known as the *trine measurement* that consists of three measurement outcomes. A standard trine measurement is composed of the outcomes

$$\Pi_0 = \frac{1 + \sigma_z}{3},$$

$$\Pi_\pm = \frac{1 - \frac{1}{2}\sigma_z \pm \frac{\sqrt{3}}{2}\sigma_x}{3} \tag{3.1.1}$$

that survey the x-z equatorial plane of the Bloch ball. Let us suppose that the corresponding measurement frequencies f_0, f_+ and f_- are consistent with a positive ML estimator, so that

$$f_0 = \text{tr}\{\widehat{\rho}_{\text{ML}}\Pi_0\},$$

$$f_\pm = \text{tr}\{\widehat{\rho}_{\text{ML}}\Pi_\pm\}. \tag{3.1.2}$$

The first equation relates f_0 with s_3, the third component of the Bloch vector $\boldsymbol{s}_{\text{ML}}$ of $\widehat{\rho}_{\text{ML}}$:

$$s_3 = 3f_0 - 1. \tag{3.1.3}$$

The last two equations relate f_+ and f_- with s_1:

$$f_\pm = \frac{1 - \frac{1}{2}s_3 \pm \frac{\sqrt{3}}{2}s_1}{3}$$

$$\Rightarrow \quad s_1 = \sqrt{3}(f_+ - f_-). \tag{3.1.4}$$

From these two relations, we conclude that the condition for $\widehat{\rho}_{\text{ML}} \geq 0$ is

$$3(f_+ - f_-)^2 + (3f_0 - 1)^2 + s_2^2 \leq 1. \tag{3.1.5}$$

We see that now, the informational incompleteness renders only s_2 arbitrary, subject to the positivity constraint, which follows from having one additional linearly independent outcome as compared to the von Neumann measurement. The family of ML estimators can thus be represented by a line that intersects the x-z equatorial plane perpendicularly.

3.1.2   *Properties of ML convex sets*

Unlike the von Neumann measurement, where all $\widehat{\rho}_{\text{ML}}$s yield f_0 and f_1 as the ML probabilities, there are measurement data that do not correspond to a legitimate estimator $\widehat{\rho}_{\text{ML}}$ for the trine measurement. For example, suppose that the trine measurement is terminated after measuring only

$N = 8$ copies of qubits, so that the measurement data collected are $\{n_0 = 6, n_+ = n_- = 1\}$. Then, the resulting measurement frequencies $\left\{ f_0 = \dfrac{3}{4}, f_+ = f_- = \dfrac{1}{8} \right\}$ do not correspond to any positive operator since $s_3 > 1$. In such a situation, the correct ML probabilities are, therefore, different from the measurement frequencies since the likelihood is to be maximized *within* the permissible space of statistical operators.

The structure of the likelihood will also be different when the POM is informationally incomplete (refer to Fig. 3.1). For the von Neumann measurement, since the ML convex set is a disc, there exists a likelihood *plateau* structure, which is also a disc, hovering above the convex set, such that the height of the plateau represents the maximal value of the likelihood. For the trine measurement, the likelihood plateau is a ridge that is situated directly above the line of ML estimators. When $D > 2$, the shape of the likelihood plateau becomes extremely complicated, owing to the complexity of its boundary that incorporates the positivity constraint for the ML estimators. See Fig. 3.1.

There are two situations in which the likelihood plateau reduces to a point. The first situation is one where the measurement frequencies result in only one positive estimator that satisfy the linear constraints $\hat{p}_j^{\mathrm{ML}} = f_j$. For the von Neumann measurement, we have seen that for $f_0 = 0, 1$, the ML convex set is reduced to a single estimator. For this POM, there exist only two ML estimators $\hat{\rho}_{\mathrm{ML}} = \dfrac{1 \pm \sigma_z}{2}$ that correspond to the sets of measurement data $\{f_0 = 1, f_1 = 0\}$ and $\{f_0 = 0, f_1 = 1\}$ respectively. The situation is even more dramatic for the trine measurement. The condition

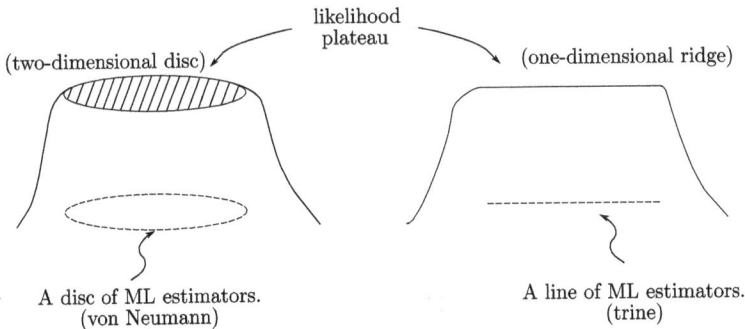

Fig. 3.1 Structure of the likelihood plateau for the single-qubit case ($D = 2$).

for which the line of ML estimators reduces to a single estimator is given by

$$3\left(f_+ - f_-\right)^2 + \left(3f_0 - 1\right)^2 = 1. \tag{3.1.6}$$

Any set of frequencies that obey Eq. (3.1.6) will result in a unique ML estimator. For this POM, the sets of measurement frequencies for which this happens are infinitely many. In practice, however, the frequencies almost never obey such stringent relations.

The second situation is one in which the measurement frequencies f_j do not correspond to any statistical operator due to the presence of data fluctuations, which happens especially often when the number of copies N is not large. This means that the entire plateau of the likelihood lies outside the state space. For a set of single-qubit measurement data $(D = 2)$ that do not correspond to any statistical operator, maximizing the likelihood within the Bloch ball always leads to a unique rank-deficient $\widehat{\rho}_{\mathrm{ML}}$ that is located on the Bloch sphere, with a set of ML probabilities that are different from the measurement frequencies.

For $D > 2$, the boundary of the state space consists of multi-dimensional corners and edges. Hence, maximizing the likelihood over such a complicated state space in the second situation may still give a convex set of ML estimators. Let us consider the three POM outcomes

$$\Pi_1 \widehat{=} \frac{1}{3}\begin{pmatrix} 1 & 0 & 0 \\ 0 & 1 & 0 \\ 0 & 0 & 0 \end{pmatrix}, \quad \Pi_2 \widehat{=} \frac{1}{3}\begin{pmatrix} 1 & 0 & 1 \\ 0 & 1 & 0 \\ 1 & 0 & 1 \end{pmatrix}, \quad \Pi_3 \widehat{=} \frac{1}{3}\begin{pmatrix} 1 & 0 & -1 \\ 0 & 1 & 0 \\ -1 & 0 & 2 \end{pmatrix} \tag{3.1.7}$$

for qutrit states $(D = 3)$ and their respective measurement frequencies $\{0.5, 0.25, 0.25\}$. With the matrix representation

$$\widehat{\rho}_{\mathrm{ML}} \widehat{=} \begin{pmatrix} \rho_{11} & \rho_{12} & \rho_{13} \\ \rho_{21} & \rho_{22} & \rho_{23} \\ \rho_{31} & \rho_{32} & \rho_{33} \end{pmatrix}, \quad \rho_{11} + \rho_{22} + \rho_{33} = 1 \tag{3.1.8}$$

for the ML estimator, the corresponding ML probabilities are

$$\widehat{p}_1^{\,\mathrm{ML}} = \frac{1}{3}(1 - \rho_{33}),$$

$$\widehat{p}_2^{\,\mathrm{ML}} = \frac{1}{3}(1 + \rho_{13} + \rho_{31}),$$

$$\widehat{p}_3^{\,\mathrm{ML}} = \frac{1}{3}(1 - \rho_{13} - \rho_{31} + \rho_{33}). \tag{3.1.9}$$

The actual ML value for this example is obtained by requiring that

$$\widehat{p}_1^{\,\text{ML}} = \frac{1}{3}(1 - \rho_{33}) = \frac{1}{2} = f_1, \qquad (3.1.10)$$

resulting in a negative diagonal element $\rho_{33} = -\frac{1}{2}$ that is not allowed for a positive $\widehat{\rho}_{\text{ML}}$. Therefore, a valid ML estimator that satisfies the extremal equation (2.1.15) must have $\rho_{33} = 0$, and the positivity constraint automatically implies that the third row and the third column that both contain ρ_{33} are zero. This conclusion is reassured by realizing that

$$R(\widehat{\rho}_{\text{ML}}) = 3\left(\frac{1}{2}\Pi_1 + \frac{1}{4}\Pi_2 + \frac{1}{4}\Pi_3\right) \stackrel{\frown}{=} \begin{pmatrix} 1 & 0 & 0 \\ 0 & 1 & 0 \\ 0 & 0 & \frac{3}{4} \end{pmatrix} \qquad (3.1.11)$$

is indeed the identity operator in the row space of $\widehat{\rho}_{\text{ML}}$.

Problem 3.1   Show that for a positive matrix, if one of its diagonal matrix elements is zero, then both the row and column that contain this zero element must be zero.

In other words, there exists a convex set of ML estimators of the form

$$\widehat{\rho}_{\text{ML}} \stackrel{\frown}{=} \begin{pmatrix} a & c & 0 \\ c^* & b & 0 \\ 0 & 0 & 0 \end{pmatrix}, \qquad a + b = 1, \quad ab \geq |c|^2 \qquad (3.1.12)$$

that satisfy (2.1.15) for these measurement frequencies, with ML probabilities all equal to one-third. These estimators reside in the *qubit subspace* of the qutrit space, where the operator-space dimension of the convex set is equal to four.

More generally, when the measurement data are inconsistent with any statistical operator, there can still be a convex set of ML estimators which give the same set of ML probabilities if there is a statistical-operator subspace that is consistent with the ML probabilities. If we denote the dimension of the operator subspace spanned by the POM outcomes — the *measurement subspace* — by D_{meas}, which is equal to the degree of linear independence $D_{\text{LI}} = D_{\text{meas}}$ of the outcomes, then, depending on the POM, there is a possibility for the dimension of the ML convex set to take values between zero and $D_{\text{unmeas}} \equiv D^2 - D_{\text{meas}}$. This phenomenon is, however, atypical in real experiments, for slight perturbations in either the measurement outcomes listed in (3.1.7) or their measurement data reduces the ML convex set to a single point. Therefore, the ML estimator obtained from the measurement data is practically unique.

3.1.3   *Measurement-subspace decomposition*

Numerical computations are often facilitated by a suitable parametrization of statistical operators. For the study of ML estimators, it is sometimes convenient to adopt a set of basis operators that characterize the D_{meas}-dimensional measurement subspace and its D_{unmeas}-dimensional complementary subspace, where the two subspaces together make up the entire positive-operator space of dimension D^2. For convenience, we want to have a set of *trace-orthonormal* basis operators that consists of D_{meas} linearly independent Hermitian operators $\{\Gamma_j\}_{j=1}^{D_{\text{meas}}}$ that span the entire measurement subspace, and D_{unmeas} other Hermitian operators $\{\Gamma_j\}_{j=D_{\text{meas}}+1}^{D^2}$ that span the complement of the measurement subspace. The ML estimator can thus be written in terms of the $D_{\text{meas}} + D_{\text{unmeas}} = D^2$ basis operators as

$$
\widehat{\rho}_{\text{ML}} = \underbrace{\sum_{j=1}^{D_{\text{meas}}} c_j^{\text{meas}} \Gamma_j}_{\equiv\, \rho_{\text{meas}}} + \underbrace{\sum_{j=D_{\text{meas}}+1}^{D^2} c_j^{\text{unmeas}} \Gamma_j}_{\equiv\, \rho_{\text{unmeas}}}, \tag{3.1.13}
$$

where ρ_{meas} resides in the measurement subspace and ρ_{unmeas} resides in its complement. The coefficients c_j^{meas} and c_j^{unmeas} are clearly real since $\text{tr}\{\Gamma_j \widehat{\rho}_{\text{ML}}\}$ are real numbers.

As we are still discussing perfect measurement outcomes, there is another interesting property for the basis operators $\Gamma_j|_{j>D_{\text{meas}}}$ that span the complement of the measurement subspace — they are *traceless* operators. To show this property, we note that the measurement outcomes Π_j must be a linear combination of only the basis operators $\Gamma_j|_{j\leq D_{\text{meas}}}$ parameterizing the measurement subspace, that is

$$
\Pi_k = \sum_{l=1}^{D_{\text{meas}}} q_{kl} \Gamma_l, \tag{3.1.14}
$$

since the outcomes themselves span the measurement subspace, of course. So,

$$
\text{tr}\left\{ \Gamma_j\Big|_{j>D_{\text{meas}}} \right\} = \text{tr}\left\{ \Gamma_j\Big|_{j>D_{\text{meas}}} \sum_{k=1}^{M} \Pi_k \right\} = 0. \tag{3.1.15}
$$

There is, thus, no trace contribution from the part ρ_{unmeas} of $\widehat{\rho}_{\text{ML}}$, and the only contribution comes from the part ρ_{meas} residing in the measurement subspace. One other important point to note is that, although $\text{tr}\{\rho_{\text{meas}}\} = 1$, ρ_{meas} is not a positive operator in general. Only the sum of ρ_{meas} and ρ_{unmeas} is positive. Having said this, we understand that there are constraints imposed on the coefficients c_j^{meas} and c_j^{unmeas} enforced by the positivity constraint. For a D-dimensional Hilbert space, there will consequently be more than one inequality that summarizes the positivity constraint on $\widehat{\rho}_{\text{ML}}$. We can, however, write down one obvious inequality that is necessarily true for all statistical operators: $\text{tr}\{\widehat{\rho}_{\text{ML}}^2\} \leq 1$. This then implies that

$$1 \geq \text{tr}\{\widehat{\rho}_{\text{ML}}^2\} = \sum_{j=1}^{D^2} \sum_{k=1}^{D^2} c_j c_k \underbrace{\text{tr}\{\Gamma_j \Gamma_k\}}_{= \delta_{j,k}} = \sum_{j=1}^{D^2} c_j^2, \qquad (3.1.16)$$

which also means that $|c_j| \leq 1$.

Problem 3.2 Suppose that $G = 1$ and ρ is a pure quantum state that resides only in the measurement subspace of a POM, show that the addition of a Hermitian operator H in the complementary subspace will always result in another nonpositive operator. How is this result consistent with the result of ρ_{unmeas}? Now, if ρ is mixed, state two *necessary* conditions on the set of eigenvalues of H for which $\rho + H$ is a quantum state.

How do we construct these basis operators numerically? A straightforward way is to use the Gram–Schmidt orthogonalization procedure on operators. To generate the basis operators that span the measurement subspace, we can always define the first operator $\widetilde{\Gamma}_1 = \Pi_1$ to be one of the measurement outcomes in the POM. Subsequently, the second operator is given by

$$\widetilde{\Gamma}_2 = \Pi_2 - \frac{\text{tr}\{\Pi_2 \widetilde{\Gamma}_1\}}{\text{tr}\{\widetilde{\Gamma}_1^2\}} \widetilde{\Gamma}_1, \qquad (3.1.17)$$

the third operator by

$$\widetilde{\Gamma}_3 = \Pi_3 - \frac{\text{tr}\{\Pi_3 \widetilde{\Gamma}_1\}}{\text{tr}\{\widetilde{\Gamma}_1^2\}} \widetilde{\Gamma}_1 - \frac{\text{tr}\{\Pi_3 \widetilde{\Gamma}_2\}}{\text{tr}\{\widetilde{\Gamma}_2^2\}} \widetilde{\Gamma}_2, \qquad (3.1.18)$$

etc., so that all D_{meas} basis operators are constructed according to the rule

$$\widetilde{\Gamma}_j = \Pi_j - \sum_{k=1}^{j-1} \frac{\text{tr}\left\{\Pi_j \widetilde{\Gamma}_k\right\}}{\text{tr}\left\{\widetilde{\Gamma}_k^2\right\}} \widetilde{\Gamma}_k. \qquad (3.1.19)$$

In this way, it is easy to see that the orthogonalization procedure generates $D_{\text{meas}} \leq M$ operators $\widetilde{\Gamma}_j$ that span the measurement subspace defined by the Π_js.

For the operators that make up ρ_{unmeas}, we can again use the orthogonalization procedure, only this time we will just use a random positive operator $\mathcal{V}_j$, which is always linearly independent of the set $\{\Pi_j\}$, for every such operator. To generate $\mathcal{V}_j$, we can adopt the parametrization $\mathcal{V}_j = \mathcal{A}_j^\dagger \mathcal{A}_j$, where $\mathcal{A}_j$ is a random complex operator which, in the computational basis, has entries that follow some pre-chosen distribution. Typically, one can choose the Gaussian distribution of zero mean and unit variance. Hence,

$$\widetilde{\Gamma}_j\big|_{j>D_{\text{meas}}} = \mathcal{V}_j - \sum_{k=1}^{j-1} \frac{\text{tr}\left\{\mathcal{V}_j \widetilde{\Gamma}_k\right\}}{\text{tr}\left\{\widetilde{\Gamma}_k^2\right\}} \widetilde{\Gamma}_k \Bigg|_{j>D_{\text{meas}}}. \qquad (3.1.20)$$

All the operators constructed this way are mutually trace-orthogonal. We can now establish the basis operators

$$\Gamma_j \equiv \frac{\widetilde{\Gamma}_j}{\sqrt{\text{tr}\left\{\widetilde{\Gamma}_j^2\right\}}} \qquad (3.1.21)$$

that are all normalized in trace.

Problem 3.3 How would one generate a complete set of D-dimensional trace-orthonormal Hermitian basis operators such that one of them is $\dfrac{1}{\sqrt{D}}$, and the rest of these operators are traceless?

With these basis operators, we can uncover an underlying structure of $\widehat{\rho}_{\text{ML}}$ that is endowed by the perfect measurement outcomes $\{\Pi_j\}$. In the convex set, all $\widehat{\rho}_{\text{ML}}$ will possess the **same** ρ_{meas}. This is because $\text{tr}\{\rho_{\text{unmeas}}\Pi_j\} = 0$ and so the resulting ML probabilities $\widehat{p}_j^{\text{ML}} = \text{tr}\{\widehat{\rho}_{\text{ML}}\Pi_j\} = \text{tr}\{\rho_{\text{meas}}\Pi_j\}$ fix the part that is in the measurement subspace. The boundary of the convex set is defined by the coefficients c_j^{unmeas} that are allowed to vary, subject to the positivity constraint on $\widehat{\rho}_{\text{ML}}$. While the dimension of the convex set is D_{unmeas}, as we knew before, the total number of linearly independent operators required to define the entire ML convex set is $D_{\text{unmeas}} + 1$. For the single-qubit case, $D = 2$, the ML convex set for the von Neumann measurement is a disc of dimension two, which is defined by three linearly independent basis operators. More generally, the convex set is defined by D_{unmeas} basis operators $\Gamma_j|_{j>D_{\text{meas}}}$ and ρ_{meas}. In other words, the ML convex set is maximally spanned by **any** set of $D_{\text{unmeas}} + 1$ linearly independent ML estimators.

3.1.4 *Maximum-likelihood–maximum-entropy (MLME) method*

3.1.4.1 *The von Neumann entropy*

In general, there exists a family of ML estimators for a given set of informationally incomplete data. In this case, one should report this entire family as the description for the source. For practical reasons, however, it is often desirable to generate a reasonable guess out of this family for statistical predictions. There is, however, **no** physical law that dictates the choice of such an estimator. In principle, there are many ways to go about it and each method requires some sort of justification. A straightforward way to generate an estimator is to associate a convex function to the convex set and choose the estimator that maximizes/minimizes this function out of the set. A popular function to pick is the *von Neumann entropy function* that is defined as

$$S(\rho) = -\text{tr}\{\rho \log \rho\}. \tag{3.1.22}$$

The corresponding ML estimator in the convex set that maximizes the entropy is the *maximum-likelihood-maximum-entropy* (MLME) estimator.

The principle of maximum entropy was extensively applied to statistical physics by Edwin Thompson Jaynes (1922–1998). When applied to state estimation, one can associate a statistical meaning to the MLME estimator. In a sense, the MLME estimation gives an estimator that represents the most conservative guess and carries the largest uncertainty quantified by $S(\rho)$, as compared to all other ML estimators in the ML convex set. In colloquial terms, we place the least amount of trust in the informationally incomplete data obtained, as far as the estimation of the unmeasured parameters of ρ_{true} is concerned. Unlike the likelihood, which is defined by the measurement subspace, where the probabilities in the argument are themselves defined only by ρ_{meas}, the entropy $S(\rho)$ is a concave function that is defined by the *entire* statistical operator. As such, it has no plateau structure since there is no ambiguity in this case. Therefore, the MLME estimator is always unique.

Problem 3.4 Show that the von Neumann entropy $S(\rho)$ is a concave function of ρ in the state space.

3.1.4.2 *The maximum-entropy estimator*

The task at hand is to maximize $S(\rho)$, subject to the positivity constraint on ρ and the ML probability constraints

$$\widehat{p}_j^{\text{ML}} = \text{tr}\{\widehat{\rho}_{\text{ML}}\Pi_j\} = \text{tr}\{\rho\,\Pi_j\}. \tag{3.1.23}$$

We can write down the Lagrange function

$$\mathcal{D}_{\text{L}}(\rho) = -\text{tr}\{\rho\log\rho\} + \sum_j \mu_j\text{tr}\{(\rho - \widehat{\rho}_{\text{ML}})\,\Pi_j\} + (\log\mu + 1)(\text{tr}\{\rho\} - 1), \tag{3.1.24}$$

for this purpose, where μ_j and $\log\mu + 1$ are the Lagrange multipliers for the respective constraints. Varying $\mathcal{D}_{\text{L}}(\rho)$ with respect to ρ, we have

$$\delta\mathcal{D}_{\text{L}}(\rho) = -\text{tr}\{\delta\rho\log\rho\} - \text{tr}\{\delta\rho\} + \sum_j \mu_j\text{tr}\{\delta\rho\,\Pi_j\}$$

$$+ (\log\mu + 1)\text{tr}\{\delta\rho\}$$

$$= \text{tr}\left\{\delta\rho\left(-\log\rho + \sum_j \mu_j\Pi_j + \log\mu\right)\right\}. \tag{3.1.25}$$

Hence, the MLME estimator for which the variation $\delta\mathcal{D}_{\text{L}}(\rho)$ is zero is of the form

$$\widehat{\rho}_{\text{MLME}} = \frac{e^{\sum_j \mu_j\Pi_j}}{\text{tr}\left\{e^{\sum_j \mu_j\Pi_j}\right\}}, \quad \mu = \frac{1}{\text{tr}\left\{e^{\sum_j \mu_j\Pi_j}\right\}}, \tag{3.1.26}$$

such that $\text{tr}\{\widehat{\rho}_{\text{MLME}}\Pi_j\} = \widehat{p}_j^{\text{ML}}$. The structure of Eq. (3.1.26) bears a nostalgic charm reminiscent of the density operator for a given canonical ensemble in statistical mechanics.

3.1.4.3 *The extremal equations*

To complete the optimization, we need to search for the Lagrange multipliers μ_j such that the ML probability constraints in Eq. (3.1.23) are obeyed. To do this, we shall optimize μ_j such that this maximum-entropy estimator maximizes the log-likelihood $\log \mathcal{L}(\{n_j\}; \rho)$. For that, we need to evaluate the form of $\delta\rho$. Since ρ is restricted to be the form of a maximum-entropy estimator,

$$
\delta\rho = \frac{\delta e^{\sum_j \mu_j \Pi_j}}{\mathrm{tr}\left\{ e^{\sum_j \mu_j \Pi_j} \right\}} - \frac{e^{\sum_j \mu_j \Pi_j}}{\mathrm{tr}\left\{ e^{\sum_j \mu_j \Pi_j} \right\}^2} \mathrm{tr}\left\{ \delta e^{\sum_j \mu_j \Pi_j} \right\}
$$

$$
= \frac{\delta e^{\sum_j \mu_j \Pi_j}}{\mathrm{tr}\left\{ e^{\sum_j \mu_j \Pi_j} \right\}} - \rho \frac{\mathrm{tr}\left\{ \delta e^{\sum_j \mu_j \Pi_j} \right\}}{\mathrm{tr}\left\{ e^{\sum_j \mu_j \Pi_j} \right\}}, \tag{3.1.27}
$$

where we need the expression for $\delta e^{\sum_j \mu_j \Pi_j}$.

This may be obtained by recalling that for any $\mathcal{B}$ with a valid $\mathcal{B}^{-1}$,

$$
\delta\mathcal{B}^{-n} = -\sum_{k=1}^{n} \mathcal{B}^{k-n-1} \delta\mathcal{B}\mathcal{B}^{-k} \quad \text{for any integer } n \geq 0, \tag{3.1.28}
$$

which is incidently part of the hints for **Problem 1.19** and **Problem 4.5**. Setting $n = 1$ gives the familiar relation $\delta\mathcal{B}^{-1} = -\mathcal{B}^{-1}\delta\mathcal{B}\mathcal{B}^{-1}$. Upon choosing $\mathcal{B}^{-n} = e^{\mathcal{A}} \equiv e^{-n\varepsilon}$, where any $\mathcal{A}$ can always be treated as a large multiple $-n\varepsilon$ of a very small operator ε, we have

$$
\delta e^{\mathcal{A}} = -\sum_{k=1}^{n} e^{(k-n-1)\varepsilon} \underbrace{\delta e^{\varepsilon}}_{\approx \delta(1+\varepsilon)} e^{-k\varepsilon}
$$

$$
= \frac{1}{n} \sum_{k=1}^{n} e^{\left(1-\frac{k}{n}\right)\mathcal{A}} \delta\mathcal{A}\, e^{\frac{k}{n}\mathcal{A}}, \tag{3.1.29}
$$

where we bear in mind that the equalities are valid as long as the limits $n \to \infty$ and $\varepsilon \to 0$ are taken. In doing so, the sum turns into an integral with the proper associations $x \equiv \frac{k}{n}$ and $dx \equiv \frac{1}{n}$,

$$
\delta e^{\mathcal{A}} = \int_0^1 dx\, e^{(1-x)\mathcal{A}} \delta\mathcal{A}\, e^{x\mathcal{A}} = \int_0^1 dx\, e^{x\mathcal{A}} \delta\mathcal{A}\, e^{(1-x)\mathcal{A}}. \tag{3.1.30}
$$

The variation of $\log \mathcal{L}(\{n_j\}; \rho)$ is then

$$\delta \log \mathcal{L}(\{n_j\}; \rho)$$

$$= N \operatorname{tr}\{R(\rho)\, \delta\rho\}$$

$$= N \operatorname{tr}\left\{ R(\rho) \left(\frac{\delta e^{\sum_j \mu_j \Pi_j}}{\operatorname{tr}\left\{ e^{\sum_j \mu_j \Pi_j} \right\}} - \frac{e^{\sum_j \mu_j \Pi_j}}{\operatorname{tr}\left\{ e^{\sum_j \mu_j \Pi_j} \right\}^2} \operatorname{tr}\left\{ \delta e^{\sum_j \mu_j \Pi_j} \right\} \right) \right\}$$

$$= \frac{N}{\operatorname{tr}\left\{ e^{\sum_j \mu_j \Pi_j} \right\}} \operatorname{tr}\left\{ [R(\rho) - 1] \int_0^1 \mathrm{d}x\, e^{(1-x)\sum_j \mu_j \Pi_j} \sum_k (\delta\mu_k \Pi_k)\, e^{x \sum_j \mu_j \Pi_j} \right\}$$

$$= N \sum_k \delta\mu_k \operatorname{tr}\left\{ \rho\, \Pi_k \left[\int_0^1 \mathrm{d}x\, e^{x \sum_j \mu_j \Pi_j} R(\rho)\, e^{-x \sum_j \mu_j \Pi_j} - 1 \right] \right\}.$$

$$(3.1.31)$$

Because of the cyclic property of the operator trace, the maximal value of $\log \mathcal{L}(\{n_j\}; \rho)$ is attained when

$$\int_0^1 \mathrm{d}x\, e^{x \sum_j \mu_j \Pi_j} R(\rho)\, e^{-x \sum_j \mu_j \Pi_j} \bigg|_{\rho = \widehat{\rho}_{\mathrm{MLME}}} = 1_{\widehat{\rho}_{\mathrm{MLME}}}, \qquad (3.1.32)$$

where $1_{\widehat{\rho}_{\mathrm{MLME}}}$ is the identity in the row space of $\widehat{\rho}_{\mathrm{MLME}}$.

Problem 3.5   Evaluate the left-hand side of Eq. (3.1.32) directly.

Problem 3.6   Show also the useful identities

$$\delta \log \mathcal{A} = \int_0^\infty \mathrm{d}t\, \frac{1}{t + \mathcal{A}}\, \delta\mathcal{A}\, \frac{1}{t + \mathcal{A}} \qquad (3.1.33)$$

and

$$\delta\sqrt{\mathcal{A}} = \int_0^\infty \frac{\mathrm{d}t}{\pi}\, \sqrt{t}\, \frac{1}{t + \mathcal{A}}\, \delta\mathcal{A}\, \frac{1}{t + \mathcal{A}} \qquad (3.1.34)$$

for any $\mathcal{A} > 0$.

3.1.4.4 *A steepest-ascent algorithm*

To solve the extremal equations in Eq. (3.1.32), we apply the steepest-ascent method by first writing down the derivative

$$\frac{\partial \log \mathcal{L}(\{n_j\};\rho)}{\partial \mu_j} = N \operatorname{tr}\left\{\rho \Pi_j \left[\int_0^1 dx\, e^{x \sum_l \mu_l \Pi_l} R(\rho) e^{-x \sum_l \mu_l \Pi_l} - 1\right]\right\},$$

(3.1.35)

which is always real.

Problem 3.7 Verify this simple fact.

Next, after defining

$$\delta\mu_j = \frac{\epsilon'}{N}\frac{\partial}{\partial \mu_j} \log \mathcal{L}(\{n_j\};\rho)$$

(3.1.36)

with a small constant $\epsilon' > 0$, we arrive at the MLME iterative algorithm involving the equations

$$\rho_{k+1} = \frac{e^{\sum_j \left[\mu_j^{(k)} + \delta\mu_j^{(k)}\right]\Pi_j}}{\operatorname{tr}\left\{e^{\sum_j \left[\mu_j^{(k)} + \delta\mu_j^{(k)}\right]\Pi_j}\right\}},$$

$$\delta\mu_j^{(k)} = \epsilon' \operatorname{tr}\left\{\rho_k \Pi_j \left[\int_0^1 dx\, e^{x \sum_l \mu_l^{(k)} \Pi_l} R(\rho_k) e^{-x \sum_l \mu_l^{(k)} \Pi_l} - 1\right]\right\}.$$

(3.1.37)

One shortcoming of this iterative algorithm is that for large dimensions, the numerical computation of the integral in Eq. (3.1.35) becomes expensive. One way to circumvent the problem is to introduce an approximation to the variation of the exponential operator:

$$\delta e^{\mathcal{A}} \approx \frac{1}{2}(e^{\mathcal{A}} \delta\mathcal{A} + \delta\mathcal{A} e^{\mathcal{A}}).$$

(3.1.38)

Consequently, the second equation in Eq. (3.1.37) reduces to

$$\delta\mu_j^{(k)} \approx \epsilon' \operatorname{tr}\left\{\rho_k \left[\frac{\Pi_j R(\rho_k) + R(\rho_k)\Pi_j}{2} - \Pi_j\right]\right\}.$$

(3.1.39)

In this way, the numerical integration can be avoided.

3.1.4.5 *Simple closed-form expressions*

There are single-qubit examples for which closed-form expressions for the MLME estimators exist. We recall that for the trine measurement, the ML convex set is a line within the Bloch ball. The measurement subspace is that of a disc defined by the measurement outcomes. It can be easily shown that the MLME estimator lies on the intersection point of the disc and the line. For the standard outcomes that we have defined in Eq. (3.1.1), the MLME estimator has the component $s_2 = 0$. Out of the many ways to show this simple fact, one may first verify that the entropy of a single-qubit state increases, as a one-to-one function, with decreasing purity.

Problem 3.8 Show that, for a single-qubit state ρ, the von Neumann entropy

$$S(\rho) = -\frac{1}{2}\left[\log\left(\frac{1-\mathcal{K}^2}{4}\right) + \mathcal{K}\log\left(\frac{1+\mathcal{K}}{1-\mathcal{K}}\right)\right], \qquad (3.1.40)$$

where $\mathcal{K} = \sqrt{2\operatorname{tr}\{\rho^2\} - 1}$ is related to the purity $\operatorname{tr}\{\rho^2\}$. Check this answer for pure states and the maximally-mixed state.

With this knowledge, the purity of an ML estimator

$$\widehat{\rho}_{\mathrm{ML}} \,\widehat{=}\, \frac{1}{2}\begin{pmatrix} 3\,\widehat{p}_0^{\,\mathrm{ML}} & \sqrt{3}(\widehat{p}_+^{\,\mathrm{ML}} - \widehat{p}_-^{\,\mathrm{ML}}) - is_2 \\ \sqrt{3}(\widehat{p}_+^{\,\mathrm{ML}} - \widehat{p}_-^{\,\mathrm{ML}}) + is_2 & 2 - 3\,\widehat{p}_0^{\,\mathrm{ML}} \end{pmatrix}, \qquad (3.1.41)$$

written in the basis of the eigenstates $|0\rangle\langle 0|$ and $|1\rangle\langle 1|$ of σ_z, is clearly minimized when $s_2 = 0$.

For a *rotated* trine measurement relative to the standard one via a unitary transformation U, that is the set $\{U\Pi_0 U^\dagger, U\Pi_\pm U^\dagger\}$, the Bloch-vector component of the MLME estimator in the rotated coordinate system that is not measured will be zero. This is clear since $\operatorname{tr}\{\widehat{\rho}_{\mathrm{ML}} U\Pi_j U^\dagger\} = \operatorname{tr}\{U^\dagger\widehat{\rho}_{\mathrm{ML}} U\Pi_j\}$ for all outcomes, so that the unitarily transformed $\widehat{\rho}_{\mathrm{ML}}$ behaves the same way to the unrotated POM outcomes as the original $\widehat{\rho}_{\mathrm{ML}}$ to the rotated outcomes.

Problem 3.9 Find the Bloch vector of the MLME estimator that gives the set of probabilities $\{\widehat{p}_1, \widehat{p}_2, \widehat{p}_3\}$ for the respective equally-distributed outcomes with Bloch vectors $\boldsymbol{b}_1 = (0, -\sin\vartheta, \cos\vartheta)^{\mathsf{T}}$, $\boldsymbol{b}_2 = \frac{1}{2}(\sqrt{3}, \sin\vartheta, -\cos\vartheta)^{\mathsf{T}}$ and $\boldsymbol{b}_3 = \frac{1}{2}(-\sqrt{3}, \sin\vartheta, -\cos\vartheta)^{\mathsf{T}}$.

For the von Neumann measurement,

$$\widehat{\rho}_{\mathrm{ML}} \,\widehat{=}\, \begin{pmatrix} f_0 & \dfrac{s_1 - is_2}{2} \\ \dfrac{s_1 + is_2}{2} & f_1 \end{pmatrix}. \tag{3.1.42}$$

By minimizing its purity, we can again show that the MLME estimator $\widehat{\rho}_{\mathrm{MLME}}$ corresponds to the components $s_1 = s_2 = 0$. That is to say, $\widehat{\rho}_{\mathrm{MLME}}$ is diagonal in the basis of the orthogonal von Neumann measurement outcomes. This result can be generalized to all Hilbert-space dimensions. For this purpose, we look at the difference of $S(\widehat{\rho}_{\mathrm{diag}})$ and $S(\widehat{\rho}_{\mathrm{ML}})$, where $\widehat{\rho}_{\mathrm{diag}}$ is the statistical operator that is diagonal in the basis of the von Neumann measurement consisting of D orthonormal projectors $|k\rangle\langle k|$, such that

$$\langle k| \widehat{\rho}_{\mathrm{diag}} |k\rangle = f_k. \tag{3.1.43}$$

Using the spectral decomposition of $\widehat{\rho}_{\mathrm{ML}}$, that is

$$\widehat{\rho}_{\mathrm{ML}} = \sum_j |\lambda_j\rangle \lambda_j \langle \lambda_j|, \tag{3.1.44}$$

and the relations

$$\langle k| \widehat{\rho}_{\mathrm{diag}} |k\rangle = \langle k| \widehat{\rho}_{\mathrm{ML}} |k\rangle$$
$$= \sum_j \lambda_j |\langle \lambda_j|k\rangle|^2, \tag{3.1.45}$$

the difference in the entropy functions becomes

$$S(\widehat{\rho}_{\mathrm{diag}}) - S(\widehat{\rho}_{\mathrm{ML}})$$

$$= \underbrace{-\sum_k \left(\sum_j \lambda_j |\langle \lambda_j|k\rangle|^2 \right) \log \left(\sum_j \lambda_j |\langle \lambda_j|k\rangle|^2 \right) + \sum_j \lambda_j \log \lambda_j}_{\displaystyle \leq \sum_k \sum_j |\langle \lambda_j|k\rangle|^2 \lambda_j \log \lambda_j}$$

$$\geq -\sum_j \lambda_j \log \lambda_j + \sum_j \lambda_j \log \lambda_j = 0, \tag{3.1.46}$$

where the inequality comes from the fact that $x \log x$ is a convex function of x, and for a convex function $\phi_{\mathrm{conv}}(x)$ and $x = \sum_j \alpha_j x_j$, such that

$\alpha_j \geq 0$ and $\sum_j \alpha_j = 1$,

$$\phi_{\text{conv}}(y) \leq \sum_j \alpha_j \phi_{\text{conv}}(x_j). \tag{3.1.47}$$

The conclusion is that for any Hilbert space of dimension D, the MLME estimator for the von Neumann measurement must be diagonal in the measurement basis.

Problem 3.10 Look for the Lagrange multipliers μ_j that make up the MLME estimator for the von Neumann measurement with the frequencies f_j.

3.2 Likelihood and entropy — imperfect detections

3.2.1 *The extremal probabilities*

So far, we have been dealing with perfect measurement outcomes: $G = 1$. Let us study the more realistic situation in experiments involving imperfect measurements where $G \leq 1$. We have looked into this matter for informationally complete data in Sec. 2.2 and found that the ML estimators obey the system of equations, in terms of the ML relative probabilities given by $\dfrac{\widehat{p}_j^{\text{ML}}}{\sum_j \widehat{p}_j^{\text{ML}}} = f_j$, where the estimated probabilities maximize the generalized log-likelihood defined by relative probabilities. We now know that for informationally complete POMs, there are enough linearly independent equations to fix the solution of $\widehat{p}_j^{\text{ML}}$, since these estimated probability must originate from a statistical operator and an informationally complete POM uniquely characterizes this operator.

When the POM is informationally incomplete, there are always not enough equations to specify all the state parameters of ρ. As a consequence, the subnormalized ML probabilities will not be uniquely specified. Just like the case for informationally complete measurements, the subnormalized ML probabilities for the outcomes have room to vary by a scalar multiple.

While the subnormalized ML probabilities are not unique for informationally incomplete POMs, the ML relative probabilities are **always** fixed, since the number of relative ML probabilities equals the number of extremal equations. Therefore, all ML probabilities will give the *same* maximal likelihood value. For each set of ML probabilities, there will in general be a convex set of ML estimators that are consistent with the probabilities.

When the measurement frequencies do not correspond to any quantum state, then for all practical situations, the positivity constraint on $\widehat{\rho}_{\mathrm{ML}}$ reduces the ML convex set to a single point. In this case, these frequencies are to be replaced by the appropriate ML relative probabilities that maximize the generalized log-likelihood in the space of positive operators. As discussed before, there is only one rank-deficient ML estimator for such measurement data.

Problem 3.11** Consider an imperfect single-qubit POM consisting of the three outcomes

$$\Pi_1' \cong \begin{pmatrix} 0.0342274 & -0.068225 + 0.0894946\,\mathrm{i} \\ -0.068225 - 0.0894946\,\mathrm{i} & 0.369994 \end{pmatrix},$$

$$\Pi_2' \cong \begin{pmatrix} 0.217291 & 0.144838 + 0.140144\,\mathrm{i} \\ 0.144838 - 0.140144\,\mathrm{i} & 0.186931 \end{pmatrix},$$

$$\Pi_3' \cong \begin{pmatrix} 0.196844 & 0.202041 - 0.000790307\,\mathrm{i} \\ 0.202041 + 0.000790307\,\mathrm{i} & 0.207378 \end{pmatrix}, \quad (3.2.1)$$

with the respective frequencies $f_1 = 0.074059$, $f_2 = 0.554196$ and $f_3 = 0.371745$. Verify numerically that, for this POM and data, the positive ML estimator is unique and rank-deficient, where the ML relative probabilities are different from the frequencies. A simple program to perform the ML algorithm is required.

3.2.2   *Multiple ML convex sets*

We can understand the structure of these ML estimators by revisiting the decomposition of $\widehat{\rho}_{\mathrm{ML}}$ into the basis operators that characterize the measurement subspace and its complementary subspace in Eq. (3.1.13). The ML probabilities can then be expressed as

$$\widehat{p}_j^{\,\mathrm{ML}} = \mathrm{tr}\{\widehat{\rho}_{\mathrm{ML}}\Pi_j'\} = \mathrm{tr}\{\rho_{\mathrm{meas}}\Pi_j'\}. \quad (3.2.2)$$

Since the $\widehat{p}_j^{\,\text{ML}}$ do not sum to unity, the trace of ρ_{meas} is no longer unity and the basis operators that span the complement of the measurement subspace are *no longer* traceless: $\text{tr}\{\rho_{\text{unmeas}}\} \neq 0$. This means that both ρ_{meas} and ρ_{unmeas} will contribute to the trace of $\widehat{\rho}_{\text{ML}}$. Moreover, since the relative probabilities are always fixed, the most general form of ρ_{meas} is given by

$$\rho_{\text{meas}} = \gamma\rho_0 \tag{3.2.3}$$

according to Eq. (2.2.27), where ρ_0 is some fixed reference positive operator such that the ML relative probabilities

$$\frac{\widehat{p}_j^{\,\text{ML}}}{\sum_k \widehat{p}_k^{\,\text{ML}}} = \frac{\text{tr}\{\rho_{\text{meas}}\Pi_j'\}}{\sum_k \text{tr}\{\rho_{\text{meas}}\Pi_k'\}} = \frac{\text{tr}\{\rho_0\Pi_j'\}}{\sum_k \text{tr}\{\rho_0\Pi_k'\}} \tag{3.2.4}$$

are fixed quantities that do not dependent on γ.

From the consequent decomposition

$$\widehat{\rho}_{\text{ML}} = \gamma\rho_0 + \rho_{\text{unmeas}}, \tag{3.2.5}$$

the maximum dimension of the ML convex set is $D^2 - D_{\text{meas}} = D_{\text{unmeas}}$, as before, and the independent state parameters that make up the dimension of the ML convex set are still the D_{unmeas} coefficients c_j^{unmeas} of the basis operators $\Gamma_j|_{j>D_{\text{meas}}}$ that define the complementary subspace. The only difference between an imperfect POM and a perfect one is that the additional operator that defines the ML convex set now varies with γ according to the unit-trace constraint on $\widehat{\rho}_{\text{ML}}$.

The general setting is as follows. Different sets of ML probabilities that give the same maximum generalized likelihood will yield different convex sets of ML estimators. This implies that we would have infinitely many MLME estimators $\widehat{\rho}_{\text{MLME}}$, one for each set of ML probabilities. We shall take the MLME estimator that possesses the *largest* entropy.

3.2.3   *A steepest-ascent algorithm*

For this, we first realize that

$$\widehat{p}_j^{\,\text{ML}} = \gamma\,\text{tr}\{\rho_0\Pi_j'\}, \tag{3.2.6}$$

where $\text{tr}\{\rho_0\Pi_j'\}$ can always be defined to be the *reference* probabilities $\widehat{p}_{j,0}^{\,\text{ML}}$. We emphasize that Eq. (3.2.6) is the most general form of $\widehat{p}_j^{\,\text{ML}}$ for imperfect detections. We can, therefore, write down the Lagrange function to be optimized:

$$\mathcal{D}_{\mathrm{L}}(\rho) = -\operatorname{tr}\{\rho \log \rho\} + \sum_{j=1}^{M} \mu_j \left(p_j - \gamma \widehat{p}_{j,0}^{\,\mathrm{ML}}\right)$$

$$+ \mu_{M+1} \left[p_{M+1} - \left(1 - \gamma \sum_{j=1}^{M} \widehat{p}_{j,0}^{\,\mathrm{ML}}\right)\right]$$

$$+ (\log \mu + 1)\left(\operatorname{tr}\{\rho\} - 1\right), \qquad (3.2.7)$$

where the third term accounts for the outcome $\Pi'_{M+1} = 1 - G$ that conserves the probabilities, which is not measured. The parameter γ is to be optimized to give an estimator with the largest entropy among the family of MLME estimators.

Not surprisingly, after a variation in γ, we get

$$\mu_{M+1} = \sum_{j=1}^{M} \beta_j \mu_j, \qquad (3.2.8)$$

where

$$\beta_j = \frac{\widehat{p}_{j,0}^{\,\mathrm{ML}}}{\sum_{k=1}^{M} \widehat{p}_{k,0}^{\,\mathrm{ML}}}. \qquad (3.2.9)$$

The optimal estimator is then given by

$$\widehat{\rho}_{\mathrm{MLME}} = \frac{e^{\sum_j \mu_j^{\mathrm{MLME}} \left(\Pi'_j + \beta_j \Pi'_{M+1}\right)}}{\operatorname{tr}\left\{e^{\sum_j \mu_j^{\mathrm{MLME}} \left(\Pi'_j + \beta_j \Pi'_{M+1}\right)}\right\}}. \qquad (3.2.10)$$

Together with Eq. (2.2.32), the algorithm for imperfect detections involves the iterative equations established previously, with Π_j replaced by $\Pi'_{j,\beta} = \Pi'_j + \beta_j \Pi'_{M+1}$ and the identity operator replaced by $\dfrac{G}{\eta^{(k)}}$,

$$\rho_{k+1} = \frac{e^{\sum_j \left(\mu_j^{(k)} + \delta\mu_j^{(k)}\right)\Pi_{j,\beta}}}{\operatorname{tr}\left\{e^{\sum_j \left(\mu_j^{(k)} + \delta\mu_j^{(k)}\right)\Pi_{j,\beta}}\right\}},$$

$$\delta\mu_j^{(k)} = \epsilon' \operatorname{tr}\left\{\rho_k \, \Pi_{j,\beta} \left(\int_0^1 dx \, e^{x \sum_l \mu_l^{(k)} \Pi'_{l,\beta}} \left[R(\rho_k) - \frac{G}{\eta^{(k)}}\right] e^{-x \sum_l \mu_l^{(k)} \Pi'_{l,\beta}}\right)\right\},$$

$$(3.2.11)$$

and the extremal equation

$$R(\widehat{\rho}_{\mathrm{MLME}})\,\widehat{\rho}_{\mathrm{MLME}} = \frac{G}{\eta_{\mathrm{MLME}}}\,\widehat{\rho}_{\mathrm{MLME}} \qquad (3.2.12)$$

that is satisfied by $\widehat{\rho}_{\mathrm{MLME}}$ is as obtained previously for any ML estimator. To speed up the numerical computation process, one can again approximate the variation of the exponential using Eq. (3.1.38), which will lead to the simplification

$$\delta\mu_j^{(k)} \approx \frac{1}{2}\mathrm{tr}\left\{\rho_k\left\{\Pi'_{j,\beta}\left[R(\rho_k) - \frac{G}{\eta^{(k)}}\right] + \left[R(\rho_k) - \frac{G}{\eta^{(k)}}\right]\Pi'_{j,\beta}\right\}\right\}.$$
$$(3.2.13)$$

To summarize, the MLME procedure for imperfect detections can be carried out in two steps:

(1) A set of reference ML probabilities $\{\widehat{p}_{j,0}^{\mathrm{ML}}\}$ is computed by purely maximizing the generalized log-likelihood for the relative probabilities.
(2) The equations for MLME are solved iteratively using these reference probabilities until the extremal equation for $\widehat{\rho}_{\mathrm{MLME}}$ is obeyed within a specified precision.

3.3   An easy trick for constrained function optimization

We discussed an approach to maximize the entropy function over the ML convex set by, instead, maximizing the likelihood over all states of the maximum-entropy form. This approach requires the computation of integrals that are, in general, quite demanding when the Hilbert-space dimension is large. Approximations to these integrals are introduced to improve the computation speed. However, as a result of these approximations, the schemes frequently compute statistical operators along directions that deviate significantly from the steepest-ascent directions. This can cause an overall slowdown in computation since the number of iterative steps required to obtain the MLME estimator within some pre-chosen accuracy is now larger. It is therefore more feasible to search for an estimator that is *close* to the actual MLME estimator using a faster algorithm that completely avoids integrals of any kind that are difficult to evaluate.

For this, we would need to rethink our strategy a little. Previously, we tried to look for the maximum-entropy estimator that maximizes the likelihood. We did this by enforcing the estimator to take the maximum-entropy form. Doing so has led to an iterative equation that involves an integral, which evaluation turns out to be the slow step for the MLME algorithm. Let us do things in a different way. We can alternatively write down a Lagrange function that involves two functions, assuming first that $\sum_j \Pi_j = 1$:

$$D_{\mathrm{L}}(\lambda; \rho) = \lambda \left(S(\rho) - S_{\mathrm{max}} \right) + \frac{1}{N} \log \mathcal{L}(\{n_j\}; \rho). \qquad (3.3.1)$$

Maximizing this Lagrange function is the same as maximizing the log-likelihood $\frac{1}{N} \log \mathcal{L}(\{n_j\}; \rho)$, subject to the constraint that the entropy function $S(\rho)$ takes the maximum value S_{max}, which is precisely our original objective. The advantage of optimizing $D_{\mathrm{L}}(\lambda; \rho)$ is that we can perform the maximization by the usual, more general parametrization of ρ that just ensures positivity and unit trace. Putting aside the issue of finding out the appropriate value of λ for the moment, we first carry out the maximization of the function $D_{\mathrm{L}}(\lambda; \rho)$, which requires the variation

$$\delta D_{\mathrm{L}}(\lambda; \rho) = -\lambda \operatorname{tr}\{\delta \rho \log \rho\} + \sum_j \frac{f_j}{p_j} \delta p_j$$

$$= \operatorname{tr}\{\delta \rho \left[-\lambda \log \rho + R(\rho) \right]\}. \qquad (3.3.2)$$

Using the standard parametrization given by Eq. (2.1.8),

$$\delta D_{\mathrm{L}}(\lambda; \rho)$$
$$= \operatorname{tr}\left\{ \left[-\lambda \log \rho + R(\rho) \right] \left(\frac{\delta A^\dagger A + A^\dagger \delta A}{\operatorname{tr}\{A^\dagger A\}} - \rho \operatorname{tr}\left\{ \frac{\delta A^\dagger A + A^\dagger \delta A}{\operatorname{tr}\{A^\dagger A\}} \right\} \right) \right\}$$
$$= \operatorname{tr}\left\{ \delta A^\dagger \frac{A\, R_\lambda(\rho)}{\operatorname{tr}\{A^\dagger A\}} + \frac{R_\lambda(\rho) A^\dagger}{\operatorname{tr}\{A^\dagger A\}} \delta A \right\}, \qquad (3.3.3)$$

with

$$R_\lambda(\rho) = R(\rho) - 1 - \lambda(\log \rho - \operatorname{tr}\{\rho \log \rho\}). \qquad (3.3.4)$$

The estimator that maximizes $D_{\mathrm{L}}(\lambda; \rho)$ now obeys the extremal equations

$$\widehat{\rho}_\lambda R_\lambda(\rho) = R_\lambda(\rho)\widehat{\rho}_\lambda = 0. \qquad (3.3.5)$$

As before, to solve this equation, we consider the steepest-ascent method and derive the iterative equations that lead to the estimator $\widehat{\rho}_\lambda$. Since the structure of the variation of the Lagrange function $\mathcal{D}_{\mathrm{L}}(\lambda; \rho)$ is the same as that for the log-likelihood, we simply write down the final equations:

$$\rho_{k+1} = \frac{\left[1 + \frac{\epsilon'}{2} R_\lambda(\rho_k)\right] \rho_k \left[1 + \frac{\epsilon'}{2} R_\lambda(\rho_k)\right]}{\mathrm{tr}\left\{\left[1 + \frac{\epsilon'}{2} R_\lambda(\rho_k)\right] \rho_k \left[1 + \frac{\epsilon'}{2} R_\lambda(\rho_k)\right]\right\}}. \tag{3.3.6}$$

We would like to make a small remark regarding the iterative equation in Eq. (3.3.6). The expression of $R_\lambda(\rho_k)$ contains the term $\log \rho_k$. In principle, for rank-deficient ρ_ks, their logarithms are *ill-defined*, since the logarithm of zero is minus infinity. Fortunately, the severity of this problem is mitigated by realizing that a rank-deficient matrix stored on a computer is only *approximately* rank-deficient, especially when such a matrix originates from a series of iterative steps that begins with an initial matrix that is full-rank. As such, this algorithm is numerically stable. For added assurance one may add a very small offset that is about one or two orders of magnitude larger than the numerical precision to completely eliminate this problem.

How should we modify the second equation in Eq. (3.3.6) for imperfect POM outcomes $\sum_j \Pi'_j < 1$? As discussed in Secs. 2.2 and 3.2, since the log-likelihood now depends on the relative probabilities, its variation yields an additional term $\dfrac{G}{\eta}$, where $\eta = \sum_j p_j < 1$. Without going through the same calculations, we simply point out that in this case, the usual $R(\rho_k)$ is to be replaced by

$$R_\lambda(\rho_k) = R(\rho_k) - \frac{G}{\eta^{(k)}} - \lambda(\log \rho_k - \mathrm{tr}\{\rho_k \log \rho_k\}). \tag{3.3.7}$$

Let us look for the appropriate value of λ such that the resulting estimator is the MLME estimator. When $\lambda = 0$, the Lagrange function $\mathcal{D}_{\mathrm{L}}(\lambda = 0; \rho)$ is just the log-likelihood divided by N. Hence, maximizing $\mathcal{D}_{\mathrm{L}}(\lambda = 0; \rho)$ with respect to ρ will give an ML estimator $\widehat{\rho}_{\mathrm{ML}}$. Owing to the informational incompleteness, there exists a convex plateau structure for $\mathcal{D}_{\mathrm{L}}(\lambda = 0; \rho)$. When λ takes a very large value, the entropy term dominates, and maximizing $\mathcal{D}_{\mathrm{L}}(\lambda \gg 1; \rho)$ will give the maximally-mixed state $\dfrac{1}{D}$ as λ approaches infinity. Naturally, when λ takes on a very small positive value, the contribution from $\lambda S(\rho)$ becomes much smaller than that from the log-likelihood term, and the change in the entropy function is only significant over the plateau region, within which the log-likelihood is maximal.

This means that, ideally, λ should be *infinitesimal* and maximizing $\mathcal{D}_{\mathrm{L}}(\lambda \to 0; \rho)$ will result in the MLME estimator $\widehat{\rho}_{\lambda \to 0} = \widehat{\rho}_{\mathrm{MLME}}$.

Alternatively, one can arrive at this conclusion upon inspecting the variation of $\mathcal{D}_\mathrm{L}(\lambda;\rho)$ with respect to ρ in more general terms:

$$\frac{\delta}{\delta\rho}\mathcal{D}_\mathrm{L}(\lambda;\rho) = \lambda\frac{\delta}{\delta\rho}S(\rho) + \frac{1}{N}\frac{\delta}{\delta\rho}\log\mathcal{L}(\{n_j\};\rho). \qquad (3.3.8)$$

Since $\dfrac{\delta}{\delta\rho}\log\mathcal{L}(\{n_j\};\rho)$ is zero on the plateau and, clearly, $\dfrac{\delta}{\delta\rho}S(\rho)$ is not zero for the ML estimators in general, it implies that λ has to approach an infinitesimal value for the derivative on the left-hand side of Eq. (3.3.8) to vanish.

We have therefore turned a constrained optimization problem into an unconstrained one. In practice, however, there is a limit to how small λ can be. If λ is too small, the change in the entropy $S(\rho)$ will be undetectable compared to the numerical precision, so that the algorithm becomes that for ML. The value of λ may be chosen such that below which there are no appreciable variations in the entropy and likelihood.

Problem 3.12 Use the constrained function optimization trick discussed in this section to minimize the function $f(x,y) = (x-1)^2 + y^3 - 2xy$ subject to the constraint $x^2 + (y-1)^2 < 1$ by deriving a steepest-ascent algorithm.

3.4 Some remarks

As we shall investigate in the next chapter, there are certain types of light sources that are described by infinite-dimensional statistical operators. With only finite amount of resources available, any quantum-state estimation protocols for these sources necessarily involve informationally incomplete data. Unlike quantum-state estimation for finite-dimensional Hilbert spaces, where the statistical operators can be stored on a computer as matrices of finite dimensions, the estimation of statistical operators that describe quantum systems of infinite-dimensional Hilbert spaces involves an additional step of choosing a suitable reconstruction subspace on which all operators are parameterized.

A naive strategy to assign a unique estimator from the measurement data would be to choose a reconstruction subspace that has the largest possible dimension such that the POM used is informationally complete, that is, the dimension D_rec of the reconstruction subspace is equal to

D_{LI}. With this reconstruction subspace, the observer can obtain a unique estimator as the data are informationally complete for this subspace. Such a strategy can often result in estimators that give highly inaccurate descriptions since it makes no reference to any other information about the source, which are important considerations that need to be taken alongside the measurement data in order to reconstruct the unknown state as accurately as possible.

Perhaps a more objective approach would be to choose a larger reconstruction subspace that is compatible with all prior information about the light source and carry out state estimation with more sophisticated methods using data that are now informationally incomplete relative to the reconstruction subspace, such as the techniques that were introduced in the preceding sections. This provides additional degrees of freedom in the state parameters to be determined and can help to improve the accuracy, say the mean squared error, for the state reconstruction with respect to the true state. The accuracy depends on the credibility of the prior information that has been taken into account when choosing the reconstruction subspace.

There are many ways of choosing a reconstruction subspace that is compatible with the prior information about the source. One straightforward way is to *truncate* the statistical operator in some pre-chosen basis and keep only the first $D_{\text{rec}} \times D_{\text{rec}}$ sector of parameters that are consistent with the prior information:

$$
\widehat{\rho} = \overbrace{\begin{pmatrix} \rho_{11} & \rho_{12} & \cdots & \rho_{1D_{\text{rec}}} \\ \rho_{21} & \rho_{22} & \cdots & \rho_{2D_{\text{rec}}} \\ \vdots & \vdots & \ddots & \vdots \\ \rho_{D_{\text{rec}}1} & \rho_{D_{\text{rec}}2} & \cdots & \rho_{D_{\text{rec}}D_{\text{rec}}} \end{pmatrix}}^{D_{\text{rec}} \times D_{\text{rec}} \text{ sector}} \left.\begin{matrix} \\ \\ \\ \\ \end{matrix}\right\} \text{truncated entries} . \tag{3.4.1}
$$

Such a truncation can also be performed in another basis and the accuracy of $\widehat{\rho}$ strongly depends on the basis in which matrix truncation is carried out. One might ask: "For a given POM and reconstruction method, what is the optimal basis for the reconstruction subspace that minimizes the mean squared error of an estimator that is averaged over all true states?" Unfortunately, this question does not have a simple straight answer because the positivity constraint imposed on estimators makes analytical studies of such problems extremely difficult.

Further reading

(1) E. T. Jaynes, Information theory and statistical mechanics, *Phys. Rev.* **106**, 620 (1957); Information theory and statistical mechanics II, *Phys. Rev.* **108**, 171 (1957); *Probability Theory: The Logic of Science*, Cambridge University Press (2003).
[Articles and reference book on the application of the maximum-entropy principle in science.]

(2) I. Bengtsson and K. Życzkowski, *Geometry of Quantum States: An Introduction to Quantum Entanglement*, Cambridge University Press (2007).
[A reference book on random operators and mathematical descriptions of the state space.]

(3) Y. S. Teo, J. Řeháček, and Z. Hradil, Informationally incomplete quantum tomography, *Quantum Measurements and Quantum Metrology* **1**, 57 (2003).
[A topical review.]

Chapter 4

Practical Aspects of State Estimation

4.1 Accuracy of unbiased state estimation

Hypothetically, in the absence of statistical fluctuation ($N = \infty$), the unique ML estimator $\widehat{\rho}_{\mathrm{ML}}$ that is constructed with an informationally complete POM is exactly equal to the true state, since the measurement frequency data f_j are just the true probabilities. However, resources in any real quantum-state tomography experiments are always limited. Therefore, in reality, the tomographic accuracy of the resulting estimator will depend on the finite number of copies N. In practice, it is useful to obtain a ballpark estimate for N required to achieve a certain desired level of reconstruction precision.

In this section, we will investigate the tomographic accuracy of the ML estimation scheme. Owing to the complexity of the resulting ML estimators, we shall only focus on situations where a sufficiently large number of copies is measured. In this regime, we would be able to obtain good approximations for the tomographic accuracy of the ML estimation scheme by working with the LIN estimators as they have well-defined mathematical structures for an analytical study. In general, the accuracy of a given estimator can be quantified in several different ways. Here, we shall take the normalized Hilbert–Schmidt distance or squared error defined by

$$\mathcal{D}_{\mathrm{H-S}}\left(\widehat{\rho}_{\mathrm{LIN}}, \rho_{\mathrm{true}}\right) = \frac{1}{2}\mathrm{tr}\left\{\left(\widehat{\rho}_{\mathrm{LIN}} - \rho_{\mathrm{true}}\right)^2\right\}, \qquad (4.1.1)$$

as a measure that quantifies the distance between $\widehat{\rho}_{\mathrm{LIN}}$ and the true state ρ_{true}. We will study the dependence of $\mathcal{D}_{\mathrm{H-S}}\left(\widehat{\rho}_{\mathrm{LIN}}, \rho_{\mathrm{true}}\right)$ on N in the large-N

limit, where the probability distribution for the unbiased estimator $\widehat{\rho}_{\text{LIN}}$,

$$p(\mathbb{D}|\widehat{\rho}_{\text{LIN}}) = \frac{1}{\sqrt{(2\pi)^{D^2} \det\{\mathbf{C}_{\text{LIN}}\}}} \exp\left(-\frac{1}{2}\left(\boldsymbol{t}_{\text{LIN}} - \boldsymbol{r}\right) \cdot \mathbf{C}_{\text{LIN}}^{-1} \cdot \left(\boldsymbol{t}_{\text{LIN}} - \boldsymbol{r}\right)\right),$$

$$(4.1.2)$$

is a Gaussian distribution centered around ρ_{true} because of the central limit theorem. Here, $\boldsymbol{t}_{\text{LIN}}$ is the column of coefficients for $\widehat{\rho}_{\text{LIN}}$. The covariance dyadic $\mathbf{C}_{\text{LIN}}$ quantifies the statistical spread of $\widehat{\rho}_{\text{LIN}}$ from ρ_{true}:

$$\mathbf{C}_{\text{LIN}} = \overline{\left(\boldsymbol{t}_{\text{LIN}} - \boldsymbol{r}\right)\left(\boldsymbol{t}_{\text{LIN}} - \boldsymbol{r}\right)^{\text{T}}}. \qquad (4.1.3)$$

We emphasize that this Gaussian distribution is only valid for the unbiased estimators $\widehat{\rho}_{\text{LIN}}$, where the positivity constraint may not be obeyed, and the uncertainty hypervolume is a hyperellipsoid that is centered at the true state.

For the unbiased estimator $\widehat{\rho}_{\text{LIN}}$, we know that there is a closed-form expression in terms of the dual operators Θ_j for an informationally complete POM: $\widehat{\rho}_{\text{LIN}} = \sum_j f_j \Theta_j$. Since we are interested in the normalized Hilbert–Schmidt distance measure that is *averaged* over all measurement data, or the mean squared error, we need

$$\overline{\mathcal{D}_{\text{H–S}}\left(\widehat{\rho}_{\text{LIN}}, \rho_{\text{true}}\right)} = \frac{1}{2}\,\text{Sp}\left\{\mathbf{C}_{\text{LIN}}\right\}, \qquad (4.1.4)$$

where the covariance dyadic

$$\mathbf{C}_{\text{LIN}} = \overline{\left(\boldsymbol{t}_{\text{LIN}} - \boldsymbol{r}\right)\left(\boldsymbol{t}_{\text{LIN}} - \boldsymbol{r}\right)^{\text{T}}}$$

$$= \overline{\boldsymbol{t}_{\text{LIN}}\,\boldsymbol{t}_{\text{LIN}}^{\text{T}}} - \overline{\boldsymbol{r}\,\boldsymbol{r}^{\text{T}}}. \qquad (4.1.5)$$

In the superket notation, the dyadic components of $\mathbf{C}_{\text{LIN}}$ are

$$\left(\mathbf{C}_{\text{LIN}}\right)_{jk} = \sum_{l,m=1}^{M} \left(\overline{f_l f_m} - p_l p_m\right) \langle\!\langle \Theta_l | \Gamma_j \rangle\!\rangle \langle\!\langle \Gamma_k | \Theta_m \rangle\!\rangle \qquad (4.1.6)$$

as a result of Eq. (1.3.1). In the present context, p_j is the simplified notation for the true probabilities of ρ_{true}.

The averaging is carried out over all measurement data that comes from a multinomial distribution, and with this in mind, we can calculate the term

$\overline{f_j f_k}$ for the set of M outcomes:

$$
\overline{f_j f_k} = \sum_{\{n_l\}} f_j f_k \, \frac{N!}{n_1! \, n_2! \, \dots \, n_M!} \, p_1^{n_1} p_2^{n_2} \dots p_M^{n_M}
$$

$$
= \frac{1}{N^2} \sum_{\{n_l\}} n_j n_k \, \frac{N!}{n_1! \, n_2! \, \dots \, n_M!} \, p_1^{n_1} p_2^{n_2} \dots p_M^{n_M}
$$

$$
= \frac{1}{N^2} \left(p_j \frac{\partial}{\partial p_j} \right) \left(p_k \frac{\partial}{\partial p_k} \right) \left. \left(\sum_{l=1}^{M} p_l \right)^{N} \right|_{\sum_l p_l = 1}
$$

$$
= \frac{1}{N} \left(p_j \frac{\partial}{\partial p_j} \right) \left. \left[p_k \left(\sum_{l=1}^{M} p_l \right)^{N-1} \right] \right|_{\sum_l p_l = 1}
$$

$$
= \frac{1}{N} [\delta_{j,k} p_k + (N-1) p_j p_k]. \tag{4.1.7}
$$

The dyadic components in Eq. (4.1.6) reduces to

$$
(\mathbf{C}_{\mathrm{LIN}})_{jk} = \frac{1}{N} \sum_{l,m=1}^{M} (\delta_{l,m} p_m - p_l p_m) \, \langle\!\langle \Theta_l | \Gamma_j \rangle\!\rangle \, \langle\!\langle \Gamma_k | \Theta_m \rangle\!\rangle
$$

$$
= \frac{1}{N} \left(\sum_{l=1}^{M} p_l \, \langle\!\langle \Theta_l | \Gamma_j \rangle\!\rangle \, \langle\!\langle \Gamma_k | \Theta_l \rangle\!\rangle - \langle\!\langle \rho_{\mathrm{true}} | \Gamma_j \rangle\!\rangle \, \langle\!\langle \Gamma_k | \rho_{\mathrm{true}} \rangle\!\rangle \right).
$$

$$
\tag{4.1.8}
$$

Using the identity in Eq. (1.2.71) for the complete set of superkets $|\Gamma_j\rangle\!\rangle$, the average distance becomes

$$
\overline{\mathcal{D}_{\mathrm{H-S}} \left(\widehat{\rho}_{\mathrm{LIN}}, \rho_{\mathrm{true}} \right)} = \frac{1}{2N} \left(\sum_{j=1}^{M} p_j \operatorname{tr}\{\Theta_j^2\} - \operatorname{tr}\{\rho_{\mathrm{true}}^2\} \right). \tag{4.1.9}
$$

Problem 4.1   Calculate $\overline{\mathcal{D}_{\mathrm{H-S}} \left(\widehat{\rho}_{\mathrm{LIN}}, \rho_{\mathrm{true}} \right)}$ for SIC POMs of dimension D using LS dual operators. Compare this answer with that of the six-outcome POM with LS dual operators for $D = 2$.

For a full-rank true state, $\hat{\rho}_{\text{LIN}}$ is always full-rank and positive for sufficiently large N. Hence, the formula in Eq. (4.1.9) quantifies the tomographic accuracies of positive ML estimators in this regime. For rank-deficient true states, since ρ_{true} is now on the boundary of the state space, there will always be a fair chance of obtaining nonpositive $\hat{\rho}_{\text{LIN}}$s that lie outside the state space regardless of the value of N. So, in general, there may be an appreciable difference between the two average distances unless N is large enough such that this difference is inconsequential. Nevertheless, the quantity $\overline{\mathcal{D}_{\text{H-S}}\left(\hat{\rho}_{\text{LIN}}, \rho_{\text{true}}\right)}$ can still provide good ballpark estimates.

Given a true state, it is possible to find the set of dual operators that attains the minimum average Hilbert–Schmidt distance $\overline{\mathcal{D}_{\text{H-S}}\left(\hat{\rho}_{\text{LIN}}, \rho_{\text{true}}\right)}$. This optimal set, like any other set of dual operators, obeys the defining property stated in Eq. (1.2.44). The relevant Lagrange function for this minimization is

$$\mathcal{D}_{\text{L}}(\{|\Theta_j\rangle\!\rangle\}) = \overline{\mathcal{D}_{\text{H-S}}\left(\hat{\rho}_{\text{LIN}}, \rho_{\text{true}}\right)}$$
$$- \sum_{l,m=1}^{D^2} \mu_{ml} \left[\left(\sum_{j=1}^{M} |\Theta_j\rangle\!\rangle \langle\!\langle\Pi_j| \right)_{lm} - \delta_{l,m} \right]. \quad (4.1.10)$$

The set of constraints accompanied by the Lagrange multipliers μ_{jk} can be written with matrix notations,

$$\sum_{l,m=1}^{D^2} \mu_{ml} \left[\left(\sum_{j=1}^{M} |\Theta_j\rangle\!\rangle \langle\!\langle\Pi_j| \right)_{lm} - \delta_{l,m} \right]$$
$$= \text{Tr}\left\{ \mathcal{M} \left(\sum_{j=1}^{M} |\Theta_j\rangle\!\rangle \langle\!\langle\Pi_j| - \mathcal{I} \right) \right\}, \quad (4.1.11)$$

where $\mathcal{M}$ is the Lagrange superoperator for the operator constraint and $\text{Tr}\{\cdot\}$ denotes the superoperator trace.

A variation on the Lagrange function $\mathcal{D}_{\text{L}}(\{|\Theta_j\rangle\!\rangle\})$ yields

$$\delta\mathcal{D}_{\text{L}}(\{|\Theta_j\rangle\!\rangle\}) = \frac{1}{2N} \sum_{j=1}^{M} p_j \left(\delta\langle\!\langle\Theta_j| \, |\Theta_j\rangle\!\rangle + \langle\!\langle\Theta_j| \, \delta|\Theta_j\rangle\!\rangle \right)$$
$$- \frac{1}{2} \sum_{j=1}^{M} \left(\delta\langle\!\langle\Theta_j| \, \mathcal{M} \, |\Pi_j\rangle\!\rangle + \langle\!\langle\Pi_j| \, \mathcal{M} \, \delta|\Theta_j\rangle\!\rangle \right), \quad (4.1.12)$$

where for convenience, we have split the operator constraint into two equal pieces. Setting this variation to zero results in the extremal equation

$$|\Theta_j\rangle\rangle \, p_j = N\mathcal{M} \, |\Pi_j\rangle\rangle. \tag{4.1.13}$$

To solve for the constant Lagrange superoperator $\mathcal{M}$, we again make use of the dual property in Eq. (1.2.44):

$$\mathcal{I} = \sum_{j=1}^{M} |\Theta_j\rangle\rangle \langle\langle\Pi_j| = N\mathcal{M} \underbrace{\sum_{j=1}^{M} \frac{|\Pi_j\rangle\rangle \langle\langle\Pi_j|}{p_j}}_{\equiv \mathcal{F}(\rho_{\text{true}})}$$

$$\Rightarrow \mathcal{M} = \frac{1}{N}\mathcal{F}(\rho_{\text{true}})^{-1}. \tag{4.1.14}$$

The optimal set of dual operators are therefore related to the corresponding POM outcomes *via* the frame superoperator $\mathcal{F}(\rho_{\text{true}})$ that also depends on the true state ρ_{true},

$$|\Theta_j\rangle\rangle = \frac{\mathcal{F}(\rho_{\text{true}})^{-1} \, |\Pi_j\rangle\rangle}{p_j}. \tag{4.1.15}$$

One can check that $\text{tr}\{\Theta_j\} = 1$ by identifying the relation

$$\mathcal{F}(\rho) \, |\rho\rangle\rangle = \sum_{j=1}^{M} \frac{|\Pi_j\rangle\rangle \langle\langle\Pi_j|\rho\rangle\rangle}{\text{tr}\{\rho\,\Pi_j\}} = \sum_{j=1}^{M} |\Pi_j\rangle\rangle = |1\rangle\rangle, \tag{4.1.16}$$

so that

$$\text{tr}\{\Theta_j\} = \frac{\langle\langle 1| \, \mathcal{F}(\rho_{\text{true}})^{-1}| \Pi_j\rangle\rangle}{p_j} = \frac{\langle\langle\rho_{\text{true}}|\Pi_j\rangle\rangle}{p_j} = 1. \tag{4.1.17}$$

For a given informationally complete POM, one can directly compute this optimal set of dual operators. However, the true state ρ_{true}, which is a necessary ingredient, is of course always unknown. To approximate these optimal dual operators, one may, for instance, replace the true probabilities with the measurement frequencies f_j, so that the operators

$$|\Theta_j\rangle\rangle = \mathcal{N} \left(\sum_{j=1}^{M} \frac{|\Pi_j\rangle\rangle \langle\langle\Pi_j|}{f_j} \right)^{-1} |\Pi_j\rangle\rangle \frac{1}{f_j} \tag{4.1.18}$$

that are now nonlinear in the measurement data, with the proper operator-trace normalization constant $\mathcal{N}$, serve as a good alternative for sufficiently large N.

> **Problem 4.2** Show that the optimal dual operators defined in both Eq. (4.1.15) and Eq. (4.1.18) are equal to the canonical dual operators in Eq. (1.2.50) for minimally complete POMs, as they should be.

To circumvent the problem of unknown true states, one may consider another measure for the tomographic accuracy that is more operational. This measure is derived by taking the average over all true states that are *unitarily equivalent*, with the state

$$\bar{\rho}^{\{U\}} = \int (dU')_{\mathrm{H}} \, U' \rho_{\mathrm{true}} U'^{\dagger}, \quad \int (dU')_{\mathrm{H}} = 1, \qquad (4.1.19)$$

defined as an integral average over all unitary transformations on ρ_{true} with respect to the *Haar* measure* $(dU')_{\mathrm{H}}$ for the group of D-dimensional unitary operators U'. This measure has the property that for a given unitary operator V,

$$(dVU')_{\mathrm{H}} = (dU')_{\mathrm{H}} = (dU'V)_{\mathrm{H}} \,. \qquad (4.1.20)$$

An example of such a measure is the rotationally-invariant measure for the single-qubit case presented in Sec. 1.2.1.1.

If we apply a unitary transformation on $\bar{\rho}^{\{U\}}$ with the unitary operator U, we have

$$
\begin{aligned}
U\bar{\rho}^{\{U\}}U^{\dagger} &= \int (dU')_{\mathrm{H}} \underbrace{UU'}_{\equiv V'} \rho_{\mathrm{true}} \underbrace{U'^{\dagger}U^{\dagger}}_{\equiv V'^{\dagger}} \\
&= \int \left(dU^{\dagger}V'\right)_{\mathrm{H}} V' \rho_{\mathrm{true}} V'^{\dagger} \\
&= \int (dV')_{\mathrm{H}} V' \rho_{\mathrm{true}} V'^{\dagger} \\
&= \bar{\rho}^{\{U\}}.
\end{aligned}
\qquad (4.1.21)
$$

---

*Alfréd Haar (1885–1933).

This implies that the average statistical operator $\bar{\rho}^{\{U\}}$ must be the maximally-mixed state $1/D$, since the equalities in (4.1.21) are true for any U, a consequence of the *Schur* lemma*. As such, an integration over the Haar measure gives the average distance

$$\overline{\mathcal{D}_{\text{H-S}}\left(\{\Pi_j\};\{\Theta_j\}\right)} = \frac{1}{2N}\left(\frac{1}{D}\sum_{j=1}^{M}\text{tr}\{\Pi_j\}\text{tr}\{\Theta_j^2\} - \text{tr}\{\rho_{\text{true}}^2\}\right), \quad (4.1.22)$$

where the purity $\text{tr}\{\rho_{\text{true}}^2\}$ may be neglected for general purposes since this value is bounded from above by one and is a small modification to the other term in the sum for moderately large dimensions.

> **Problem 4.3** Find the optimal set of dual operators that achieves a minimum $\overline{\mathcal{D}_{\text{H-S}}\left(\{\Pi_j\};\{\Theta_j\}\right)}$ for LIN estimators.

Another quantity of practical interest is the measure of the statistical fluctuation of the mean squared error between $\hat{\rho}_{\text{LIN}}$ and ρ_{true}. This quantity is the variance of $\mathcal{D}_{\text{H-S}}\left(\hat{\rho}_{\text{LIN}}, \rho_{\text{true}}\right)$ and is defined to be

$$\text{Var}\left\{\mathcal{D}_{\text{H-S}}\left(\hat{\rho}_{\text{LIN}}, \rho_{\text{true}}\right)\right\} = \frac{1}{4}\overline{\text{tr}\left\{\left(\hat{\rho}_{\text{LIN}} - \rho_{\text{true}}\right)^2\right\}^2} - \frac{1}{4}\overline{\text{tr}\left\{\left(\hat{\rho}_{\text{LIN}} - \rho_{\text{true}}\right)^2\right\}}^2$$

$$= \frac{1}{4}\overline{\left(t_{\text{LIN}} - r\right)^4} - \frac{1}{4}\text{Sp}\left\{\mathbf{C}_{\text{LIN}}\right\}^2. \quad (4.1.23)$$

For large N, the first term can be expressed in terms of the Gaussian distribution of $\hat{\rho}_{\text{LIN}}$ about ρ_{true} as

$$\overline{\left(t_{\text{LIN}} - r\right)^4} = \frac{1}{\sqrt{(2\pi)^{D^2}\text{Det}\{\mathbf{C}_{\text{LIN}}\}}}\int (dt)(t_{\text{LIN}} - r)^4$$

$$\times \exp\left(-\frac{1}{2}(t_{\text{LIN}} - r)^\mathsf{T}\mathbf{C}_{\text{LIN}}^{-1}(t_{\text{LIN}} - r)\right). \quad (4.1.24)$$

To evaluate the integral in Eq. (4.1.24), we first look at the general integral

$$\int (d\boldsymbol{x})\left(\boldsymbol{x}^\mathsf{T}\boldsymbol{x}\right)^2 e^{-\frac{1}{2}\boldsymbol{x}^\mathsf{T}\mathbf{A}\boldsymbol{x}}, \quad (4.1.25)$$

---

*Issai Schur (1875–1941).

where the real dyadic $\mathbf{A} \geq 0$ and $\boldsymbol{x} = (x_1 \ x_2 \ \ldots \ x_D)^{\mathrm{T}}$ is a real column of dimension κ. It is easy to figure out that the integral can be evaluated by thinking of the integrand as a cascade of dyadic derivatives of $\mathbf{A}$. So to facilitate subsequent calculations, let us understand the action of such a dyadic derivative. We may do so by starting with the variation

$$\delta \left(\boldsymbol{x}^{\mathrm{T}} \mathbf{A} \boldsymbol{x} \right) = \delta \operatorname{Sp} \{ \boldsymbol{x} \boldsymbol{x}^{\mathrm{T}} \mathbf{A} \}$$
$$= \operatorname{Sp} \{ \boldsymbol{x} \boldsymbol{x}^{\mathrm{T}} \delta \mathbf{A} \}, \qquad (4.1.26)$$

which allows us to make the identification

$$\frac{\delta}{\delta \mathbf{A}} \left(\boldsymbol{x}^{\mathrm{T}} \mathbf{A} \boldsymbol{x} \right) = \boldsymbol{x} \boldsymbol{x}^{\mathrm{T}}. \qquad (4.1.27)$$

For the integral evaluation, we would also need the variation on $\operatorname{Det}\{\mathbf{A}\}$. For this, it may be more convenient to consider the logarithmic $\log(\operatorname{Det}\{\mathbf{A}\})$, and since $\log(\operatorname{Det}\{\mathbf{A}\}) = \operatorname{tr}\{\log \mathbf{A}\}$,

$$\delta \log(\operatorname{Det}\{\mathbf{A}\}) = \delta \operatorname{tr}\{\log \mathbf{A}\} = \operatorname{tr}\{\mathbf{A}^{-1}\delta \mathbf{A}\}, \qquad (4.1.28)$$

or

$$\frac{\delta}{\delta \mathbf{A}} \operatorname{Det}\{\mathbf{A}\} = \operatorname{Det}\{\mathbf{A}\} \, \mathbf{A}^{-1}. \qquad (4.1.29)$$

The integral in (4.1.25) is therefore given by

$$\int (\mathrm{d}\boldsymbol{x}) \, (\boldsymbol{x}^{\mathrm{T}} \boldsymbol{x})^2 \, \mathrm{e}^{-\frac{1}{2} \boldsymbol{x}^{\mathrm{T}} \mathbf{A} \boldsymbol{x}}$$
$$= -2 \int (\mathrm{d}\boldsymbol{x}) \boldsymbol{x}^{\mathrm{T}} \boldsymbol{x} \operatorname{Sp} \left\{ \frac{\delta}{\delta \mathbf{A}} \, \mathrm{e}^{-\frac{1}{2} \boldsymbol{x}^{\mathrm{T}} \mathbf{A} \boldsymbol{x}} \right\}$$
$$= 4 \operatorname{Sp} \left\{ \frac{\delta}{\delta \mathbf{A}} \operatorname{Sp} \left\{ \frac{\delta}{\delta \mathbf{A}} \int (\mathrm{d}\boldsymbol{x}) \, \mathrm{e}^{-\frac{1}{2} \boldsymbol{x}^{\mathrm{T}} \mathbf{A} \boldsymbol{x}} \right\} \right\}$$
$$= 4 \operatorname{Sp} \left\{ \frac{\delta}{\delta \mathbf{A}} \operatorname{Sp} \left\{ \frac{\delta}{\delta \mathbf{A}} \sqrt{\frac{(2\pi)^\kappa}{\operatorname{Det}\{\mathbf{A}\}}} \right\} \right\}, \qquad (4.1.30)$$

where we have made use of the identity

$$\int (d\boldsymbol{x})\, e^{-\boldsymbol{x}^{\mathrm{T}}\mathbf{A}\boldsymbol{x}+\boldsymbol{b}^{\mathrm{T}}\boldsymbol{x}} = \sqrt{\frac{\pi^{\kappa}}{\mathrm{Det}\{\mathbf{A}\}}}\, e^{\frac{1}{4}\boldsymbol{b}^{\mathrm{T}}\mathbf{A}^{-1}\boldsymbol{b}} \qquad (4.1.31)$$

for any real positive dyadic $\mathbf{A}$ and complex column $\boldsymbol{b}$.

Problem 4.4 Prove this identity using the well-known result

$$\int dy\, e^{-ay^2+by} = \sqrt{\frac{\pi}{a}}\, e^{\frac{b^2}{4a}}, \quad \mathrm{Re}\,\{a\} \geq 0 \qquad (4.1.32)$$

for one-dimensional Gaussian integrals.

The cascade of dyadic derivatives can now be evaluated accordingly as

$$4\,\mathrm{Sp}\left\{\frac{\delta}{\delta\mathbf{A}}\,\mathrm{Sp}\left\{\frac{\delta}{\delta\mathbf{A}}\sqrt{\frac{(2\pi)^{\kappa}}{\mathrm{Det}\{\mathbf{A}\}}}\right\}\right\}$$

$$= -2\sqrt{(2\pi)^{\kappa}}\,\mathrm{Sp}\left\{\frac{\delta}{\delta\mathbf{A}}\left(\mathrm{Det}\{\mathbf{A}\}^{-\frac{1}{2}}\,\mathrm{Sp}\{\mathbf{A}^{-1}\}\right)\right\}$$

$$= -2\sqrt{(2\pi)^{\kappa}}\,\mathrm{Sp}\left\{-\frac{1}{2}\mathrm{Det}\{\mathbf{A}\}^{-\frac{1}{2}}\mathbf{A}^{-1}\,\mathrm{Sp}\{\mathbf{A}^{-1}\} - \mathrm{Det}\{\mathbf{A}\}^{-\frac{1}{2}}\mathbf{A}^{-2}\right\}$$

$$= \sqrt{\frac{(2\pi)^{\kappa}}{\mathrm{Det}\{\mathbf{A}\}}}\left(\mathrm{Sp}\{\mathbf{A}^{-1}\}^2 + 2\,\mathrm{Sp}\{\mathbf{A}^{-2}\}\right), \qquad (4.1.33)$$

after which we arrive at the result

$$\int (d\boldsymbol{x})\,(\boldsymbol{x}^{\mathrm{T}}\boldsymbol{x})^2\, e^{-\frac{1}{2}\boldsymbol{x}^{\mathrm{T}}\mathbf{A}\boldsymbol{x}} = \sqrt{\frac{(2\pi)^{\kappa}}{\mathrm{Det}\{\mathbf{A}\}}}\left(\mathrm{Sp}\{\mathbf{A}^{-1}\}^2 + 2\,\mathrm{Sp}\{\mathbf{A}^{-2}\}\right).$$
$$\qquad (4.1.34)$$

The variance of the squared error can then be calculated to be

$$\mathrm{Var}\,\{\mathcal{D}_{\mathrm{H-S}}\,(\widehat{\rho}_{\mathrm{LIN}},\rho_{\mathrm{true}})\} = \frac{1}{2}\,\mathrm{Sp}\{\mathbf{C}_{\mathrm{LIN}}^2\}, \qquad (4.1.35)$$

in contrast with the expression of the mean squared error given in Eq. (4.1.4). Thus, not surprisingly, the variance goes as the *square* of the reciprocal of N, where the term $\mathrm{Sp}\left\{\mathbf{C}_{\mathrm{LIN}}^2\right\}$ now includes higher moments of the measurement frequencies, $\overline{f_j f_k f_l}$, $\overline{f_j f_k f_l f_m}$, etc. The square of the covariance dyadic $\mathbf{C}_{\mathrm{LIN}}$ has dyadic components

$$
\begin{aligned}
\left(\mathbf{C}_{\mathrm{LIN}}^2\right)_{mn} = \frac{1}{N^2} \Bigg[& \sum_{l,l'=1}^{M} p_l\, p_{l'} \, \langle\!\langle \Theta_l | \Gamma_m \rangle\!\rangle \, \langle\!\langle \Gamma_n | \Theta_{l'} \rangle\!\rangle \, \mathrm{tr}\{\Theta_l \Theta_{l'}\} \\
& - 2 \sum_{l=1}^{M} p_l \, \langle\!\langle \Theta_l | \Gamma_m \rangle\!\rangle \, \langle\!\langle \Gamma_n | \rho_{\mathrm{true}} \rangle\!\rangle \, \mathrm{tr}\{\rho_{\mathrm{true}} \Theta_l\} \\
& + \langle\!\langle \rho_{\mathrm{true}} | \Gamma_m \rangle\!\rangle \, \langle\!\langle \Gamma_n | \rho_{\mathrm{true}} \rangle\!\rangle \, \mathrm{tr}\{\rho_{\mathrm{true}}^2\} \Bigg].
\end{aligned}
\tag{4.1.36}
$$

This leads to the expression

$$
\begin{aligned}
& \mathrm{Var}\left\{\mathcal{D}_{\mathrm{H-S}}\left(\widehat{\rho}_{\mathrm{LIN}}, \rho_{\mathrm{true}}\right)\right\} \\
& = \frac{1}{2\,N^2} \left(\sum_{l,l'=1}^{M} p_l\, p_{l'} \mathrm{tr}\{\Theta_l \Theta_{l'}\}^2 - 2 \sum_{l=1}^{M} p_l \, \mathrm{tr}\{\rho_{\mathrm{true}} \Theta_l\}^2 + \mathrm{tr}\{\rho_{\mathrm{true}}^2\}^2 \right)
\end{aligned}
\tag{4.1.37}
$$

for the variance. Equation (4.1.37) involves higher order terms of ρ_{true} in a nontrivial manner. One can still perform an average over all unitarily equivalent states to arrive at an operational expression. Alternatively, one may estimate the statistical fluctuation of the mean squared error by replacing ρ_{true} in Eq. (4.1.37) with a statistical operator that best represents the quantum state of the source the observer had initially intended to prepare, say an estimator obtained by measuring sufficiently large number of copies.

There exists another figure of merit one can use to quantify the accuracy and statistical fluctuation for the ML scheme, or any other estimation strategy for that matter, namely the *trace-class distance* defined as

$$
\mathcal{D}_{\mathrm{tr}}\left(\widehat{\rho}_{\mathrm{ML}}, \rho_{\mathrm{true}}\right) = \frac{1}{2}\,\mathrm{tr}\{|\widehat{\rho}_{\mathrm{ML}} - \rho_{\mathrm{true}}|\},
\tag{4.1.38}
$$

where $|\mathcal{A}| = \sqrt{\mathcal{A}^\dagger \mathcal{A}}$ for a given complex operator $\mathcal{A}$. The evaluation of the average trace-class distance involves the square root operation that is in general complicated to handle and requires more sophisticated approximation techniques that are beyond the scope of this survey.

Problem 4.5 Given that t represents a column of parameters distributed according to the Gaussian distribution

$$p(t; \mu, \mathbf{C}) = \frac{1}{\sqrt{(2\pi)^M \det\{\mathbf{C}\}}} \exp\left(-\frac{1}{2}\,(t - \mu) \cdot \mathbf{C}^{-1} \cdot (t - \mu)\right) \tag{4.1.39}$$

governed by the mean column μ and the covariance dyadic $\mathbf{C}$, show that the Fisher information dyadic for this distribution is

$$\mathbf{F}_1(a) = -\overline{\frac{\partial}{\partial a}\frac{\partial}{\partial a}} \log p(t; \mu, \mathbf{C})$$

$$= \frac{\partial \mu}{\partial a} \cdot \mathbf{C}^{-1} \cdot \frac{\partial \mu}{\partial a} + \frac{1}{2}\mathrm{Sp}\left\{\mathbf{C}^{-1} \cdot \frac{\partial \mathbf{C}}{\partial a} \cdot \mathbf{C}^{-1} \cdot \frac{\partial \mathbf{C}}{\partial a}\right\}, \tag{4.1.40}$$

where the dyadic trace acts only on the product of $\mathbf{C}$s and $\mathbf{C}^{-1}$s, not on the column derivatives with respect to a.

Problem 4.6 Equation (1.2.35) in **Problem 1.3** is useful for defining dyadic derivatives. For constant dyadics $\mathbf{A}$ and $\mathbf{B}$,

$$\delta|\mathbf{AXB}\rangle\rangle = |\mathbf{A}\delta\mathbf{XB}\rangle\rangle = \mathbf{A} \otimes \mathbf{B}^{\mathsf{T}}|\delta\mathbf{X}\rangle\rangle, \tag{4.1.41}$$

and we may define the dyadic derivative

$$\frac{\delta}{\delta\mathbf{X}}(\mathbf{AXB}) = \frac{\delta}{\delta|\mathbf{X}\rangle\rangle}|\mathbf{AXB}\rangle\rangle = \mathbf{A} \otimes \mathbf{B}^{\mathsf{T}}. \tag{4.1.42}$$

Use this prescription to systematically establish the relations

$$\frac{\delta}{\delta\mathbf{A}}\mathbf{A}^{-1} = -\mathbf{A}^{-1} \otimes \mathbf{A}^{\mathsf{T}-1}, \tag{4.1.43}$$

$$\frac{\delta}{\delta x}(xx^{\mathsf{T}}) = \mathbf{1} \otimes x + x \otimes \mathbf{1}, \tag{4.1.44}$$

for a full-rank dyadics $\mathbf{A}$ and column x.

4.2 Experimental schemes

In this second section, we shall look at some experimental schemes of common measurements. Throughout this section, we shall focus on *photonic* quantum systems that are produced by a light source, such as a laser.

4.2.1 *Discrete-variable measurements — photon polarization*

The simplest degree of freedom that can be conveniently manipulated is the photon polarization degree of freedom, which consists of the two possible orthogonal polarizations $|\text{H}\rangle$ and $|\text{V}\rangle$. Any polarization ket is then a superposition of these two possibilities. Therefore, the Hilbert space for such a qubit[*] is two-dimensional. The measurement of polarization states is thus a *discrete-variable* measurement since the outcomes are discrete. A light source that is appropriate for such polarization measurements is a single-photon source, that is, a source that ideally produces a single photon at any one time.

4.2.1.1 *Six-outcome measurement*

The first measurement that we shall consider is the familiar six-outcome informationally complete POM that was introduced in Sec. 2.1. These are the six rank-one outcomes $(1 \pm \sigma_x)/6$, $(1 \pm \sigma_y)/6$ and $(1 \pm \sigma_z)/6$ that detect the polarization degree of freedom of single photons. For this experimental scheme (refer to Fig. 4.1), we need *beam splitters* (BS), *polarizing* beam

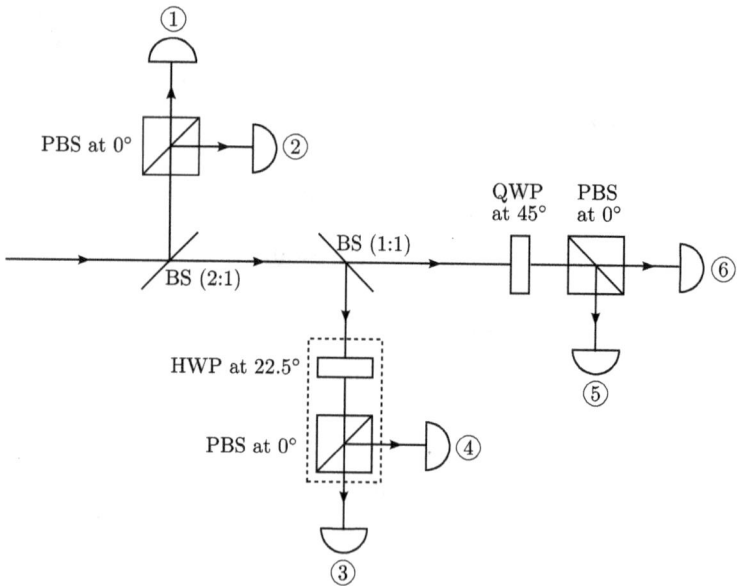

Fig. 4.1 Schematic setup for the six-outcome measurement.

---

[*]A qubit always refers to a quantum-mechanical degree of freedom which, in this case, is the polarization degree of freedom.

splitters (PBS) that fully transmits photons with horizontal polarization $|\mathrm{H}\rangle$ and reflects photons with vertical polarization $|\mathrm{V}\rangle$, a *half-wave* plate (HWP) that rotates the polarization of linearly-polarized light, and a *quarter-wave* plate (QWP) that converts linearly-polarized light to circularly-polarized light and *vice versa*. Here, we use the convention

$$|\mathrm{H}\rangle \mathrel{\hat=} \begin{pmatrix} 1 \\ 0 \end{pmatrix} \quad \text{and} \quad |\mathrm{V}\rangle \mathrel{\hat=} \begin{pmatrix} 0 \\ 1 \end{pmatrix}. \tag{4.2.1}$$

An incoming photon, upon hitting a BS that is capable of transmitting two-thirds of light intensity, has a one-third chance of being re-emitted into the reflected path and be detected by the photodetectors ① and ② after a PBS that is positioned with its optical axis at $0°$ with the pre-chosen "horizontal" direction. The detectors ① and ② measure the respective expectation values $\langle |\mathrm{H}\rangle \langle \mathrm{H}| /3\rangle = \langle (1 + \sigma_z)/6\rangle$ and $\langle |\mathrm{V}\rangle \langle \mathrm{V}| /3\rangle = \langle (1 - \sigma_z)/6\rangle$.

A photon traveling in the transmitted path encounters another BS that equally transmits and reflects light to two other separate paths. In practice, the components in the boxed region may be simplified to a PBS positioned at $45°$. In any case, the action of the HWP at an angle θ can be described by the unitary matrix

$$\mathbf{M}_{\mathrm{HWP}} \left(\theta \right) = \begin{pmatrix} \cos(2\theta) & \sin(2\theta) \\ \sin(2\theta) & -\cos(2\theta) \end{pmatrix}, \tag{4.2.2}$$

which naturally introduces a relative phase of π between the two orthogonal polarization states, a characteristic feature of the HWP. The state measured by ③ is thus described by the ket $|+\rangle$ that is given by

$$|+\rangle \mathrel{\hat=} \mathbf{M}_{\mathrm{HWP}} \left(\frac{\pi}{8} \right) \begin{pmatrix} 1 \\ 0 \end{pmatrix} = \frac{1}{\sqrt{2}} \begin{pmatrix} 1 & 1 \\ 1 & -1 \end{pmatrix} \begin{pmatrix} 1 \\ 0 \end{pmatrix} = \frac{1}{\sqrt{2}} \begin{pmatrix} 1 \\ 1 \end{pmatrix}, \tag{4.2.3}$$

where $\langle |+\rangle \langle +| /3\rangle = \langle (1 + \sigma_x)/6\rangle$ indeed. By the same token, the expectation value $\langle |-\rangle \langle -| /3\rangle = \langle (1 - \sigma_x)/6\rangle$ associated with the ket $|-\rangle\!\rangle$ is measured with detector ④.

Finally, for the path to detectors ⑤ and ⑥, the unitary matrix describing the action of the QWP at the angle θ' is

$$\mathbf{M}_{\mathrm{QWP}} \left(\theta' \right) = \begin{pmatrix} (\cos\theta')^2 + \mathrm{i}\,(\sin\theta')^2 & \sqrt{2}\mathrm{e}^{-\mathrm{i}\frac{\pi}{4}} \sin\theta' \cos\theta' \\ \sqrt{2}\mathrm{e}^{-\mathrm{i}\frac{\pi}{4}} \sin\theta' \cos\theta' & (\sin\theta')^2 + \mathrm{i}\,(\cos\theta')^2 \end{pmatrix}. \tag{4.2.4}$$

With this, the state that is measured by detector ⑤ is then

$$|\mathrm{L}\rangle \,\widehat{=}\, \mathbf{M}_{\mathrm{QWP}}\left(\frac{\pi}{4}\right)\begin{pmatrix}0\\1\end{pmatrix} = \frac{\mathrm{e}^{\mathrm{i}\frac{\pi}{4}}}{\sqrt{2}}\begin{pmatrix}1 & -\mathrm{i}\\-\mathrm{i} & 1\end{pmatrix}\begin{pmatrix}0\\1\end{pmatrix} = \frac{\mathrm{e}^{\mathrm{i}\frac{\pi}{4}}}{\sqrt{2}}\begin{pmatrix}-\mathrm{i}\\1\end{pmatrix}, \qquad (4.2.5)$$

such that $\langle|\mathrm{L}\rangle\langle\mathrm{L}|/3\rangle = \langle(1+\sigma_y)/6\rangle$. The measurement data collected by ⑥ is correspondingly $\langle|\mathrm{R}\rangle\langle\mathrm{R}|/3\rangle = \langle(1-\sigma_y)/6\rangle$, so that, in the end, all the six outcomes coming from three rotated von Neumann measurements are realized.

4.2.1.2   *Trine measurement*

The next POM implementation that we are going to discuss is the standard trine measurement comprising the three outcomes $(1+\sigma_z)/3$, $(1\pm\sqrt{3}\sigma_x/2 - \sigma_z/2)/3$, which are mutually nonorthogonal. To realize this measurement, we shall make use of the *Naimark* theorem*, which instructs us to perform a POM by using von Neumann measurements on a larger Hilbert space.

To prepare such a measurement scheme (refer to Fig. 4.2), an additional common optical component that we need is the *partially polarizing beam splitter* (PPBS) that transmits or reflects light of different polarizations in fixed arbitrary ratios.

The action of the PPBS turns out to be important in setting up a larger Hilbert space to realize a general POM. We label the PPBS with the ratio $(r_\mathrm{H}{:}r_\mathrm{V})$ of real reflection amplitudes, where the polarization state $|\mathrm{H}\rangle\langle\mathrm{H}|$ is reflected with probability r_H^2, and the state $|\mathrm{V}\rangle\langle\mathrm{V}|$ is reflected with

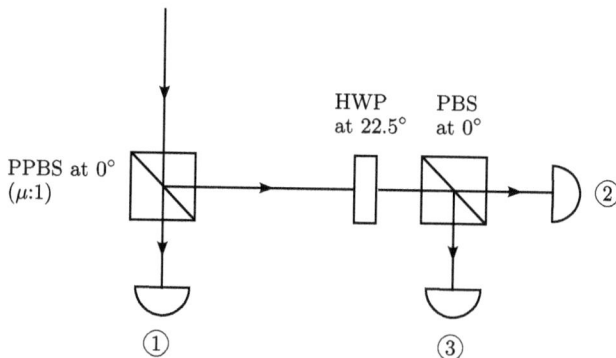

Fig. 4.2   Schematic setup for the trine measurement.

---

*Mark Aronovich Naimark (1909–1978).

probability r_V^2. Let us also define the respective transmission amplitudes t_H and t_V for $|\mathrm{H}\rangle\langle\mathrm{H}|$ and $|\mathrm{V}\rangle\langle\mathrm{V}|$, with

$$t_\mathrm{H}^2 + r_\mathrm{H}^2 = 1 = t_\mathrm{V}^2 + r_\mathrm{V}^2. \tag{4.2.6}$$

The output of the PPBS gives two paths, the transmitted path and the reflected path. In the transmitted path, the action of the PPBS on an incoming quantum state can be described with the unitary matrix

$$\mathbf{M}_\mathrm{PPBS}\left(t_\mathrm{H}, t_\mathrm{V}\right) = \begin{pmatrix} t_\mathrm{H} & 0 \\ 0 & t_\mathrm{V} \end{pmatrix}. \tag{4.2.7}$$

In the reflected path, the relevant matrix is $\mathbf{M}_\mathrm{PPBS}\left(r_\mathrm{H}, r_\mathrm{V}\right)$. Throughout the subsequent discussions, we shall assume that there are no polarization switching effects from the PPBS, so that the off-diagonal elements of the PPBS matrix are all zero.

The PPBS that is chosen for this scheme now provides two additional spatial degrees of freedom for the incoming photons by reflecting the horizontal polarization state with a certain amplitude μ and fully reflects the vertical polarization state. The transmitted path therefore leads to the measurement of the outcome $(1 - \mu^2)(1 + \sigma_z)/2$ at detector ①. It is clear that the value of μ should be $\pm 1/\sqrt{3}$. We need to check, however, if this value indeed gives us the required measurement outcomes in the reflected path. This can be done by multiplying the matrices for the respective optical components in this path.

For instance, the outcome measured by detector ② is represented by

$$\left[\mathbf{M}_\mathrm{PPBS}(\mu, 1)\, \mathbf{M}_\mathrm{HWP}\left(\frac{\pi}{8}\right)\right] \begin{pmatrix} 1 \\ 0 \end{pmatrix} \begin{pmatrix} 1 & 0 \end{pmatrix} \left[\mathbf{M}_\mathrm{HWP}\left(\frac{\pi}{8}\right) \mathbf{M}_\mathrm{PPBS}(\mu, 1)\right]$$

$$= \begin{pmatrix} \mu & 0 \\ 0 & 1 \end{pmatrix} \left[\frac{1}{\sqrt{2}} \begin{pmatrix} 1 & 1 \\ 1 & -1 \end{pmatrix}\right] \begin{pmatrix} 1 \\ 0 \end{pmatrix} \begin{pmatrix} 1 & 0 \end{pmatrix} \left[\frac{1}{\sqrt{2}} \begin{pmatrix} 1 & 1 \\ 1 & -1 \end{pmatrix}\right] \begin{pmatrix} \mu & 0 \\ 0 & 1 \end{pmatrix}$$

$$= \frac{1}{2} \begin{pmatrix} \mu^2 & \mu \\ \mu & 1 \end{pmatrix}. \tag{4.2.8}$$

The corresponding outcome measured by detector ③ is

$$\frac{1}{2} \begin{pmatrix} \mu^2 & -\mu \\ -\mu & 1 \end{pmatrix}. \tag{4.2.9}$$

Upon comparing with the desired trine measurement outcomes

$$\frac{1 \pm \frac{\sqrt{3}}{2}\sigma_x - \frac{1}{2}\sigma_z}{3} \,\hat{=}\, \frac{1}{3}\begin{pmatrix} \frac{1}{2} & \pm\frac{\sqrt{3}}{2} \\ \pm\frac{\sqrt{3}}{2} & \frac{3}{2} \end{pmatrix}$$

$$= \frac{1}{2}\begin{pmatrix} \frac{1}{3} & \pm\frac{1}{\sqrt{3}} \\ \pm\frac{1}{\sqrt{3}} & 1 \end{pmatrix} \qquad (4.2.10)$$

that we would like to measure, we confirm that $\mu = \pm 1/\sqrt{3}$. In this way, the trine measurement is implemented using von Neumann measurements with the PBS on a larger Hilbert space supplied by the PPBS of a suitable specification.

4.2.1.3   *Tetrahedron measurement*

We can take this idea one step forward and try to implement another informationally complete POM that contains exactly four outcomes with a symmetry property. This POM is the tetrahedron measurement, which exemplifying set of Bloch vectors $\{a_j\}$ and their property are listed in Eq. (2.1.2) and Eq. (2.1.3). The scheme (refer to Fig. 4.3) is in fact very

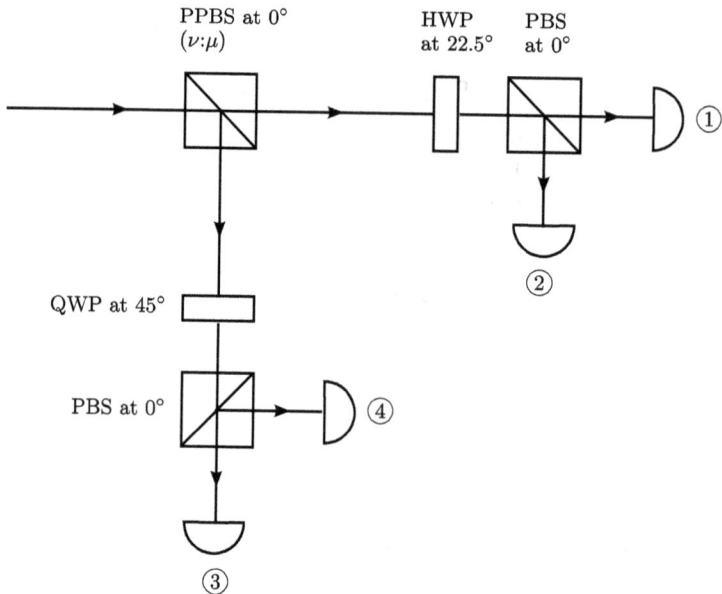

Fig. 4.3   Schematic setup for the tetrahedron measurement.

similar to that for the trine measurement, with an additional PBS and a QWP aligned at 45°.

This time, to reduce the number of parameters that define the PPBS, we shall suppose that the reflection amplitudes ν and μ satisfy the relation $\mu^2 + \nu^2 = 1$. In the transmitted path, the two outcomes detected are represented by

$$① : \quad \left[\mathbf{M}_{\text{PPBS}}(\mu, \nu) \, \mathbf{M}_{\text{HWP}} \left(\frac{\pi}{8} \right) \right] \begin{pmatrix} 1 \\ 0 \end{pmatrix} (1 \;\; 0) \left[\mathbf{M}_{\text{HWP}} \left(\frac{\pi}{8} \right) \mathbf{M}_{\text{PPBS}}(\mu, \nu) \right]$$

$$= \frac{1}{2} \begin{pmatrix} \mu^2 & \mu\nu \\ \mu\nu & \nu^2 \end{pmatrix},$$

$$② : \quad \left[\mathbf{M}_{\text{PPBS}}(\mu, \nu) \, \mathbf{M}_{\text{HWP}} \left(\frac{\pi}{8} \right) \right] \begin{pmatrix} 0 \\ 1 \end{pmatrix} (0 \;\; 1) \left[\mathbf{M}_{\text{HWP}} \left(\frac{\pi}{8} \right) \mathbf{M}_{\text{PPBS}}(\mu, \nu) \right]$$

$$= \frac{1}{2} \begin{pmatrix} \mu^2 & -\mu\nu \\ -\mu\nu & \nu^2 \end{pmatrix}. \tag{4.2.11}$$

In the reflected path, they are represented by

$$③ : \quad \left[\mathbf{M}_{\text{PPBS}}(\nu, \mu) \, \mathbf{M}_{\text{QWP}} \left(\frac{\pi}{4} \right) \right] \begin{pmatrix} 1 \\ 0 \end{pmatrix} (1 \;\; 0) \left[\mathbf{M}_{\text{QWP}} \left(\frac{\pi}{4} \right)^* \mathbf{M}_{\text{PPBS}}(\nu, \mu) \right]$$

$$= \frac{1}{2} \begin{pmatrix} \nu^2 & i\mu\nu \\ -i\mu\nu & \mu^2 \end{pmatrix},$$

$$④ : \quad \left[\mathbf{M}_{\text{PPBS}}(\nu, \mu) \, \mathbf{M}_{\text{QWP}} \left(\frac{\pi}{4} \right) \right] \begin{pmatrix} 0 \\ 1 \end{pmatrix} (0 \;\; 1) \left[\mathbf{M}_{\text{QWP}} \left(\frac{\pi}{4} \right)^* \mathbf{M}_{\text{PPBS}}(\nu, \mu) \right]$$

$$= \frac{1}{2} \begin{pmatrix} \nu^2 & -i\mu\nu \\ i\mu\nu & \mu^2 \end{pmatrix}. \tag{4.2.12}$$

We can learn more about these outcomes by extracting their Bloch vectors. Examining the intended outcome to be measured by detector ① and denoting its Bloch vector by $\boldsymbol{a}_1$ with coefficients $a_j^{(1)}$,

$$\frac{1}{2} \begin{pmatrix} \mu^2 & \mu\nu \\ \mu\nu & \nu^2 \end{pmatrix} = \frac{1}{4} \begin{pmatrix} 1 + a_3^{(1)} & a_1^{(1)} - i a_2^{(1)} \\ a_1^{(1)} + i a_2^{(1)} & 1 - a_3^{(1)} \end{pmatrix}$$

$$\Rightarrow \boldsymbol{a}_1 \hat{=} \begin{pmatrix} 2\mu\nu \\ 0 \\ \mu^2 - \nu^2 \end{pmatrix}. \tag{4.2.13}$$

All the other Bloch vectors can also be extracted in this way,

$$a_2 \hat{=} \begin{pmatrix} -2\mu\nu \\ 0 \\ \mu^2 - \nu^2 \end{pmatrix}, \quad a_3 \hat{=} \begin{pmatrix} 0 \\ 2\mu\nu \\ \nu^2 - \mu^2 \end{pmatrix}, \quad a_4 \hat{=} \begin{pmatrix} 0 \\ -2\mu\nu \\ \nu^2 - \mu^2 \end{pmatrix}. \quad (4.2.14)$$

Invoking the symmetry property described by Eq. (2.1.3), which is obeyed by all of these Bloch vectors, we arrive at the equations

$$(\mu^2 - \nu^2)^2 = \frac{1}{3},$$

$$4\mu^2\nu^2 + (\mu^2 - \nu^2)^2 = 1, \quad (4.2.15)$$

from which we get

$$\nu^2 = \frac{1}{6\mu^2},$$

$$36\mu^8 - 24\mu^4 + 1 = 0. \quad (4.2.16)$$

These yield

$$\mu^4 = \frac{1}{3} \pm \frac{1}{2\sqrt{3}} \quad (4.2.17)$$

or

$$\mu^2 = \sqrt{\frac{1}{3} \pm \frac{1}{2\sqrt{3}}} = \frac{1}{2}\left(1 \pm \frac{1}{\sqrt{3}}\right), \quad (4.2.18)$$

for which we may choose the pair

$$\mu = \sqrt{\frac{1}{2}\left(1 + \frac{1}{\sqrt{3}}\right)} \approx 0.888 \quad (\mu^2 \approx 0.789),$$

$$\nu = \sqrt{\frac{1}{2}\left(1 - \frac{1}{\sqrt{3}}\right)} \approx 0.459 \quad (\nu^2 \approx 0.211) \quad (4.2.19)$$

as one possible specification for the PPBS.

Larger quantum systems that involve more than one pair of polarization degree of freedom, two-photon systems for instance, can also be measured with the help of single-photon measurements schemes, such as the ones that we have studied. To estimate the quantum state of a source that produces two photons, for example, one can set up two tetrahedron measurements, one for each photon, to carry out informationally complete quantum-state tomography on the two-photon systems. This measurement is *separable*, since the corresponding POM outcomes are derived from tensor products of single-qubit outcomes. Such a tensor-product measurement can easily be generalized to quantum systems that are composed of any number of single-photon subsystems.

4.2.2 *Discrete-variable measurements — photon counting and its difficulties*

4.2.2.1 *Photon-number statistics*

In the previous sections, we have introduced polarization measurements for single-photon sources. One can perform a different kind of measurement on a multi-photon source that produces more than one photon at one go — *photon counting*. In the present scope, we shall restrict all discussions hereafter to the situation in which all photons produced by the source have exactly the same set of properties (polarization, frequency, wave vector, *etc.*), that is a *single-mode* photonic source.

In probing the photon-number statistics of incoming photons, expectation values of the photon-number states are measured. These photon-number states are the *Fock** states that form a basis spanning the infinite-dimensional Hilbert space,

$$\sum_{n=0}^{\infty} |n\rangle \langle n| = 1. \tag{4.2.20}$$

The expectation values of the Fock states, or the probability of detecting n photons, are related to a statistical operator $\rho_{\text{true}} \equiv \rho_{\text{src}}$ that describes the quantum signal from the source by

$$p(n) = \langle |n\rangle \langle n| \rangle_{\text{src}} = \langle n| \rho_{\text{src}} |n\rangle. \tag{4.2.21}$$

Ideally, the expectation values $\langle |n\rangle \langle n| \rangle_{\text{src}}$ can be measured by a photodetector that can identify the number of photons it detects. Such a detector is called a *photon-number resolving detector* (PNRD). As the number of photons n produced by the source at one go is not conserved, ρ_{src} is infinite-dimensional in general.

4.2.2.2 *Characteristic functions*

The probability of registering n photons can be calculated using a standard formalism in statistics, and since this formalism will be important for a later discussion, let us take some time to review it.

---

*Vladimir Aleksandrovich Fock (1898–1974).

Consider a Hermitian observable $Y = Y^\dagger = \sum_y |y\rangle\, y\, \langle y|$ that has distinct integer eigenvalues y. Given the state ρ_{src}, the expectation value

$$\left\langle e^{i\frac{2\pi l}{L}Y} \right\rangle_{\text{src}} = \text{tr}\left\{ \rho_{\text{src}}\, e^{i\frac{2\pi l}{L}Y} \right\}$$

$$= \sum_{y'=-\infty}^{\infty} p(y')\, e^{i\frac{2\pi l}{L}y'}, \qquad (4.2.22)$$

where l is an integer that takes values $0, 1, \ldots, L-1$.

We would like to extract the probability $p(y)$ of measuring the eigenvalue y. This can be done by summing both sides of the equation over the variable l with appropriate weights, that is

$$\sum_{l=0}^{L-1} e^{-i\frac{2\pi l}{L}y} \left\langle e^{i\frac{2\pi l}{L}Y} \right\rangle_{\text{src}} = \sum_{y'=-\infty}^{\infty} p(y') \underbrace{\sum_{l=0}^{L-1} e^{i\frac{2\pi l}{L}(y'-y)}}_{= L\,\delta_{y',y}}$$

$$= L \sum_{y'=-\infty}^{\infty} p(y')\,\delta_{y',y}$$

$$= L\, p(y), \qquad (4.2.23)$$

or

$$p(y) = \frac{1}{L} \sum_{l=0}^{L-1} \left\langle e^{i\frac{2\pi l}{L}(Y-y)} \right\rangle_{\text{src}}. \qquad (4.2.24)$$

Since

$$\frac{1}{L} \sum_{l=0}^{L-1} e^{i\frac{2\pi l}{L}(Y-y)} = \delta_{Y,y} = |y\rangle\langle y|, \qquad (4.2.25)$$

where $|y\rangle\langle y|$ are the orthogonal eigenstates of Y for the eigenvalues y, we have

$$p(y) = \langle |y\rangle\langle y| \rangle, \qquad (4.2.26)$$

of course.

From Eq. (4.2.24), we note that the probability distribution $p(y)$ is in fact a *discrete Fourier* transform of the expectation value of the exponential operator $e^{i\frac{2\pi l}{L}Y}$. This expectation value is the *characteristic function* of Y and completely characterizes the probability distribution $p(y)$. The distribution $p(y)$ is therefore a linear combination of all moments of Y: $\langle Y \rangle$, $\langle Y^2 \rangle$, $\langle Y^3 \rangle$, *etc.*

Returning to our photon-counting scenario, the probability of measuring n photons with a PNRD is given by

$$p(n) = \langle |n\rangle \langle n| \rangle = \frac{1}{L} \sum_{l=0}^{L-1} \left\langle e^{i\frac{2\pi l}{L}\left(A^\dagger A - n\right)} \right\rangle_{\text{src}}, \qquad (4.2.27)$$

where $A^\dagger A$, composed of the *annihilation operator* A and its adjoint, is the *number operator* that has eigenstates $|n\rangle \langle n|$ with eigenvalues n.

4.2.2.3 *The physics of photodetection*

Measuring these photon-number probabilities seems straightforward with PNRDs. Unfortunately, it is not possible to count photons in practice. The typical photodetectors available, which work at temperatures ranging from 200K to 273K with reasonable efficiencies, are based on the technology of *avalanche photodiodes* (APDs). Since they are very common, it is worth side-tracking a little to understand some important features of an APD.

A simple schematic for an APD is that of a *p-n junction*, a junction made by joining a p-type material with an n-type material (refer to Fig. 4.4). A p-type material consists of a semiconductor substrate, silicon (Si) (Group IV) for instance, that is doped with an impurity that results in vacancies when its atoms are bonded with the substrate. Usually, a Group III element, for example, like boron (B), with one less valence electron compared to Si, is used. These vacancies introduce additional energy states that have only slightly higher energies than the valence band of Si, so that the valence electrons can be excited into the energy states, thereby creating positively-charged carriers known as *holes* in the valence band. On the other hand, an n-type material results from doping the Si substrate with, for instance, a Group V element like phosphorous (P), which supplies one additional loosely-bound valence electron per dopant atom. Additional electronic energy states that are occupied, with energies that are

---

*Jean-Baptiste Joseph Fourier (1768–1830).

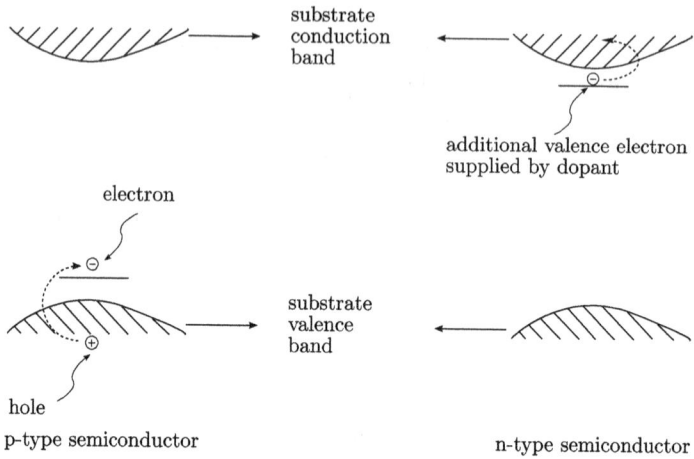

Fig. 4.4 Energy-level and band diagrams of doped semiconductors.

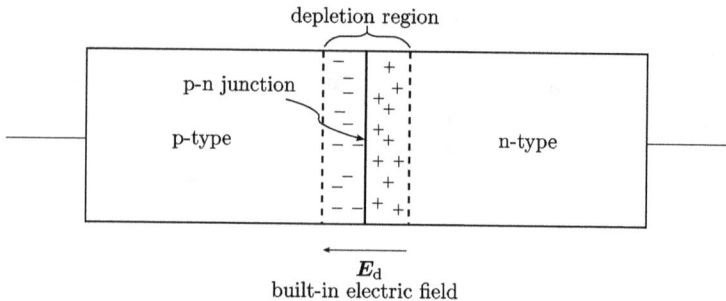

Fig. 4.5 A typical p-n junction.

slightly lower than the conduction band of Si, are consequently introduced, allowing electrons from these states to excite into the conduction band.

Therefore, p-type materials have *excess holes*, in direct contrast to n-type materials which have *excess electrons*. When a junction is made by combining a p-type material with an n-type material, the excess holes from the p-type region will diffuse into the n-type region, and at the same time, the excess electrons in the n-type region will diffuse into the p-type region. The diffused (minority) holes and electrons *recombine* at the respective regions, resulting in a layer around the p-n junction called the *depletion region* (refer to Fig. 4.5). Since the minority carriers recombine with the majority carriers in each region, this depletion region consists of net space charges with no mobile charge carriers.

As the net space charges in both regions are being created by diffusion and recombination events, an electric field $\boldsymbol{E}_d$ is generated across the depletion region. The width of the depletion region can widen if an external *reverse-bias* voltage V_r is supplied to the APD, where the positive external voltage is applied to the n-type material. Doing so causes the excess electrons in the n-type region to drift to the positive terminal, and excess holes in the p-type region to drift to the negative terminal, thus widening the depletion region. As a consequence, the magnitude of $\boldsymbol{E}_d$ increases.

Beyond a certain critical magnitude of $\boldsymbol{E}_d$, the depletion region of the APD will be suitable for the detection of light signals. If an incoming light beam shines on the APD, the electron-hole pairs that are generated in the depletion region will accelerate to high speeds, due to the strong electric field $\boldsymbol{E}_d$, and will be able to impart high energies to the region to excite new charge carriers. This process, also known as *impact ionization*, can yield a huge amount of charge carriers, thus amplifying the signal detection output. Such an *avalanche event* is exploited by APDs to detect light.

There are typically two modes of operation for an APD; the linear mode and the Geiger mode (refer to Fig. 4.6). In the linear mode, the reverse-bias voltage is set to a value below the breakdown voltage V_b. In this regime, the photocurrent that is amplified by an avalanche event is approximately *proportional* to the intensity of an incoming light beam. In this mode of operation, the amplification of the signal is only high enough to detect signals of moderate strengths, since the electric field $\boldsymbol{E}_d$ is not very strong. In other words, the APD cannot detect weak signal pulses in this regime, and "photon counting" is not possible.

To detect weak pulses, the Geiger mode of operation is commonly used. In this mode, the reverse-bias voltage is set to a value V_r that is higher than V_b. In this case, we define the *excess bias* $V_{ex} = V_r - V_b \geq 0$. To prevent spontaneous breakdown of the detector under this mode of operation, the external circuit usually limits the current to the order of microamperes. An incoming weak signal pulse that may contain more than one single-photon pulse can now trigger significant avalanche photocurrents of the order of milliamperes within tens of a picosecond, thus saturating the photocurrent of the APD. For this, an APD running in this mode is a *single-photon avalanche diode* (SPAD), which, on hindsight, should more appropriately be named as the *pulse-resolved avalanche diode*.

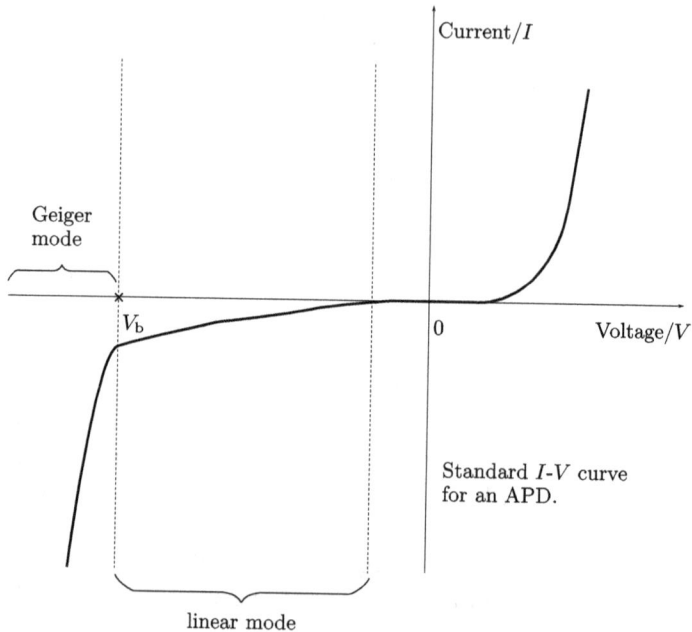

Fig. 4.6 Characteristic current-voltage behavior and modes of operation of a conventional APD.

The response of the photocurrent to the signal pulse is nonlinear and the resulting sharp photocurrent waveform is prone to statistical fluctuations, since an avalanche event produces random amounts of photocurrents. Also, the photocurrent saturates almost immediately as the signal pulse is detected by the SPAD. As such, while the SPAD is able to detect signal pulses one by one, it is not possible to "count" the number of photons in each pulse with it.

There are other imperfections that complicate matters. First, the detection efficiency of an APD strongly depends on the wavelength of the signal. Typical materials such as Si and indium gallium arsenide/indium phosphide (InGaAs/InP) structures can detect signals in the visible spectrum and infra-red spectrum respectively. Photodetection outside the appropriate range results in extremely low detection efficiency. Second, there also exist dark counts that may arise when charge carriers are excited by thermal fluctuations. Once these charge carriers are excited, for sufficiently large E_d, they can contribute to spurious photocurrents that do not originate from the signal.

These dark carriers can also arise from a phenomenon known as *afterpulsing*. During an avalanche event, some energetic charge carriers may be reabsorbed by the APD to occupy empty energy states, which are usually high in energy that are otherwise inaccessible to low-energy carriers. Examples of such states are those with energies in the bandgap. These states typically originate from defects in the APD that are commonly found in InGaAs/InP structures, for instance, and have finite relaxation time. The de-excitations of these trapped carriers are completely random. When these trapped carriers are released, they can initiate a separate avalanche event, which is commonly referred to as an afterpulse, in the absence of the external light source, increasing the noise of the photocurrents. Consequently, the dark-count rates will be much higher for the SPAD as compared to the linear mode APD, since the total electric field in the depletion region is much stronger.

To ensure the functionality of the SPAD, it is necessary to *quench* the detector photocurrent after an avalanche event so that it can be used to perform a subsequent photodetection (refer to Fig. 4.7). This is done by lowering the reverse-bias voltage from V_r to some value V_{off} that is below V_b in magnitude, such that the SPAD is now in the linear mode of operation and negligible photocurrents can be generated from weak signal pulses. At this stage, the SPAD is switched off and no pulses are detected. After this short period of time, also known as the *hold-off time*, the reverse-bias voltage is then returned to the operational level V_r and the SPAD is switched on for signal detection. This quenching procedure is typically carried out either by a control circuit that actively arms and disarms the detector at regular intervals, or a circuit consisting of current-modulating components to passively adjust the input voltage of the detector according to the circuit relaxation time.*

To reduce dark-count rates, the SPAD may be turned off for a longer period of time. This, however, limits the operation time of the SPAD, which decreases the number of signal pulses that are detected per unit time. A stable low-temperature environment can also help to minimize the dark-count rates. A recent design of an SPAD, for example, that is based on InGaAs/InP, working at 225K with an excess bias of 5V, can achieve a dark-count rate of about 10^5 copies per second with a hold-off time of a few microseconds for an "on" period of 20ns.†

---

*Refer to A. Gallivanoni, *IEEE Trans. Nucl. Sci.* **57**, 3815 (2010) for a progress report.
†Refer to A. Tosi *et al.*, *IEEE J. Quantum Electron.* **48**, 1227 (2012).

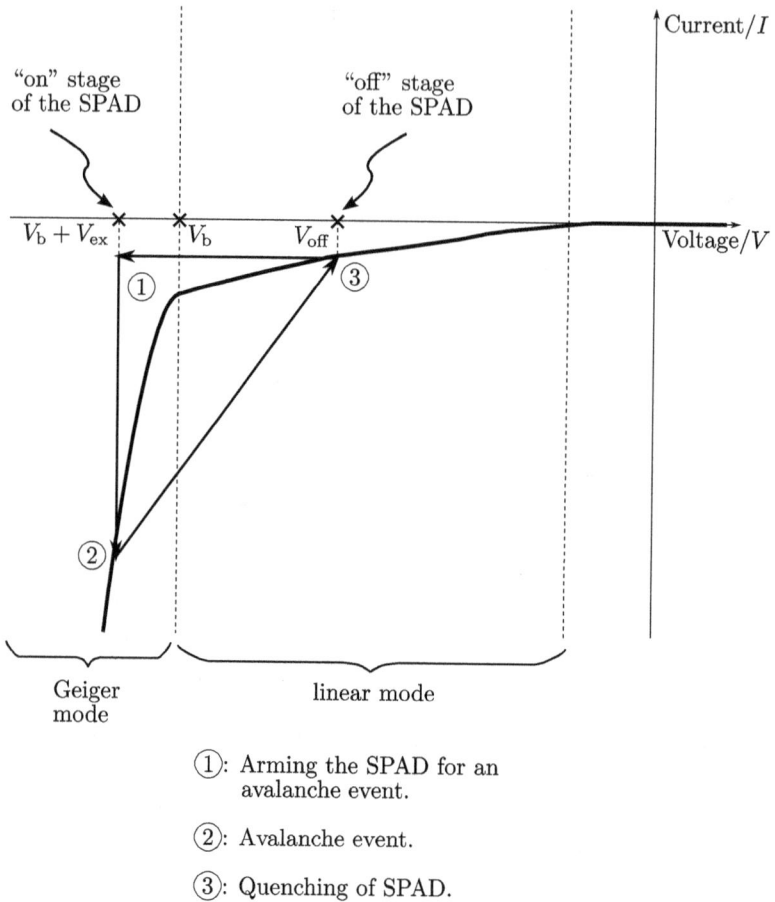

① : Arming the SPAD for an
     avalanche event.

② : Avalanche event.

③ : Quenching of SPAD.

Fig. 4.7   Modes of operation of a typical SPAD.

4.2.2.4   *Indirect photon counting*

Despite the technical difficulties in measuring pure Fock states, it is still possible to probe the photon-number statistics for a given source using only conventional APDs. This can be done by indirect photon counting.

In other words, we would need the expectation values of measurement outcomes that completely characterize the D_{rec} diagonal elements of the state ρ_{src} in the Fock basis, where D_{rec} refers to a cutoff dimension for the reconstruction subspace under certain assumptions about the properties of the source.

In fact, the feasible option would be a POM that is composed of *mixtures* of Fock states

$$\Pi_j = \sum_{n=0}^{D_{\text{rec}}-1} |n\rangle\, c_{jn}\, \langle n|, \qquad (4.2.28)$$

where the coefficients $c_{jk} \geq 0$ are such that the POM $\{\Pi_j\}$ corresponds to $D_{\text{LI}} = D_{\text{rec}}$ for estimating the D_{rec} diagonal elements of ρ_{src}. The corresponding true probabilities are then related to linear combinations of the expectation values $\langle |n\rangle\langle n| \rangle$ according to the equation

$$p_j^{\text{true}} = \sum_{n=0}^{D_{\text{rec}}-1} c_{jk}\, \langle |n\rangle\langle n| \rangle. \qquad (4.2.29)$$

In practice, such a POM can be realized by means of a *multi-port device* that splits a signal pulse into multiple pulses and distribute them to the respective output ports (refer to Fig. 4.8). With this device, it is possible to split a pulse that contains more than one single-photon pulse into multiple weaker pulses and measure the output coincidences with $\mathcal{K}$ SPAD detectors that either idle or respond to an incoming signal by registering it as a "click".

Alternatively, the $\mathcal{K}$ ports may be replaced by waveguides, for instance an optical fiber, that split the signal pulse into several weak pulses, such that every pulse is time-delayed with respect to the rest. These delayed pulses are then recombined and detected by an SPAD detector. The latter

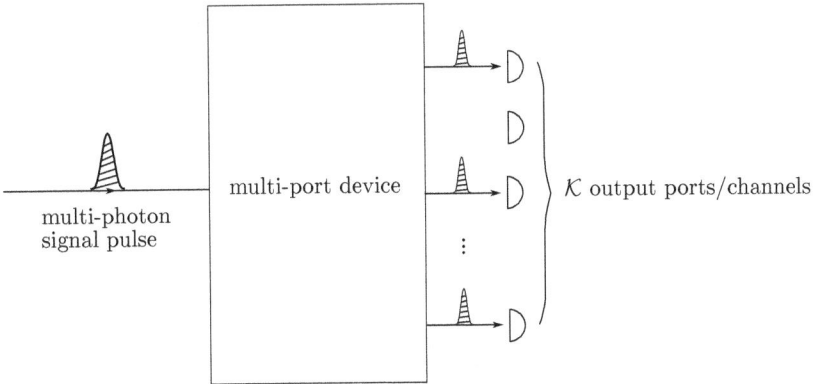

Fig. 4.8 Multi-photon pulse measurement with a multi-port device.

approach is a *time-multiplexed detection* and for this to work, the delay time has to be at least larger than the hold-off time of the detector.

Strictly speaking, this $\mathcal{K}$-port device measures the photon-"click" statistics of the source — each photodetection represents a detection of a weak pulse, or a "click". To approximately connect this with photon numbers, one needs to assume that the multiple pulses that come from the splitting of the original signal pulse are, with *high* probability, single-photon pulses. Now, the physical situation is that the incoming pulse contains photons up to a certain critical number, above which the probability of detecting such a pulse is negligible. The aforementioned assumption would therefore require $\mathcal{K}$ to be at least this critical number, since a device with $\mathcal{K}$ physical ports can detect at most $\mathcal{K}$ single-photon pulses simultaneously. In practice, an observer can never be sure what this critical number should be, but is able to estimate, instead, the mean number of photons per pulse. Therefore, as a general rule, the number of ports $\mathcal{K}$ should be much larger than this mean to be on the safe side.

As a simple example, we will look for the POM outcomes that correspond to a two-port device ($\mathcal{K} = 2$). Let T_1 and T_2 be the transmission coefficients for the respective ports, where the transmission coefficient T_k is the product of the transmissivity during the splitting of the signal pulse and the efficiency of the photodetector for the output port k, such that $T_1 + T_2 < 1$.

Since the photodetectors now act as binary detectors, that is each detector either idles or registers a "click", we can denote these two events by "0" and "1" respectively. Such a detection is also known as an "on/off" detection. As there are altogether two ports, the number of different configurations is given by $2^2 = 4$.

We first calculate the probability for the "00" configuration. The probability that the two detectors register no "click"s is given by $1-T_1-T_2$.* Then, the probability p_{00} for the "00" configuration, which includes the detection of single-photons up to the number $D_{\text{rec}} - 1$, is given by

$$p_{00} = \sum_{n=0}^{D_{\text{rec}}-1} (1 - T_1 - T_2)^n \, |n\rangle \langle n|. \qquad (4.2.30)$$

---

*This is analogous to an imperfect BS that can absorb photons with probability $A = 1 - T - R$, where T and R are respectively its transmission and reflection coefficients.

For the probability p_{01} of obtaining the "01" configuration, this is equal to $p_{0\forall} - p_{00}$, where the symbol "$\forall$" stands for "for all binary numbers" and

$$p_{0\forall} = \sum_{n=0}^{D_{\mathrm{rec}}-1} (1-T_1)^n \langle |n\rangle \langle n| \rangle. \tag{4.2.31}$$

It follows that

$$p_{01} = \sum_{n=0}^{D_{\mathrm{rec}}-1} [(1-T_1)^n - (1-T_1-T_2)^n] \langle |n\rangle \langle n| \rangle. \tag{4.2.32}$$

Similarly, for the "10" configuration, we have $p_{10} = p_{\forall 0} - p_{00}$ with

$$p_{\forall 0} = \sum_{n=0}^{D_{\mathrm{rec}}-1} (1-T_2)^n \langle |n\rangle \langle n| \rangle, \tag{4.2.33}$$

so that

$$p_{10} = \sum_{n=0}^{D_{\mathrm{rec}}-1} [(1-T_2)^n - (1-T_1-T_2)^n] \langle |n\rangle \langle n| \rangle. \tag{4.2.34}$$

The final probability is then $p_{11} = 1 - p_{00} - p_{01} - p_{10}$.

The four POM outcomes for the two-port device are therefore given by

$$\Pi_1 = \sum_{n=0}^{D_{\mathrm{rec}}-1} |n\rangle (1-T_1-T_2)^n \langle n|,$$

$$\Pi_2 = \sum_{n=0}^{D_{\mathrm{rec}}-1} |n\rangle [(1-T_1)^n - (1-T_1-T_2)^n] \langle n|,$$

$$\Pi_3 = \sum_{n=0}^{D_{\mathrm{rec}}-1} |n\rangle [(1-T_2)^n - (1-T_1-T_2)^n] \langle n|,$$

$$\Pi_4 = 1 - \Pi_1 - \Pi_2 - \Pi_3. \tag{4.2.35}$$

More generally, the set of $2^{\mathcal{K}}$ POM outcomes for a $\mathcal{K}$-port device can be obtained in this manner.

Problem 4.7** As a familiarization exercise, work out all the POM outcomes for a three-port device by hand, with transmission coefficients $T_1 + T_2 + T_3 < 1$. Try to write a computer program to compute these outcomes for any $\mathcal{K}$-port device with transmission coefficients $\sum_{k=1}^{\mathcal{K}} T_k < 1$. For the latter exercise, verify the program for $\mathcal{K} = 2$ and $\mathcal{K} = 3$.

4.2.3 *Continuous-variable measurements*

4.2.3.1 *Displaced Fock-state measurement*

Since the POM outcomes that are obtained through the multi-port device only characterize the photon-number statistics of the source, we cannot estimate the entire statistical operator ρ_{src}, for its off-diagonal entries in the Fock basis are not probed by these outcomes. To estimate these off-diagonal elements, we need to introduce unitary transformations to the POM so that its outcomes are no longer all diagonal in the Fock basis and for this, it is convenient to adopt another description of the photonic source.

We can think of the Fock state $|n\rangle \langle n|$ as being created from the *vacuum state* $|0\rangle \langle 0|$. Mathematically,

$$|n\rangle \langle n| = A^{\dagger^n} |0\rangle \frac{1}{n!} \langle 0| A^n. \qquad (4.2.36)$$

Here, we have made use of the relation

$$A^{\dagger} |n\rangle = |n+1\rangle \sqrt{n+1} \qquad (4.2.37)$$

for the photonic *creation operator* $A^{\dagger}$, where $[A, A^{\dagger}] = 1$ applies. We can therefore attribute a ladder operator A to one particular *mode* of light, and the Fock states for this mode of light are generated by this operator and its adjoint.

As a quick recap, the eigenstates of A are known as the *coherent states* $|\alpha\rangle \langle \alpha^*|$, with α being a complex eigenvalue of A. These normalized, yet *nonorthogonal*, states are related to the Fock states by the equation

$$|\alpha\rangle = e^{-\frac{1}{2}|\alpha|^2} \sum_{n'=0}^{\infty} |n'\rangle \frac{\alpha^{n'}}{\sqrt{n'!}}. \qquad (4.2.38)$$

Note also that both A and $A^\dagger$ has no eigenbras and eigenkets respectively. The reader is strongly encouraged to attempt **Problems 4.8–4.10** on the properties of the ladder operators and their coherent states as a revision.

Problem 4.8 Starting with the commutator $[A, A^\dagger] = 1$ for the ladder operators A and $A^\dagger$ and the eigenvalue equations $A^\dagger A |n\rangle = |n\rangle n$ and $A |\alpha\rangle = |\alpha\rangle \alpha$, refresh your memory by confirming the following results:

(1) Equation (4.2.37) and

$$A |n\rangle = |n - 1\rangle \sqrt{n}; \qquad (4.2.39)$$

(2) Equation (4.2.38);
(3) The inner product relation

$$\langle \alpha^* | \alpha' \rangle = e^{-\frac{1}{2}(|\alpha|^2 + |\alpha'|^2) + \alpha^* \alpha'} \qquad (4.2.40)$$

between the coherent kets $|\alpha\rangle$ and $|\alpha'\rangle$.

Problem 4.9 Establish that the wave function $\langle x | \alpha \rangle$ for the eigenket $|\alpha\rangle$ of A is given by

$$\langle x | \alpha \rangle = \frac{1}{\pi^{\frac{1}{4}}} e^{-\frac{1}{2}|\alpha|^2 - \frac{1}{2}\alpha^2 - \frac{1}{2}x^2 + \sqrt{2}\alpha x}. \qquad (4.2.41)$$

Problem 4.10 Show that A has no eigenbras and $A^\dagger$ has no eigenkets.

With this description for the signal, there exists a natural unitary operator for the infinite-dimensional Hilbert space — the *displacement operator* given by

$$D(\alpha) = e^{\alpha A^\dagger - \alpha^* A} = D(-\alpha)^\dagger. \qquad (4.2.42)$$

Hereafter, the notation $D(\alpha)$ should be understood as a shorthand for the function $D(\alpha, \alpha^*)$ of α and α^*. This is the unitary operator that displaces the ket $|\alpha'\rangle$ to the ket $|\alpha' + \alpha\rangle$. To demonstrate this, we first remember that if the two operators $\mathcal{A}$ and $\mathcal{B}$ give a commutator $[\mathcal{A}, \mathcal{B}]$ such that $[\mathcal{A}, [\mathcal{A}, \mathcal{B}]] = [\mathcal{B}, [\mathcal{A}, \mathcal{B}]] = 0$, the *Baker–Campbell–Hausdorff** (BCH) relation states that

$$e^{\mathcal{A}+\mathcal{B}} = e^{\mathcal{A}} e^{\mathcal{B}} e^{-\frac{1}{2}[\mathcal{A},\mathcal{B}]}. \tag{4.2.43}$$

According to this formula, $D(\alpha)$ can therefore be disentangled into

$$D(\alpha) = e^{-\frac{1}{2}|\alpha|^2} e^{\alpha A^\dagger} e^{-\alpha^* A}, \tag{4.2.44}$$

and it then follows that

$$D(\alpha) |\alpha'\rangle = e^{\alpha A^\dagger} |\alpha'\rangle e^{-\frac{1}{2}|\alpha|^2 - \alpha^*\alpha'}. \tag{4.2.45}$$

To evaluate $e^{\alpha A^\dagger} |\alpha'\rangle$, we need the result of $A^\dagger |\alpha'\rangle$ and this can be found by noting that, with Eq. (4.2.38),

$$\langle n| A^\dagger |\alpha'\rangle = \sqrt{n} \, \langle n-1|\alpha'\rangle$$

$$= \sqrt{n} \, e^{-\frac{1}{2}|\alpha'|^2} \frac{\alpha'^{n-1}}{\sqrt{(n-1)!}}$$

$$= e^{-\frac{1}{2}|\alpha'|^2} \frac{\partial}{\partial \alpha'} \left(\frac{\alpha'^n}{\sqrt{n!}} \right)$$

$$= e^{-\frac{1}{2}|\alpha'|^2} \frac{\partial}{\partial \alpha'} \left(\langle n|\alpha'\rangle \, e^{\frac{1}{2}|\alpha'|^2} \right)$$

$$\Rightarrow \quad A^\dagger \left(|\alpha'\rangle \, e^{\frac{1}{2}|\alpha'|^2} \right) = \frac{\partial}{\partial \alpha'} \left(|\alpha'\rangle \, e^{\frac{1}{2}|\alpha'|^2} \right). \tag{4.2.46}$$

Hence, the ket $|\alpha'\rangle \, e^{\frac{1}{2}|\alpha'|^2}$ is an *entire* function, a function that is *analytic* or complex differentiable over the whole complex plane, whereas $|\alpha'\rangle$ is not.

---

*Henry Frederick Baker (1866–1956), John Edward Campbell (1862–1924), and Felix Hausdorff (1868–1942).

Problem 4.11 If $f(x,p) = u(x,p) + iv(x,p)$ is entire, where $u(x,p)$ and $v(x,p)$ are real functions of the real variables x and p, it must satisfy the following *Cauchy–Riemann** equations:

$$\frac{\partial u}{\partial x} = \frac{\partial v}{\partial p},$$

$$\frac{\partial u}{\partial p} = -\frac{\partial v}{\partial x}. \tag{4.2.47}$$

With this fact, confirm the above remark on the kets $|\alpha'\rangle\, e^{\frac{1}{2}|\alpha'|^2}$ and $|\alpha'\rangle$.

Completing the calculation,

$$D(\alpha)\left(|\alpha'\rangle\, e^{\frac{1}{2}|\alpha'|^2}\right) = e^{\alpha\frac{\partial}{\partial\alpha'}}\left(|\alpha'\rangle\, e^{\frac{1}{2}|\alpha'|^2}\right) e^{-\frac{1}{2}|\alpha|^2 - \alpha^*\alpha'}$$

$$= |\alpha' + \alpha\rangle\, e^{\frac{1}{2}|\alpha'+\alpha|^2 - \frac{1}{2}|\alpha|^2 - \alpha^*\alpha'}$$

$$= |\alpha' + \alpha\rangle\, e^{\frac{1}{2}|\alpha'|^2 + \frac{1}{2}(\alpha\alpha'^* - \alpha^*\alpha')}, \tag{4.2.48}$$

or

$$D(\alpha)|\alpha'\rangle = |\alpha' + \alpha\rangle\, e^{i\,\mathrm{Im}\{\alpha\alpha'^*\}}. \tag{4.2.49}$$

This implies that the statements

$$D(\beta)D(\alpha)|\alpha'\rangle = D(\beta)|\alpha'+\alpha\rangle\, e^{i\,\mathrm{Im}\{\alpha\alpha'^*\}}$$

$$= |\alpha+\beta+\alpha'\rangle\, e^{i\,\mathrm{Im}\{\beta(\alpha'^*+\alpha^*)\}} e^{i\,\mathrm{Im}\{\alpha\alpha'^*\}}$$

$$= D(\alpha+\beta)|\alpha'\rangle\, e^{i\,\mathrm{Im}\{\beta\alpha^*\}} \tag{4.2.50}$$

are true for all $|\alpha'\rangle$, or

$$D(\beta)D(\alpha) = D(\alpha+\beta)\, e^{i\,\mathrm{Im}\{\beta\alpha^*\}}. \tag{4.2.51}$$

The deceptively simple relation

$$D(\alpha)|0\rangle = |\alpha\rangle, \tag{4.2.52}$$

which is a consequence of the general expression in Eq. (4.2.49), is in fact a very useful formula to bear in mind.

---

* *Baron* Augustin-Louis Cauchy (1789–1857) and Georg Friedrich Bernhard Riemann (1826–1866).

We can apply the displacement operator $D(\alpha)$ on the set of orthonormal kets $|n\rangle$ to unitarily transform them into another complete set of orthonormal kets. The new kets $D(\alpha)|n\rangle$ are linearly independent of $\{|n\rangle\}$, and can therefore characterize the off-diagonal entries of ρ_{src}. Just like $|\alpha'\rangle$, the action of $D(\alpha)$ on $|n\rangle$ displaces $|n\rangle$ in the complex plane. It is for this reason that $D(\alpha)|n\rangle\langle n|D(-\alpha)$ are called the *displaced Fock states*.

For completeness, we shall illustrate this displacement by writing the Fock kets $|n\rangle$ in terms of the unnormalized coherent kets $|\alpha'\rangle = |\alpha'\rangle\, e^{\frac{1}{2}|\alpha'|^2}$. Beginning with Eq. (4.2.38), we multiply both sides of this equation by $\alpha'^{-(n+1)}$, with $n > -1$, and integrate the equation along the unit circle centered at the origin of the complex plane:

$$\sum_{n'=0}^{\infty} \frac{|n'\rangle}{\sqrt{n'!}} \underset{\substack{\text{unit}\\\text{circle}}}{\oint} d\alpha'\, \alpha'^{n'-n-1} = \underset{\substack{\text{unit}\\\text{circle}}}{\oint} d\alpha'\, \frac{|\alpha'\rangle}{\alpha'^{n+1}}. \qquad (4.2.53)$$

On the left-hand side, the complex contour integral is evaluated using the *Cauchy residue theorem* given by

$$\underset{\substack{\text{unit}\\\text{circle}}}{\oint} d\alpha'\, \frac{f(\alpha')}{\alpha'^{n+1}} = 2\pi i \operatorname{Res}\left(\frac{f(\alpha')}{\alpha'^{n+1}}, \{\alpha' = 0, n+1\}\right), \qquad (4.2.54)$$

where $f(\alpha')$ is an analytic function of α' and "Res" refers to the *residue* of the integrand for the $(n+1)$th order pole at $\alpha' = 0$, which can be computed with the formula

$$\operatorname{Res}\left(\frac{f(\alpha')}{\alpha'^{n+1}}, \{\alpha' = 0, n+1\}\right) = \frac{1}{n!}\left(\frac{d}{d\alpha'}\right)^n f(\alpha')\Bigg|_{\alpha'=0}. \qquad (4.2.55)$$

With $f(\alpha') = \alpha'^{n'}$,

$$\operatorname{Res}\left(\frac{\alpha'^{n'}}{\alpha'^{n+1}}, \{\alpha' = 0, n+1\}\right) = \frac{1}{n!}\left(\frac{d}{d\alpha'}\right)^n \alpha'^{n'}\Bigg|_{\alpha'=0}$$

$$= \frac{n'!}{n!(n'-n)!}\, \alpha'^{n'-n}\Bigg|_{\alpha'=0}$$

$$= \delta_{n',n}, \qquad (4.2.56)$$

where we have used the identity

$$\left(\frac{\mathrm{d}}{\mathrm{d}x}\right)^{\mu} x^{\nu} = \frac{\nu!}{(\nu - \mu)!} x^{\nu - \mu}, \tag{4.2.57}$$

which is defined for any μ and ν and consequently gives

$$\oint_{\substack{\text{unit} \\ \text{circle}}} \mathrm{d}\alpha' \, \alpha'^{n'-n-1} = 2\pi i \, \delta_{n',n}. \tag{4.2.58}$$

Finally, we have the expression

$$|n\rangle = \frac{\sqrt{n!}}{2\pi i} \oint_{\substack{\text{unit} \\ \text{circle}}} \mathrm{d}\alpha' \, \frac{|\alpha'\rangle}{\alpha'^{n+1}} \tag{4.2.59}$$

for the Fock kets in terms of the kets $|\alpha'\rangle$. So, the Fock ket is in fact a superposition of coherent kets along the unit circle,* centered at the origin of the complex plane.

This tells us that there is indeed a displacement for all the components of $|n\rangle$ in the complex plane after an application of $D(\alpha)$, inasmuch as

$$
\begin{aligned}
D(\alpha)\,|n\rangle &= \frac{\sqrt{n!}}{2\pi i} \oint_{\substack{\text{unit} \\ \text{circle}}} \mathrm{d}\alpha' \, \frac{|\alpha' + \alpha\rangle\, e^{\frac{1}{2}|\alpha'|^2 + i\,\mathrm{Im}\{\alpha\alpha'^*\}}}{\alpha'^{n+1}} \\[2mm]
&= \frac{\sqrt{n!}}{2\pi i} \oint_{\substack{\text{unit} \\ \text{circle}}} \mathrm{d}\alpha' \, \frac{|\alpha' + \alpha\rangle\, e^{\frac{1}{2}|\alpha' + \alpha|^2 - \frac{1}{2}|\alpha|^2 - \alpha^*\alpha'}}{\alpha'^{n+1}} \quad (\alpha'' = \alpha' + \alpha) \\[2mm]
&= \frac{\sqrt{n!}}{2\pi i}\, e^{\frac{1}{2}|\alpha|^2} \oint_{\substack{\text{unit circle} \\ \text{at } \alpha''=\alpha}} \mathrm{d}\alpha'' \, \frac{|\alpha''\rangle\, e^{-\alpha^*\alpha''}}{(\alpha'' - \alpha)^{n+1}}, \tag{4.2.60}
\end{aligned}
$$

that is, the entire superposition is deformed and translated by the complex amplitude α.

Problem 4.12   Perform a simple consistency check on the validity of Eq. (4.2.59) by inserting it into the right-hand side of Eq. (4.2.38).

---

*In the theory of contour integration, all closed contours that surround the relevant pole(s) give exactly the same answers. The unit-circle contour is simply a convenient representative of these equivalent contours.

Problem 4.13 Verify that

$$\int \frac{dx\,dp}{2\pi} |\alpha\rangle \langle\alpha^*| = 1 \qquad (4.2.61)$$

for the parametrization $\alpha = (x + ip)/\sqrt{2}$, which says that the coherent states are complete. These states have, in fact, *more than one* completeness relation. They are said to be *overcomplete* and the equivalent mathematical statement for this is given by

$$\int \frac{dx\,dp}{2\pi} \frac{|\alpha'\rangle \langle\alpha^*|}{\langle\alpha^*|\alpha'\rangle} = 1, \qquad (4.2.62)$$

where the dimensionless real variables x and p appropriately parameterize the complex variables α and α'. Confirm this relation with the parametrization $\alpha' = x$ and $\alpha = ip$.

One can implement an experimental scheme to measure these displaced Fock states (refer to Fig. 4.9) using just a BS, with a transmission amplitude t and a reflection amplitude r such that $|t|^2 + |r|^2 = 1$. A mode of light

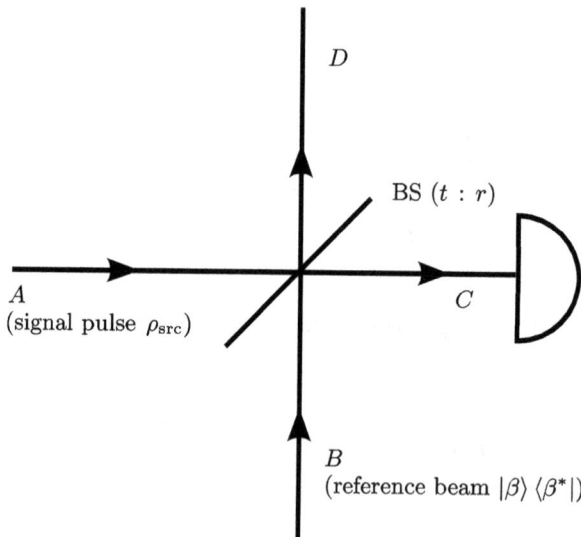

Fig. 4.9   Schematic setup for the displaced Fock-state measurement.

from the source, associated with the signal mode A, and a coherent state $|\beta\rangle \langle \beta^*|$ with $\beta = |\beta|e^{i\vartheta}$, associated with mode B, are fed to the BS, after which a photodetector is used to collect the transmitted light. This reference coherent state is generated by the electronic signal oscillator that is part of a *local oscillator* set-up that stabilizes the reference coherent state. In the language of the ladder operators, the BS can be regarded as a unitary matrix

$$\mathbf{U}_{\text{BS}}(t,r) = \begin{pmatrix} t & r \\ r^* & -t^* \end{pmatrix} \tag{4.2.63}$$

that turns a column of light modes into a new column of light modes:

$$\begin{pmatrix} C \\ D \end{pmatrix} = \mathbf{U}_{\text{BS}}(t,r) \begin{pmatrix} A \\ B \end{pmatrix}. \tag{4.2.64}$$

Without loss of generality, we may define the amplitudes $t = \tilde{t}$ and $r = \tilde{r}$, where $\tilde{t}$ and $\tilde{r}$ are now real. The photodetector collects signals of mode C in the form of photocurrents, which are proportional to the expectation value

$$\langle C^\dagger C \rangle = \text{tr}\{\rho_{\text{src}} \otimes |\beta\rangle \langle \beta^*| C^\dagger C\} \tag{4.2.65}$$

of the number operator $C^\dagger C$ for this mode. Using the specification of the BS, this number operator is given by

$$C^\dagger C = \left(\tilde{t}A^\dagger + \tilde{r}B^\dagger\right)\left(\tilde{t}A + \tilde{r}B\right)$$

$$= \tilde{t}^2 \left(A^\dagger + \frac{\tilde{r}}{\tilde{t}}B^\dagger\right)\left(A + \frac{\tilde{r}}{\tilde{t}}B\right). \tag{4.2.66}$$

To look for the probability of measuring the eigenvalue n_c of $C^\dagger C$ (the number of photons in mode C), we make use of the concept of the characteristic function introduced earlier. Let us calculate the characteristic function $\left\langle e^{i\frac{2\pi l}{L}C^\dagger C} \right\rangle$. We can first expand this exponential term into the series

$$\left\langle e^{i\frac{2\pi l}{L}C^\dagger C} \right\rangle = \sum_{k=0}^{\infty} \frac{1}{k!} \left(\frac{2\pi i l}{L}\right)^k \left\langle (C^\dagger C)^k \right\rangle. \tag{4.2.67}$$

The next task is to evaluate the moments $\left\langle (C^\dagger C)^k \right\rangle$. To simplify the subsequent expressions, we define $M_y \equiv A + yB$ and $M_y^\dagger \equiv A^\dagger + yB^\dagger$, with $y = \tilde{r}/\tilde{t}$ in our case, that obey the commutation relation

$$
\begin{aligned}
\left[M_y, M_y^\dagger\right] &= \left[A + yB, A^\dagger + yB^\dagger\right] \\
&= \underbrace{\left[A, A^\dagger\right]}_{= 1} + y^2 \underbrace{\left[B, B^\dagger\right]}_{= 1} \\
&= 1 + y^2,
\end{aligned}
\tag{4.2.68}
$$

which reduces to $\left[A, A^\dagger\right] = 1$ when $y = 0$, obviously. At this point, we will make an assumption about $\tilde{r}$. We shall take the limit $\tilde{r} \to 0$ and $|\beta| \to \infty$ such that $\tilde{r}|\beta|$ remains a finite value. In this limit, the terms that involve $\tilde{r}$ of second-order and beyond vanish, so that $\left[M_y, M_y^\dagger\right] = 1$.

The expectation values

$$
\begin{aligned}
\left\langle \left(C^\dagger C\right)^k \right\rangle &= \mathrm{tr}\left\{ \left(\rho_{\mathrm{src}} \otimes |\beta\rangle \langle \beta^*|\right) \left(C^\dagger C\right)^k \right\} \\
&= \mathrm{tr}\left\{ \left(\rho_{\mathrm{src}} \otimes |\beta\rangle \langle \beta^*|\right) \left(M_y^\dagger M_y\right)^k \right\},
\end{aligned}
\tag{4.2.69}
$$

which involve partial traces with the coherent state $|\beta\rangle \langle \beta^*|$, can be simplified by ordering the operators in the term $\left(M_y^\dagger M_y\right)^k$ in such a way that all $M_y^\dagger$s are to the *left* of all M_ys. In other words, we need to *normally order* these operators. For $k = 1$, the result

$$
\begin{aligned}
\left\langle M_y^\dagger M_y \right\rangle &= \mathrm{tr}\left\{ \left(\rho_{\mathrm{src}} \otimes |\beta\rangle \langle \beta^*|\right) M_y^\dagger M_y \right\} \\
&= \left\langle \left(A^\dagger + y\beta^*\right)\left(A + y\beta\right) \right\rangle_{\mathrm{src}}
\end{aligned}
\tag{4.2.70}
$$

follows immediately since $M_y^\dagger$ is already on the left of M_y. For the term that corresponds to $k = 2$, we have

$$
\begin{aligned}
\left\langle \left(M_y^\dagger M_y\right)^2 \right\rangle &= \mathrm{tr}\left\{ \left(\rho_{\mathrm{src}} \otimes |\beta\rangle \langle \beta^*|\right) \left(M_y^\dagger M_y M_y^\dagger M_y\right) \right\} \\
&= \mathrm{tr}\left\{ \left(\rho_{\mathrm{src}} \otimes |\beta\rangle \langle \beta^*|\right) \left(M_y^{\dagger 2} M_y^2 + M_y^\dagger M_y\right) \right\} \\
&= \mathrm{tr}\left\{ \rho_{\mathrm{src}} \left[\left(A^\dagger + y\beta^*\right)^2 \left(A + y\beta\right)^2 + \left(A^\dagger + y\beta^*\right)\left(A + y\beta\right) \right] \right\} \\
&= \left\langle \left[\left(A^\dagger + y\beta^*\right)\left(A + y\beta\right) \right]^2 \right\rangle_{\mathrm{src}}.
\end{aligned}
\tag{4.2.71}
$$

Without performing any more calculations, it is straightforward to realize that

$$
\left\langle \left(M_y^\dagger M_y\right)^k \right\rangle = \left\langle \left[\left(A^\dagger + y\beta^*\right)\left(A + y\beta\right) \right]^k \right\rangle_{\mathrm{src}},
\tag{4.2.72}
$$

since all terms in every product always contain linear combinations of the ladder operators A and B that are of the same adjointness. So,

$$\left\langle e^{i\frac{2\pi l}{L}C^\dagger C}\right\rangle = \left\langle e^{i\frac{2\pi l}{L}\left(A^\dagger-\alpha^*\right)(A-\alpha)}\right\rangle_{\text{src}}, \tag{4.2.73}$$

where $\alpha = -\tilde{r}\beta$.

To rewrite Eq. (4.2.73) into a more recognizable form, we first look at the result of sandwiching the ladder operator A with two displacement operators of an infinitesimal parameter $\delta\alpha$:

$$
\begin{aligned}
D(\delta\alpha)\,A\,D(-\delta\alpha) &= \left(1 + \delta\alpha A^\dagger - \delta\alpha^* A\right)A\left(1 - \delta\alpha A^\dagger + \delta\alpha^* A\right) \\
&= A + \delta\alpha A^\dagger A - \delta\alpha^* A^2 - \delta\alpha A A^\dagger + \delta\alpha^* A^2 \\
&= A - \delta\alpha. \tag{4.2.74}
\end{aligned}
$$

Hence, applying an infinitesimal displacement transformation on A shifts it by an infinitesimal multiple of the identity, and with enough displacements, we have

$$D(\alpha)\,A\,D(-\alpha) = A - \alpha. \tag{4.2.75}$$

Therefore

$$D(\alpha)F(A, A^\dagger)D(-\alpha) = F(A - \alpha, A^\dagger - \alpha^*) \tag{4.2.76}$$

for any function $F(A, A^\dagger)$ of A and $A^\dagger$, and the characteristic function is given by

$$\left\langle e^{i\frac{2\pi l}{L}C^\dagger C}\right\rangle = \left\langle D(\alpha)\,e^{i\frac{2\pi l}{L}A^\dagger A}\,D(-\alpha)\right\rangle_{\text{src}}. \tag{4.2.77}$$

From this, the probability for measuring the discrete eigenvalues n_c is given as

$$
\begin{aligned}
p(n_c) &= \frac{1}{L}\sum_{l=0}^{L-1}\left\langle D(\alpha)\,e^{i\frac{2\pi l}{L}\left(A^\dagger A - n_c\right)}\,D(-\alpha)\right\rangle_{\text{src}} \\
&= \left\langle D(\alpha)\,|n = n_c\rangle\,\langle n = n_c|\,D(-\alpha)\right\rangle_{\text{src}}. \tag{4.2.78}
\end{aligned}
$$

Note that the measurement outcomes $D(\alpha)\,|n\rangle\,\langle n|\,D(-\alpha)$ are now labeled by the continuous complex variable α. Measurements of this kind are also known as *continuous-variable measurements*.

4.2.3.2 *Quadrature-eigenstate measurement*

There is an alternative experimental scheme (refer to Fig. 4.10) to tap into continuous degrees of freedom in order to estimate the off-diagonal elements of ρ_{src}, and that is to consider solely the BS set-up, only this time we collect photocurrents from both arms of the BS output. The photocurrents that are proportional to $\langle C^\dagger C \rangle$ and $\langle D^\dagger D \rangle$ are first detected by the respective photodetectors. After which, the two registered measurement data are digitally subtracted. Suppose that the BS ratio is exactly 1 : 1 $(\tilde{t} = \tilde{r} = 1/\sqrt{2})$, where the transmission and reflection amplitudes are real, we have

$$
\begin{aligned}
I_{\text{diff}} &\equiv \frac{C^\dagger C - D^\dagger D}{\sqrt{2}\,|\beta|} \\
&= \frac{1}{\sqrt{2}\,|\beta|}\left[\left(\frac{A^\dagger + B^\dagger}{\sqrt{2}}\right)\left(\frac{A + B}{\sqrt{2}}\right) - \left(\frac{A^\dagger - B^\dagger}{\sqrt{2}}\right)\left(\frac{A - B}{\sqrt{2}}\right)\right] \\
&= \frac{1}{\sqrt{2}\,|\beta|}\left(A^\dagger B + B^\dagger A\right),
\end{aligned}
\tag{4.2.79}
$$

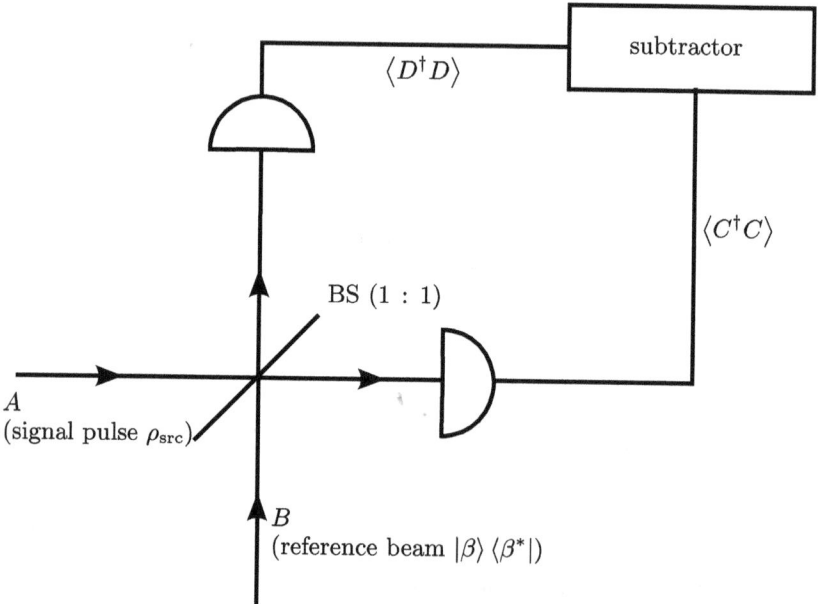

Fig. 4.10 Schematic setup for the quadrature-eigenstate measurement.

where the scaling factor $1/(\sqrt{2}\,|\beta|)$ can easily be introduced electronically. As usual, we are interested in the characteristic function

$$\left\langle e^{ikI_{\text{diff}}}\right\rangle = \text{tr}\left\{\rho_{\text{src}} \otimes |\beta\rangle\,\langle\beta^*|\,e^{i\frac{k}{\sqrt{2}\,|\beta|}\left(A^\dagger B + B^\dagger A\right)}\right\}. \qquad (4.2.80)$$

Upon expanding the characteristic function into its power series

$$\left\langle e^{ikI_{\text{diff}}}\right\rangle = \sum_{l=0}^{\infty} \frac{1}{l!}\left(\frac{ik}{\sqrt{2}\,|\beta|}\right)^l \left\langle\left(A^\dagger B + B^\dagger A\right)^l\right\rangle, \qquad (4.2.81)$$

it is then possible to evaluate the expectation value of each term in the expansion.

Evaluating these expectation values requires us to normally order the operators of every order in the power series. To do this, we further expand the operators in the lth term into a binomial expansion. For commuting objects x and y, the lth-order binomial $(x+y)^l$ for commuting variables x and y exhibits the usual expansion

$$(x+y)^l = \sum_{m=0}^{l}\binom{l}{m} x^{l-m} y^m, \qquad (4.2.82)$$

where the combinatorial factor counts the number of possible orderings to multiply $l-m$ xs and m ys together, which are *all* equal. If the variables x and y are now replaced by the noncommuting operators $A^\dagger B$ and $B^\dagger A$, the lth-order binomial then expands into sums of products of $A^\dagger B$s and $B^\dagger A$s in all different orders. This is known as *Weyl* ordering*, or symmetric ordering. Formally, such a symmetrical sum of products may be defined as

$$\left[\left(A^\dagger B\right)^m \left(B^\dagger A\right)^n\right]_{\text{WS}} = \frac{\{\text{Sum over all products of } m\ A^\dagger B\text{s and } n\ B^\dagger A\text{s.}\}}{\binom{m+n}{n}},$$
$$(4.2.83)$$

where we take this opportunity to introduce the bracket notation $[O^m O'^n]_{\text{WS}}$ denotes the *Weyl-ordered form* of the product of $m\ O$ operators and $n\ O'$ operators that are noncommuting. This bracket notation should not be confused with the Weyl ordering procedure on this product. This is best

---

*Hermann Klaus Hugo Weyl (1885–1955).

illustrated using an example with $m = 2$, $n = 1$, $O = A$ and $O' = A^\dagger$. Then Weyl ordering the product $A^2 A^\dagger$ gives

$$
\begin{aligned}
A^2 A^\dagger &= \frac{1}{3} A^2 A^\dagger + \frac{2}{3} A^2 A^\dagger \\
&= \frac{1}{3} A^2 A^\dagger + \frac{2}{3} \left(A A^\dagger A + A \right) \\
&= \frac{1}{3} \left(A^2 A^\dagger + A A^\dagger A \right) + \frac{1}{3} \left(A^\dagger A^2 + A \right) + \frac{2}{3} A \\
&= \frac{1}{3} \left(A^2 A^\dagger + A A^\dagger A + A^\dagger A^2 \right) + A \\
&= \left[A^2 A^\dagger \right]_{\text{WS}} + A.
\end{aligned}
\tag{4.2.84}
$$

Using this bracket notation, it is straightforward to see that the binomial expansion for the operators is given by

$$
\left(A^\dagger B + B^\dagger A \right)^l = \sum_{m=0}^{l} \binom{l}{m} \left[\left(A^\dagger B \right)^{l-m} \left(B^\dagger A \right)^m \right]_{\text{WS}}.
\tag{4.2.85}
$$

The expectation value of $\left(A^\dagger B + B^\dagger A \right)^l$ can now be calculated by noting that since B commutes with A, each Weyl ordered term in the binomial expansion for a given m can be written as a factored product of functions of mode A and mode B. For instance, the second-order binomial

$$
\left(A^\dagger B + B^\dagger A \right)^2 = \left[\left(A^\dagger B \right)^2 \right]_{\text{WS}} + 2 \left[\left(A^\dagger B \right) \left(B^\dagger A \right) \right]_{\text{WS}} + \left[\left(B^\dagger A \right)^2 \right]_{\text{WS}}
\tag{4.2.86}
$$

involves the three Weyl-ordered forms

$$
\left[\left(A^\dagger B \right)^2 \right]_{\text{WS}} = \left(A^\dagger B \right)^2,
\tag{4.2.87}
$$

$$
\left[\left(B^\dagger A \right)^2 \right]_{\text{WS}} = \left(B^\dagger A \right)^2,
\tag{4.2.88}
$$

and

$$
\begin{aligned}
\left[\left(A^\dagger B \right) \left(B^\dagger A \right) \right]_{\text{WS}} &= \frac{1}{2} \left(A^\dagger B B^\dagger A + B^\dagger A A^\dagger B \right) \\
&= \frac{1}{2} \left(A^\dagger A B B^\dagger + A A^\dagger B^\dagger B \right).
\end{aligned}
\tag{4.2.89}
$$

Hence, the expectation value of the second-order binomial is readily obtained by normal ordering the ladder operators for mode B. The first two Weyl-ordered forms are already properly ordered in B and $B^\dagger$, and the third one can be straightforwardly ordered to give

$$\left[\left(A^\dagger B\right)\left(B^\dagger A\right)\right]_{\mathrm{WS}} = \frac{1}{2}\left(A^\dagger A + A A^\dagger\right) B^\dagger B + \frac{1}{2} A^\dagger A$$

$$= \frac{1}{2}\left(A^\dagger A + A A^\dagger\right)\left[BB^\dagger\right]_{\mathrm{N}} + \{\text{fringe term}\}, \quad (4.2.90)$$

where we introduce another bracket notation for the *normal-ordered form** of $BB^\dagger$. Again, this is not to be understood as performing normal ordering on $BB^\dagger$, for

$$BB^\dagger = B^\dagger B + 1 = \left[BB^\dagger\right]_{\mathrm{N}} + 1. \qquad (4.2.91)$$

It is pertinent to observe that after normal ordering is performed on mode B, the fringe term is at most two orders lower in the ladder operators of mode B to the Weyl-ordered form. All other terms in the second-order binomial of Eq. (4.2.86) contain products of exactly two ladder operators of mode B, apart from the fringe term that arises from normal ordering. Thus, after taking the trace with $|\beta\rangle\langle\beta^*|$, we find that these terms give a factor of $|\beta|^2$ that precisely cancels the $1/|\beta|^2$ prefactor in the second-order term in Eq. (4.2.81).

Problem 4.14   Show that

$$\left[F(A^\dagger A)\right]_{\mathrm{N}} = F\left(\frac{\mathrm{d}}{\mathrm{d}x}\right) x^{A^\dagger A}\bigg|_{x=1}, \qquad (4.2.92)$$

which states that the normal-ordered form of an operator function of $A^\dagger A$ is yet another operator function of $A^\dagger A$. With this, establish the identities

$$\left[e^{-\lambda A^\dagger A}\right]_{\mathrm{N}} = (1-\lambda)^{A^\dagger A}, \qquad (4.2.93)$$

$$A^{\dagger\, n} A^n = \frac{(A^\dagger A)!}{(A^\dagger A - n)!} \quad \text{for integer } n. \qquad (4.2.94)$$

---

*A common notation for the normal-ordered form is $: B^m B^{\dagger\, n} : = B^{\dagger\, n} B^m$. Here, the bracket notation is adopted to distinguish between different types of ordered forms.

The same goes for the higher-order binomials

$$\left(A^\dagger B + B^\dagger A\right)^l = \sum_{m=0}^{l} \binom{l}{m} \left[A^{\dagger\,l-m} A^m\right]_{\text{WS}} \left[B^{l-m} B^{\dagger\,m}\right]_{\text{N}} + \{\text{fringe terms}\},$$

(4.2.95)

so that

$$\langle\beta^*|\left(A^\dagger B + B^\dagger A\right)^l|\beta\rangle = |\beta|^l \left(A^\dagger e^{i\vartheta} + A e^{-i\vartheta}\right)^l + O\left(|\beta|^{l-2}\right).$$ (4.2.96)

From here, let us impose the condition that the intensity of the local oscillator is much larger than the mean number of photons the source contains, taking the limit $\langle A^\dagger A\rangle_{\text{src}} \ll |\beta|^2 \to \infty$. In this limit, the fringe terms do not contribute to the characteristic function, and we finally get

$$\left\langle e^{ikI_{\text{diff}}}\right\rangle = \left\langle e^{i\frac{k}{\sqrt{2}}\left(A e^{-i\vartheta} + A^\dagger e^{i\vartheta}\right)}\right\rangle_{\text{src}}$$

$$= \left\langle e^{ikX_\vartheta}\right\rangle_{\text{src}},$$ (4.2.97)

where X_ϑ is the *quadrature operator* of phase ϑ. So, in the high-energy limit, the BS set-up realizes a POM consisting of a complete set of eigenstates of X_ϑ — the quadrature eigenstates.

That X_ϑ is derived from unitarily rotating the usual *position quadrature operator* X can be seen by first noting the usual definition

$$X = \frac{A + A^\dagger}{\sqrt{2}}$$ (4.2.98)

for the position operator and, next, recalling the identities

$$F(A^\dagger A)\, A = A\, F(A^\dagger A - 1),$$

$$F(A^\dagger A)\, A^\dagger = A^\dagger\, F(A^\dagger A + 1).$$ (4.2.99)

Problem 4.15   Show these identities.

An application of the unitary operator $e^{i\vartheta A^\dagger A}$ on X thus gives

$$e^{i\vartheta A^\dagger A}\, X\, e^{-i\vartheta A^\dagger A} = e^{i\vartheta A^\dagger A}\left(\frac{A + A^\dagger}{\sqrt{2}}\right) e^{-i\vartheta A^\dagger A}$$

$$= \frac{1}{\sqrt{2}}\left[A e^{i\vartheta\left(A^\dagger A - 1\right)} + A^\dagger e^{i\vartheta\left(A^\dagger A + 1\right)}\right] e^{-i\vartheta A^\dagger A}$$

$$= \frac{1}{\sqrt{2}} \left(A e^{-i\vartheta} + A^\dagger e^{i\vartheta} \right)$$

$$= X_\vartheta. \tag{4.2.100}$$

This means that the familiar position $X = X_0$ and momentum $P = X_{\frac{\pi}{2}}$ quadratures are just two special cases of the general operator X_ϑ. A given pair of quadrature operators X_ϑ and $X_{\vartheta'}$ exhibit the commutation relation

$$[X_\vartheta, X_{\vartheta'}] = \left[\frac{1}{\sqrt{2}} \left(A e^{-i\vartheta} + A^\dagger e^{i\vartheta} \right), \frac{1}{\sqrt{2}} \left(A e^{-i\vartheta'} + A^\dagger e^{i\vartheta'} \right) \right]$$

$$= \frac{1}{2} \left[e^{i(\vartheta'-\vartheta)} - e^{-i(\vartheta'-\vartheta)} \right] = i \sin(\vartheta' - \vartheta), \tag{4.2.101}$$

where the right-hand side is maximized in magnitude for *any* pair of quadrature operators such that $|\vartheta' - \vartheta|$ is an odd-integer multiple of $\pi/2$. In terms of the standard position and momentum operators,

$$X_\vartheta = \frac{1}{\sqrt{2}} \left(A e^{-i\vartheta} + A^\dagger e^{i\vartheta} \right)$$

$$= \frac{1}{\sqrt{2}} \left(\frac{X + iP}{\sqrt{2}} e^{-i\vartheta} + \frac{X - iP}{\sqrt{2}} e^{i\vartheta} \right)$$

$$= \left(X \frac{e^{i\vartheta} + e^{-i\vartheta}}{2} + P \frac{e^{i\vartheta} - e^{-i\vartheta}}{2i} \right)$$

$$= X \cos\vartheta + P \sin\vartheta, \tag{4.2.102}$$

where $0 \le \vartheta < \pi$.

The eigenvalues x_ϑ of X_ϑ are, therefore, continuous since they are derived from the eigenvalues of X:

$$\int dx_\vartheta \, |x_\vartheta\rangle x_\vartheta \langle x_\vartheta| = X_\vartheta = e^{i\vartheta A^\dagger A} X e^{-i\vartheta A^\dagger A}$$

$$= \int dx \left(e^{i\vartheta A^\dagger A} |x\rangle \right) x \left(\langle x| e^{-i\vartheta A^\dagger A} \right)$$

$$\Rightarrow \begin{cases} |x_\vartheta\rangle = e^{i\vartheta A^\dagger A} |x\rangle, \\ x_\vartheta = x. \end{cases} \tag{4.2.103}$$

The characteristic function $\left\langle e^{ikX_\vartheta} \right\rangle_{\mathrm{src}}$ then defines the probability distribution for the continuous set of eigenvalues x_ϑ. The continuous analogue of the relationship between the probability distribution and its characteristic function is defined by the *continuous Fourier Transform*:

$$
\begin{aligned}
p(x_\vartheta) &= \int \frac{\mathrm{d}k}{2\pi} \left\langle e^{ik(X_\vartheta - x_\vartheta)} \right\rangle_{\mathrm{src}} \\
&= \left\langle \delta\left(X_\vartheta - x_\vartheta\right) \right\rangle_{\mathrm{src}} \\
&= \left\langle |x_\vartheta\rangle \langle x_\vartheta| \right\rangle_{\mathrm{src}}.
\end{aligned}
\tag{4.2.104}
$$

This means that if we allow the coherent state intensity to increase to infinity, the resulting probability distribution that is measured by the set-up is that for the eigenstates of the quadrature operator X_ϑ, where the angle ϑ is defined by the phase of the local oscillator. This entire procedure is known as *homodyne detection*.

As far as the physical principle is concerned, we have shown that the measurement frequencies corresponding to the probability distribution of the quadrature eigenstates can be obtained by measuring the observable I_{diff} with a BS set up that is fed with a signal state and a reference coherent state of *infinite* intensity. There is, however, an important remark on these quadrature eigenstates. Being eigenstates of the continuous-variable operator X_ϑ, they normalize to a delta function for a fixed angle ϑ, that is $\langle x_\vartheta | x'_\vartheta \rangle = \delta(x'_\vartheta - x_\vartheta)$, and are therefore unphysical projectors. This is reminiscent of the unphysical nature of position eigenstates $|x\rangle \langle x|$ that are a result of an overidealization of the physical situation in which we are unable to prepare a quantum system in a particular position eigenstate, as this requires infinite resources that we can never hope to have. For the quadrature projectors, this would require an infinite amount of energy from the local oscillator. Hence, in reality, the eigenstates $|x_\vartheta\rangle \langle x_\vartheta|$ can never be realized. It cannot be overemphasized that the inability to prepare a quantum system in one of these eigenstates is tied to its fundamental physical limitation, not instrumental capabilities.

Let us see what happens when the photodetectors are not perfect, which is the situation in any real experiment. Suppose the efficiencies of the two photodetectors are equal to η_{L}. For such imperfections, we can model the effects on the detected modes with an imaginary BS that is mode dependent and distributes some of the incoming photons to imaginary *vacuum modes*, which are modes prepared in the vacuum state $|0\rangle \langle 0|$. These are summarized by the relations

$$C \rightarrow C' = \sqrt{\eta_{\mathrm{L}}}\, C + \sqrt{1 - \eta_{\mathrm{L}}}\, C_{\mathrm{L}}\,,$$

$$D \rightarrow D' = \sqrt{\eta_{\mathrm{L}}}\, D + \sqrt{1 - \eta_{\mathrm{L}}}\, D_{\mathrm{L}}\,. \tag{4.2.105}$$

Here, the operators C_{L} and D_{L} are the respective vacuum modes that describe losses for modes C and D, which are the original modes of light without photon-loss imperfections. The relations in Eq. (4.2.105) account for the reductions in photocurrents according to the respective efficiencies, so that

$$\langle C'^{\dagger} C' \rangle = \mathrm{tr}\Big\{ \rho_{\mathrm{src}} \otimes |\beta\rangle\,\langle \beta^{*}| \otimes \underbrace{|0\rangle\,\langle 0|}_{\text{eigenstate of } C_{\mathrm{L}}/D_{\mathrm{L}}} \big(C'^{\dagger} C' \big) \Big\}$$

$$= \eta_{\mathrm{L}} \left\langle C^{\dagger} C \right\rangle_{\mathrm{src}}, \tag{4.2.106}$$

as well as

$$\left\langle D'^{\dagger} D' \right\rangle = \eta_{\mathrm{L}} \left\langle D^{\dagger} D \right\rangle_{\mathrm{src}}. \tag{4.2.107}$$

We can define

$$I'_{\mathrm{diff}} = \frac{C'^{\dagger} C' - D'^{\dagger} D'}{\sqrt{2}\,|\beta|\eta_{\mathrm{L}}} \tag{4.2.108}$$

for the scaled difference in the number operators for modes C and D that is equal to

$$\begin{aligned}
I'_{\mathrm{diff}} &= \frac{1}{\sqrt{2}\,|\beta|\eta_{\mathrm{L}}} \Big[\eta_{\mathrm{L}} \left(C^{\dagger} C - D^{\dagger} D \right) + \sqrt{\eta_{\mathrm{L}}(1 - \eta_{\mathrm{L}})} \\
&\quad \times \left(C^{\dagger} C_{\mathrm{L}} + C_{\mathrm{L}}^{\dagger} C - D^{\dagger} D_{\mathrm{L}} - D_{\mathrm{L}}^{\dagger} D \right) + (1 - \eta_{\mathrm{L}}) \left(C_{\mathrm{L}}^{\dagger} C_{\mathrm{L}} - D_{\mathrm{L}}^{\dagger} D_{\mathrm{L}} \right) \Big] \\
&\approx \frac{1}{\sqrt{2}\,|\beta|\eta_{\mathrm{L}}} \Big[\eta_{\mathrm{L}} \left(A^{\dagger} B + B^{\dagger} A \right) + \sqrt{\frac{\eta_{\mathrm{L}}(1 - \eta_{\mathrm{L}})}{2}} \\
&\quad \times \left(B^{\dagger} C_{\mathrm{L}} + C_{\mathrm{L}}^{\dagger} B + B^{\dagger} D_{\mathrm{L}} + D_{\mathrm{L}}^{\dagger} B \right) \Big] \\
&= \frac{1}{\sqrt{2 \eta_{\mathrm{L}}}\,|\beta|} \Bigg\{ \left[\sqrt{\eta_{\mathrm{L}}} A^{\dagger} + \sqrt{\frac{1 - \eta_{\mathrm{L}}}{2}} \left(C_{\mathrm{L}}^{\dagger} + D_{\mathrm{L}}^{\dagger} \right) \right] B \\
&\quad + B^{\dagger} \left[\sqrt{\eta_{\mathrm{L}}} A + \sqrt{\frac{1 - \eta_{\mathrm{L}}}{2}} \left(C_{\mathrm{L}} + D_{\mathrm{L}} \right) \right] \Bigg\}, \tag{4.2.109}
\end{aligned}$$

where we have consistently kept terms with the local-oscillator mode B only, as all other terms do not lead to any contribution to the characteristic function of I'_{diff} in the high-energy limit.

Problem 4.16 Explain why this is so.

Without going through the same calculation, we simply define

$$A' = \sqrt{\eta_{\text{L}}}A + \sqrt{\frac{1-\eta_{\text{L}}}{2}}\,(C_{\text{L}} + D_{\text{L}}), \qquad (4.2.110)$$

which behaves exactly like A in terms of the commutation algebra:

$$[A', A'^\dagger] = \eta_{\text{L}}\,[A, A^\dagger] + \frac{1-\eta_{\text{L}}}{2}\,[C_{\text{L}} + D_{\text{L}}, C_{\text{L}}^\dagger + D_{\text{L}}^\dagger] = 1,$$
$$[A', B] = [A', B^\dagger] = 0. \qquad (4.2.111)$$

With this, we immediately write down the final observable that is measured, after taking the high-energy limit $|\beta|^2 \gg \langle A^\dagger A\rangle_{\text{src}}$, as

$$W_\vartheta = \frac{A'e^{-i\vartheta} + A'^\dagger e^{i\vartheta}}{\sqrt{2\,\eta_{\text{L}}}}$$
$$= \frac{1}{\sqrt{2}}\left\{\left[A + \sqrt{\frac{1-\eta_{\text{L}}}{2\,\eta_{\text{L}}}}\,(C_{\text{L}} + D_{\text{L}})\right]e^{-i\vartheta}\right.$$
$$\left. + \left[A^\dagger + \sqrt{\frac{1-\eta_{\text{L}}}{2\,\eta_{\text{L}}}}\,(C_{\text{L}}^\dagger + D_{\text{L}}^\dagger)\right]e^{i\vartheta}\right\}$$
$$= X_\vartheta + \frac{1}{2}\sqrt{\frac{1-\eta_{\text{L}}}{\eta_{\text{L}}}}\,[e^{-i\vartheta}(C_{\text{L}} + D_{\text{L}}) + e^{i\vartheta}(C_{\text{L}}^\dagger + D_{\text{L}}^\dagger)]. \qquad (4.2.112)$$

Since the vacuum modes are independent of the signal mode in X_ϑ, the characteristic function $\langle e^{ikW_\vartheta}\rangle$, with the expectation value taken with respect to the tensor product of ρ_{src} and two copies of the vacuum state for the modes C_{L} and D_{L} (each denoted by $|0\rangle\langle 0|$ in short), can be disentangled as

$$
\begin{aligned}
\left\langle e^{ikW_\vartheta} \right\rangle &= \left\langle e^{ikX_\vartheta} \, e^{ik\sqrt{\frac{1-\eta_L}{4\,\eta_L}}\left(C_L e^{-i\vartheta}+C_L^\dagger e^{i\vartheta}\right)} \, e^{ik\sqrt{\frac{1-\eta_L}{4\,\eta_L}}\left(D_L e^{-i\vartheta}+D_L^\dagger e^{i\vartheta}\right)} \right\rangle \\
&= \left\langle e^{ikX_\vartheta} \right\rangle_{\mathrm{src}} \left\langle 0 \right| e^{ik\sqrt{\frac{1-\eta_L}{4\,\eta_L}}C_L^\dagger e^{i\vartheta}} \, e^{-\frac{k^2}{2}\left(\frac{1-\eta_L}{4\,\eta_L}\right)} \, e^{ik\sqrt{\frac{1-\eta_L}{4\,\eta_L}}C_L e^{-i\vartheta}} \left| 0 \right\rangle \\
&\quad \times \left\langle 0 \right| e^{ik\sqrt{\frac{1-\eta_L}{4\,\eta_L}}D_L^\dagger e^{i\vartheta}} \, e^{-\frac{k^2}{2}\left(\frac{1-\eta_L}{4\,\eta_L}\right)} \, e^{ik\sqrt{\frac{1-\eta_L}{4\,\eta_L}}D_L e^{-i\vartheta}} \left| 0 \right\rangle \\
&= \left\langle e^{ikX_\vartheta} \right\rangle_{\mathrm{src}} e^{-k^2\left(\frac{1-\eta_L}{4\,\eta_L}\right)}.
\end{aligned}
\tag{4.2.113}
$$

The probability distribution $p(w_\vartheta)$ for the eigenvalues w_ϑ of the observable W_ϑ is thus given as

$$
\begin{aligned}
p(w_\vartheta) &= \int \frac{dk}{2\pi} \left\langle e^{ik(W_\vartheta - w_\vartheta)} \right\rangle_{\mathrm{src}} \\
&= \int \frac{dk}{2\pi} \left\langle e^{ik(X_\vartheta - w_\vartheta)} \, e^{-k^2\left(\frac{1-\eta_L}{4\,\eta_L}\right)} \right\rangle_{\mathrm{src}},
\end{aligned}
\tag{4.2.114}
$$

and with the help of the identity in Eq. (4.1.32), the Gaussian integral in Eq. (4.2.114) can be easily evaluated, yielding

$$
\begin{aligned}
p(w_\vartheta) &= \left\langle \frac{1}{\sqrt{2\pi\sigma_{\eta_L}^2}} e^{-\frac{1}{2\sigma_{\eta_L}^2}(X_\vartheta - w_\vartheta)^2} \right\rangle_{\mathrm{src}} \qquad \left(\sigma_{\eta_L}^2 = \frac{1-\eta_L}{2\,\eta_L}\right) \\
&= \left\langle \int dx'_\vartheta \, \left| x'_\vartheta \right\rangle \left[\frac{1}{\sqrt{2\pi\sigma_{\eta_L}^2}} e^{-\frac{1}{2\sigma_{\eta_L}^2}(x'_\vartheta - w_\vartheta)^2} \right] \left\langle x'_\vartheta \right| \right\rangle_{\mathrm{src}}.
\end{aligned}
\tag{4.2.115}
$$

The outcomes that are measured with two photodetectors of efficiency η_L is, as it turns out, a *mixture* of quadrature eigenstates weighted by a Gaussian distribution. Thus, realistic photodetectors inevitably result in a line-width for the otherwise infinitely narrow delta-function distribution.

It is useful to represent the continuous-variable kets $|x_\vartheta\rangle$ with vectors in a discrete orthonormal basis for numerical computation. We shall consider the Fock basis as the computational basis. From Eq. (4.2.103), we have

$$\langle n | x_\vartheta \rangle = \langle n | e^{i\vartheta A^\dagger A} | x \rangle \Big|_{x=x_\vartheta}$$

$$= e^{in\vartheta} \langle n | x \rangle \Big|_{x=x_\vartheta}$$

$$= \frac{1}{\pi^{\frac{1}{4}} \sqrt{2^n \, n!}} e^{in\vartheta} e^{-\frac{1}{2} x_\vartheta^2} H_n(x_\vartheta), \qquad (4.2.116)$$

where $H_n(x)$ is the *Hermite* polynomial in x of degree n and we have used the expression

$$\langle x | n \rangle = \frac{1}{\pi^{\frac{1}{4}} \sqrt{2^n \, n!}} e^{-\frac{1}{2} x^2} H_n(x) = \langle n | x \rangle \qquad (4.2.117)$$

for the eigenstate wave function of a quantum harmonic oscillator. With this, the matrix elements of $|x_\vartheta\rangle \langle x_\vartheta|$ in the computational basis is given by

$$\langle n' | x_\vartheta \rangle \langle x_\vartheta | n \rangle = \frac{1}{\sqrt{\pi \, 2^{n'+n} \, n'! \, n!}} e^{i(n'-n)\vartheta} e^{-x_\vartheta^2} H_{n'}(x_\vartheta) H_n(x_\vartheta)$$

$$(4.2.118)$$

in terms of the kets $|n\rangle$ and $|n'\rangle$.

We may also express these quadrature eigenstates in the position or momentum basis to obtain the corresponding wave functions. We start by looking for the position wave function $\langle x' | x_\vartheta \rangle$. It follows that

$$\langle x' | x_\vartheta \rangle = \langle x' | e^{i\vartheta A^\dagger A} | x \rangle \Big|_{x=x_\vartheta}$$

$$= \sum_{n=0}^{\infty} \langle x' | n \rangle \langle n | e^{i\vartheta A^\dagger A} | x \rangle \Big|_{x=x_\vartheta}$$

$$= \sum_{n=0}^{\infty} e^{in\vartheta} \langle x' | n \rangle \langle n | x_\vartheta \rangle$$

$$= \frac{1}{\sqrt{\pi}} e^{-\frac{1}{2}\left(x_\vartheta^2 + x'^2\right)} \sum_{n=0}^{\infty} \frac{e^{in\vartheta}}{2^n \, n!} H_n(x_\vartheta) H_n(x'), \qquad (4.2.119)$$

---

*Charles Hermite (1822–1901).

in which we need to evaluate the sum of products of Hermite polynomials. To continue the calculation, we recall the definition

$$H_n(x) = (-1)^n e^{x^2} \left(\frac{d}{dx}\right)^n e^{-x^2}, \qquad (4.2.120)$$

which is also known as a *Rodrigues* formula* for $H_n(x)$. It is convenient to express Eq. (4.2.120) in terms of a Gaussian integral by using the identity

$$e^{-x^2} = \frac{1}{\sqrt{4\pi}} \int dy\, e^{-\frac{1}{4}y^2 + ixy}, \qquad (4.2.121)$$

thus turning it into

$$H_n(x) = \frac{(-i)^n}{\sqrt{4\pi}} e^{x^2} \int dy\, y^n\, e^{-\frac{1}{4}y^2 + ixy}. \qquad (4.2.122)$$

So, the sum

$$\sum_{n=0}^{\infty} \frac{u^n}{n!} H_n(x_\vartheta) H_n(x'), \qquad (4.2.123)$$

where $u = e^{i\vartheta}/2$ in this case, can be evaluated as a series of Gaussian integrals,

$$\sum_{n=0}^{\infty} \frac{u^n}{n!} H_n(x_\vartheta) H_n(x')$$

$$= \frac{1}{4\pi} e^{x_\vartheta^2 + x'^2} \sum_{n=0}^{\infty} \frac{(-u)^n}{n!} \int dy\, y^n\, e^{-\frac{1}{4}y^2 + ix_\vartheta y} \int dy'\, y'^n\, e^{-\frac{1}{4}y'^2 + ix'y'}$$

$$= \frac{1}{4\pi} e^{x_\vartheta^2 + x'^2} \int dy\, dy'\, e^{-\frac{1}{4}y^2 + ix_\vartheta y}\, e^{-\frac{1}{4}y'^2 + ix'y'} \underbrace{\sum_{n=0}^{\infty} \frac{(-uyy')^n}{n!}}_{= e^{-uyy'}}$$

$$= \frac{1}{4\pi} e^{x_\vartheta^2 + x'^2} \int dy\, e^{-\frac{1}{4}y^2 + ix_\vartheta y} \underbrace{\int dy'\, e^{-\frac{1}{4}y'^2 + (ix' - uy)y'}}_{= \sqrt{4\pi}\, e^{(ix' - uy)^2}}$$

---

*Benjamin Olinde Rodrigues (1795–1851).

$$= \frac{1}{\sqrt{4\pi}} \, e^{x_\vartheta^2} \int dy \, e^{-\left(\frac{1}{4}-u^2\right)y^2 + i\left(x_\vartheta - 2\,x'u\right)y}$$

$$= \frac{1}{\sqrt{1-4\,u^2}} \, e^{x_\vartheta^2} \, e^{-\frac{(x_\vartheta - 2\,x'u)^2}{1-4\,u^2}}$$

$$= \frac{1}{\sqrt{1-4\,u^2}} \, e^{-\frac{4\,u^2}{1-4\,u^2}\left(x_\vartheta^2 + x'^2\right)} \, e^{\frac{4\,x_\vartheta x'u}{1-4\,u^2}} . \tag{4.2.124}$$

Problem 4.17 From the Rodrigues formula in Eq. (4.2.120), derive the identity

$$\sum_{n=0}^{\infty} \frac{t^n}{n!} \, H_n(x) = e^{2xt - t^2}, \tag{4.2.125}$$

where the exponential function on the right-hand side is called the *generating function* for the Hermite polynomials $H_n(x)$. Hence, show that

$$\int dx \, e^{-x^2} H_m(x) H_n(x) = 2^n \, n! \, \sqrt{\pi} \, \delta_{m,n}, \tag{4.2.126}$$

which illustrates the orthogonality of the Hermite polynomials.

After substituting the variable u with the original value, we have the terms

$$\sqrt{1-4\,u^2} = \sqrt{1 - e^{2\,i\vartheta}} = \sqrt{-2\,i\,e^{i\vartheta}\sin\vartheta},$$

$$\frac{4\,u}{1-4\,u^2} = \frac{2\,e^{i\vartheta}}{1 - e^{2\,i\vartheta}} = \frac{i}{\sin\vartheta}, \tag{4.2.127}$$

and

$$\frac{4\,u^2}{1-4\,u^2} = \frac{i}{2}\frac{e^{i\vartheta}}{\sin\vartheta}, \tag{4.2.128}$$

all of which constitute the wave function

$$\langle x'|x_\vartheta\rangle = \frac{1}{\sqrt{-2\pi i\, e^{i\vartheta}\sin\vartheta}} \, e^{-\frac{i}{2}\cot\vartheta\left(x_\vartheta^2 + x'^2\right) + \frac{i x_\vartheta x'}{\sin\vartheta}} . \tag{4.2.129}$$

A simple check that

$$\langle x'|x_{\frac{\pi}{2}}\rangle = \frac{1}{\sqrt{2\pi}} \, e^{i x' x_{\frac{\pi}{2}}} = \langle x'|p\rangle \tag{4.2.130}$$

tells us that Eq. (4.2.129) is consistent with the usual x'–p transformation function. The corresponding momentum wave function $\langle p'|x_\vartheta\rangle$ can be found by taking the Fourier transform of Eq. (4.2.129).

Problem 4.18   Show that the momentum wave function for $|x_\vartheta\rangle$ is

$$\langle p'|x_\vartheta\rangle = \frac{1}{\sqrt{2\pi\,\mathrm{e}^{\mathrm{i}\vartheta}\cos\vartheta}}\,\mathrm{e}^{\frac{\mathrm{i}}{2}\tan\vartheta(x_\vartheta^2+p'^2)-\frac{\mathrm{i}x_\vartheta p'}{\cos\vartheta}}. \tag{4.2.131}$$

Problem 4.19   Demonstrate that Eq. (4.2.129) and Eq. (4.2.131) are consistent with the relations

$$\langle x'|p'\rangle = \int \mathrm{d}x_\vartheta\, \langle x'|x_\vartheta\rangle\langle x_\vartheta|p'\rangle = \frac{1}{\sqrt{2\pi}}\,\mathrm{e}^{\mathrm{i}x'p'},$$

$$\langle x'|x''\rangle = \int \mathrm{d}x_\vartheta\, \langle x'|x_\vartheta\rangle\langle x_\vartheta|x''\rangle = \delta(x'-x''), \tag{4.2.132}$$

$$\langle p'|p''\rangle = \int \mathrm{d}x_\vartheta\, \langle p'|x_\vartheta\rangle\langle x_\vartheta|p''\rangle = \delta(p'-p''),$$

by evaluating the integrals.

Problem 4.20   Derive the transformation function

$$\langle x_\vartheta|x'_{\vartheta'}\rangle = \frac{\mathrm{e}^{\frac{\mathrm{i}}{2}(\vartheta-\vartheta')}}{\sqrt{2\pi\mathrm{i}\sin(\vartheta-\vartheta')}}\,\mathrm{e}^{\frac{\mathrm{i}}{2\sin(\vartheta-\vartheta')}\left[(x_\vartheta^2+x'^2_{\vartheta'})\cos(\vartheta-\vartheta')-2x_\vartheta x'_{\vartheta'}\right]}, \tag{4.2.133}$$

and obtain the obvious transformation function $\langle x_\vartheta|x'_\vartheta\rangle$ by recalling that

$$\lim_{\gamma\to 0}\left\{\frac{1}{\sqrt{2\pi\gamma}}\,\mathrm{e}^{-\frac{x^2}{2\gamma}}\right\} = \delta(x), \quad \mathrm{Re}\{\gamma\} \geq 0. \tag{4.2.134}$$

Also, obtain the transformation function $\langle x_\vartheta|x'_{\vartheta-\frac{\pi}{2}}\rangle$ from Eq. (4.2.133).

Problem 4.21 By defining the *Fourier operator* $\mathcal{F}_k$ as

$$\mathcal{F}_k\{g(x)\} = \int \frac{\mathrm{d}x}{\sqrt{2\pi}}\, \mathrm{e}^{-ikx}\, g(x) \qquad (4.2.135)$$

for a function $g(x)$, compute the action of $\mathcal{F}_k$ on $g(x)$ for $g(x) = g(n;x) = \langle x|n\rangle$, which is the wave function of a quantum harmonic oscillator that is stated in Eq. (4.2.117). Hence, write down the eigenvalues and eigenfunctions of $\mathcal{F}_k$ and state the nature of this operator.

Further reading

(1) S. L. Braunstein, How large a sample is needed for the maximum likelihood estimator to be approximately Gaussian?, *J. Phys. A: Math. Gen.* **25**, 3813 (1992).
   [An article that gives timely warning to callous users of the central limit theorem.]
(2) A. I. Lvovsky and M. G. Raymer, Continuous-variable optical quantum-state tomography, *Rev. Mod. Phys.* **81**, 299 (2009).
   [A topical review.]
(3) H. Zhu and B.-G. Englert, Quantum state tomography with fully symmetric measurements and product measurements, *Phys. Rev. A* **84**, 022327 (2011).
   [A comparison of tomographic accuracies of linear-inversion estimators for different quantum measurements with different distance measures; a combination of statistics and random-matrix theory.]
(4) H. Zhu, Quantum state estimation with informationally overcomplete measurements, *Phys. Rev. A* **90**, 012115 (2014).
   [More on optimal estimators with respect to the Hilbert–Schmidt distance.]

Chapter 5

Quasi-Probability Distributions

5.1 Distribution functions

So far, we have always represented data with statistical operators, which are very often used for expectation value predictions. In the context of continuous-variable quantum-state estimation, we know, from the previous chapter as well as our experience in intermediate-level quantum mechanics courses, that quantum states of light can be characterized by observables with continuous eigenvalues. Examples of such observables are, for instance, the quadrature operators discussed in Sec. 4.2.3.2. These continuous-variable observables provide a natural alternative approach to representing the measurement data, besides the usual statistical-operator representation that we are very familiar with.

This equivalent representation is the *distribution-function* representation, which translates the measurement data into values of a real function over a *phase space*. As an introduction, we restrict our discussion to only single-mode photonic sources, and for these sources, the phase space is a plot of the position (x) and momentum (p) values that is analogous to the classical phase space for a single particle. Here, the values x and p do not refer to the respective dynamical variables of any particle. Rather, they are simply labels assigned to the distribution function so that it *resembles* a classical probability distribution function, and are, respectively, the eigenvalues of the X and P quadrature operators.

5.1.1   *Wigner function*

5.1.1.1   *Definitions and properties*

The first distribution function of interest is the *Wigner* function*, which is defined as

$$\rho_{\mathrm{W}}(x,p) = \int \mathrm{d}y\, \mathrm{e}^{\mathrm{i}py} \left\langle x - \tfrac{y}{2} \middle| \rho \middle| x + \tfrac{y}{2} \right\rangle \qquad (5.1.1)$$

for a state ρ. We can rewrite $\rho_{\mathrm{W}}(x,p)$ as the expectation value of an operator $W_{x,p}$, which we call the *Wigner operator*, inasmuch as

$$\rho_{\mathrm{W}}(x,p) = \mathrm{tr}\{\rho\, W_{x,p}\},$$

$$W_{x,p} = \int \mathrm{d}y\, \left| x + \tfrac{y}{2} \right\rangle \mathrm{e}^{\mathrm{i}py} \left\langle x - \tfrac{y}{2} \right|. \qquad (5.1.2)$$

The Wigner function yields *marginal probability distributions* in the variable x or p when one of the two variables is integrated over. By recalling the familiar identities

$$\int \mathrm{d}x\, |x\rangle \langle x| = \int \mathrm{d}p\, |p\rangle \langle p| = 1,$$

$$\langle x|\, \mathrm{e}^{\mathrm{i}x'P} = \langle x + x'|, \qquad (5.1.3)$$

$$\mathrm{e}^{\mathrm{i}p'X}\, |p\rangle = |p + p'\rangle,$$

and using Eq. (5.1.2), it follows that

$$\int \mathrm{d}x\, \rho_{\mathrm{W}}(x,p) = \mathrm{tr}\left\{ \rho \int \mathrm{d}y \int \mathrm{d}x\, \left| x + \tfrac{y}{2} \right\rangle \mathrm{e}^{\mathrm{i}py} \left\langle x - \tfrac{y}{2} \right| \right\}$$

$$= \mathrm{tr}\left\{ \rho \int \mathrm{d}y\, \mathrm{e}^{\mathrm{i}y(p-P)} \right\}$$

$$= 2\pi\, \mathrm{tr}\{\rho\, \delta\,(p - P)\}$$

$$= 2\pi\, \mathrm{tr}\{\rho\, |p\rangle \langle p|\}$$

$$= 2\pi\, \langle p|\, \rho\, |p\rangle, \qquad (5.1.4)$$

---

*Eugene Paul Wigner (1902–1995).

as well as

$$\int \frac{dp}{2\pi} \rho_{\rm w}(x,p) = {\rm tr}\left\{ \rho \int dy \, |x + \tfrac{y}{2}\rangle \langle x - \tfrac{y}{2}| \int \frac{dp}{2\pi} e^{ipy} \right\}$$

$$= {\rm tr}\left\{ \rho \int dy \, |x + \tfrac{y}{2}\rangle \, \delta(y) \, \langle x - \tfrac{y}{2}| \right\}$$

$$= {\rm tr}\{ \rho \, |x\rangle \langle x| \}$$

$$= \langle x| \rho |x\rangle. \tag{5.1.5}$$

So, the Wigner function $\rho_{\rm w}(x,p)$ really places x and p on equal footing, although at first sight, they appear to be asymmetrically assigned. From the above marginal distributions, we see that $\rho_{\rm w}(x,p)$ does indeed behave *like* a normalized probability distribution function of x and p as

$$\int \frac{dx \, dp}{2\pi} \rho_{\rm w}(x,p) = 1, \tag{5.1.6}$$

where $dx \, dp/(2\pi)$ is the natural integration measure for the phase space.

As an operator that corresponds to a distribution function, the Wigner operator $W_{x,p}$ must be Hermitian and this is easily confirmed by taking the adjoint of $W_{x,p}$,

$$W_{x,p}^{\dagger} = \int dy \, |x - \tfrac{y}{2}\rangle e^{-ipy} \langle x + \tfrac{y}{2}| \qquad (y' = -y)$$

$$= \int dy' \, |x + \tfrac{y'}{2}\rangle e^{ipy'} \langle x - \tfrac{y'}{2}|$$

$$= W_{x,p}. \tag{5.1.7}$$

For a statistical operator ρ, $\rho_{\rm w}(x,p)$ is a bounded function, and without going through elaborate calculations just yet, we can obtain these bounds by inspecting the eigenvalues of the Wigner operator. Upon squaring $W_{x,p}$, we find that

$$W_{x,p}^2 = \int dy \, |x + \tfrac{y}{2}\rangle e^{ipy} \langle x - \tfrac{y}{2}| \int dy' \, |x + \tfrac{y'}{2}\rangle e^{ipy'} \langle x - \tfrac{y'}{2}|$$

$$= \int dy \int dy' \, e^{ip(y+y')} \, |x + \tfrac{y}{2}\rangle \underbrace{\langle x - \tfrac{y}{2}|x + \tfrac{y'}{2}\rangle}_{= \delta\left(\frac{y+y'}{2}\right)} \langle x - \tfrac{y'}{2}|$$

$$= 2 \int dy \ |x + \tfrac{y}{2}\rangle \langle x + \tfrac{y}{2}| \qquad \left(y' = x + \frac{y}{2}\right)$$

$$= 4. \tag{5.1.8}$$

This means that $W_{x,p}/2$ is a *Hermitian unitary operator* with eigenvalues ± 1. Since $W_{x,p}$ is Hermitian, it possesses a complete set of orthonormal eigenstates. We can then express this operator in its spectral decomposition,

$$W_{x,p} = 2\,W_+^{\{x,p\}} - 2\,W_-^{\{x,p\}}, \tag{5.1.9}$$

where $W_+^{\{x,p\}}$ and $W_-^{\{x,p\}}$ are the projectors corresponding to the eigenvalues ± 2 for $W_{x,p}$. It is now easy to find an upper bound for $\rho_{\mathrm{w}}(x,p)$, since the positivity constraint for ρ implies that

$$\begin{aligned}
\rho_{\mathrm{w}}(x,p) &= \mathrm{tr}\{\rho\, W_{x,p}\} \\
&= 2\left[\mathrm{tr}\!\left\{\rho\, W_+^{\{x,p\}}\right\} - \mathrm{tr}\!\left\{\rho\, W_-^{\{x,p\}}\right\}\right] \\
&\leq 2\left[\mathrm{tr}\!\left\{\rho\, W_+^{\{x,p\}}\right\} + \mathrm{tr}\!\left\{\rho\, W_-^{\{x,p\}}\right\}\right] \\
&= 2.
\end{aligned} \tag{5.1.10}$$

In a similar fashion, a lower bound for $\rho_{\mathrm{w}}(x,p)$ is obtained from the following calculation:

$$\begin{aligned}
\rho_{\mathrm{w}}(x,p) &= \mathrm{tr}\{\rho\, W_{x,p}\} \\
&= 2\left[\mathrm{tr}\!\left\{\rho\, W_+^{\{x,p\}}\right\} - \mathrm{tr}\!\left\{\rho\, W_-^{\{x,p\}}\right\}\right] \\
&\geq -2\,\mathrm{tr}\!\left\{\rho\, W_-^{\{x,p\}}\right\} \\
&\geq -2\left\|W_-^{\{x,p\}}\right\|_2 \\
&= -2.
\end{aligned} \tag{5.1.11}$$

We therefore find that

$$-2 \leq \rho_{\mathrm{w}}(x,p) \leq 2 \quad \text{for any state } \rho. \tag{5.1.12}$$

This result supports the possibility that $\rho_{\mathrm{w}}(x,p)$ can take negative values since, after all, $\rho_{\mathrm{w}}(x,p)$ is the expectation value of a Hermitian operator. One can verify that the bounds for $\rho_{\mathrm{w}}(x,p)$ are indeed attainable.

As an example, let us consider the special case in which $x = p = 0$. The resulting Wigner operator *at the origin* is calculated to be

$$
\begin{aligned}
W_{0,0} &= \int dy \left| \tfrac{y}{2} \right\rangle \left\langle -\tfrac{y}{2} \right| \\
&= \int dy \left| \tfrac{y}{2} \right\rangle \sum_{n=0}^{\infty} \underbrace{\left\langle -\tfrac{y}{2} \middle| n \right\rangle}_{= (-1)^n \left\langle \tfrac{y}{2} \middle| n \right\rangle} \langle n| \\
&= 2 \sum_{n=0}^{\infty} |n\rangle \, (-1)^n \, \langle n| , \qquad\qquad (5.1.13)
\end{aligned}
$$

where we have used the property

$$
H_n(-y) = (-1)^n \, H_n(y) \qquad\qquad (5.1.14)
$$

for the Hermite polynomial that is present in the eigenstate wave function $\langle y | n \rangle$ for the quantum harmonic oscillator. By defining the *parity operator* to be

$$
\sum_{n=0}^{\infty} |n\rangle \, (-1)^n \, \langle n| = (-1)^{A^\dagger A}, \qquad\qquad (5.1.15)
$$

we have

$$
W_{0,0} = 2 \, (-1)^{A^\dagger A}. \qquad\qquad (5.1.16)
$$

We now realize that the eigenstates for $W_{0,0}$ are just the Fock states $|n\rangle \langle n|$! Since the eigenvalue 2 is associated with the **even-numbered** Fock states, and the eigenvalue -2 is associated with the **odd-numbered** Fock states, that is

$$
W_{0,0} = 2 \underbrace{\sum_{\text{even } n} |n\rangle \langle n|}_{= W_+^{\{0,0\}}} - 2 \underbrace{\sum_{\text{odd } n} |n\rangle \langle n|}_{= W_-^{\{0,0\}}}, \qquad\qquad (5.1.17)
$$

the lower and upper bounds for $\rho_W(0,0)$ can now be shown to be tight. To attain the upper bound, one can choose ρ to be any even-numbered Fock state. On the other hand, any odd-numbered Fock state will achieve the lower bound. So, it turns out that the Wigner function $\rho_W(\alpha, \alpha^*)$ is not a proper probability distribution in general, as this function can take negative values. It is for this reason that $\rho_{\rm w}(\alpha, \alpha^*)$ is often called a *quasi-probability distribution*.

What about the eigenstates of $W_{x,p}$ for arbitrary x and p? This can also be calculated with slightly more work. Let us revisit the definition for $W_{x,p}$, from which

$$W_{x,p} = \int dy \, |x + \tfrac{y}{2}\rangle \, e^{ipy} \, \langle x - \tfrac{y}{2}|$$

$$= e^{-ixP} \int dy \, |\tfrac{y}{2}\rangle \, e^{ipy} \, \langle -\tfrac{y}{2}| \, e^{ixP}$$

$$= e^{-ixP} e^{ipX} \int dy \, |\tfrac{y}{2}\rangle \langle -\tfrac{y}{2}| \, e^{-ipX} e^{ixP}$$

$$= 2\, e^{-ixP} e^{ipX} (-1)^{A^\dagger A} e^{-ipX} e^{ixP}. \tag{5.1.18}$$

After combining the exponentials

$$e^{-ixP} e^{ipX} = e^{i(pX - xP)} e^{-\frac{i}{2} xp}, \tag{5.1.19}$$

into a single exponential operator using the BCH relation in Eq. (4.2.43), we have

$$W_{x,p} = 2\, e^{i(pX - xP)} (-1)^{A^\dagger A} e^{-i(pX - xP)}. \tag{5.1.20}$$

The above expression all the more shows the symmetry in the variables x and p in the Wigner operator $W_{x,p}$ and, since they are no longer separated, we shall introduce a complex variable α that combines these two variables into one as $\alpha = (x + ip)/\sqrt{2}$. Let us also incorporate the ladder operators A and $A^\dagger$ that relate the X and P quadrature operators with the identities

$$X = \frac{A + A^\dagger}{\sqrt{2}},$$

$$P = \frac{A - A^\dagger}{i\sqrt{2}}. \tag{5.1.21}$$

As a result, $pX - xP$ now turns into

$$pX - xP = \left(\frac{\alpha - \alpha^*}{i\sqrt{2}}\right)\left(\frac{A + A^\dagger}{\sqrt{2}}\right) - \left(\frac{\alpha + \alpha^*}{\sqrt{2}}\right)\left(\frac{A - A^\dagger}{i\sqrt{2}}\right)$$

$$= \frac{1}{2i}\left(2\alpha A^\dagger - 2\alpha^* A\right)$$

$$= \frac{1}{i}\left(\alpha A^\dagger - \alpha^* A\right), \tag{5.1.22}$$

so that, finally, we arrive at the Fock representation

$$W_\alpha \equiv W_{x,p} = 2D(\alpha)(-1)^{A^\dagger A} D(-\alpha) = 2(-1)^{(A^\dagger - \alpha^*)(A - \alpha)},$$

$$\rho_W(\alpha, \alpha^*) \equiv \rho_W(x, p) = 2 \operatorname{tr}\left\{\rho(-1)^{(A^\dagger - \alpha^*)(A - \alpha)}\right\}. \qquad (5.1.23)$$

Therefore, as one may have suspected, the eigenstates of W_α are indeed the displaced Fock states $D(\alpha)|n\rangle \langle n| D(-\alpha)$.

Very generally, the Wigner function $F_W(x, p)$ is defined for any operator F by simply replacing ρ with F in Eq. (5.1.1). The function $\rho_W(x, p)$ then serves as a weight distribution for computing the phase-space average $F_W(x, p)$. This phase-space average is, in fact, related to a familiar quantity, and to find out what that is, we proceed to evaluate this average:

$$\int \frac{dx\, dp}{2\pi} \rho_W(x, p) F_W(x, p)$$

$$= \int \frac{dx\, dp}{2\pi} \operatorname{tr}\left\{ F \int dy\, \left|x + \tfrac{y}{2}\right\rangle e^{ipy} \left\langle x - \tfrac{y}{2}\right| \int dy'\, e^{ipy'} \left\langle x - \tfrac{y'}{2}\right| \rho \left|x + \tfrac{y'}{2}\right\rangle \right\}$$

$$= \int dy \int dy' \int dx\, \operatorname{tr}\left\{ F\, \left|x + \tfrac{y}{2}\right\rangle \left\langle x - \tfrac{y'}{2}\right| \rho \left|x + \tfrac{y'}{2}\right\rangle \left\langle x - \tfrac{y}{2}\right| \right\}$$

$$\times \underbrace{\int \frac{dp}{2\pi} e^{ip(y+y')}}_{= \delta(y + y')}$$

$$= \int dy \int dx\, \operatorname{tr}\{ F\, \left|x + \tfrac{y}{2}\right\rangle \left\langle x + \tfrac{y}{2}\right| \rho \left|x - \tfrac{y}{2}\right\rangle \left\langle x - \tfrac{y}{2}\right| \} \qquad \begin{pmatrix} u = x + \tfrac{y}{2} \\ v = x - \tfrac{y}{2} \end{pmatrix}$$

$$= \int du \int dv\, \operatorname{tr}\{ F\, |u\rangle \langle u| \rho |v\rangle \langle v| \}$$

$$= \operatorname{tr}\{\rho F\} = \langle F \rangle. \qquad (5.1.24)$$

In summary, we have

$$\operatorname{tr}\{\rho F\} = \int \frac{dx\, dp}{2\pi} \rho_W(x, p) F_W(x, p), \qquad (5.1.25)$$

which is an extremely useful relation when dealing with calculations involving Wigner functions and expectation values.

Problem 5.1   Show that

$$\text{tr}\{W_{x,p}\,W_{x',p'}\} = 2\pi\,\delta\,(x-x')\,\delta\,(p-p'). \tag{5.1.26}$$

What can we say about the set of all Wigner operators? Finally, express a statistical operator in terms of its Wigner function.

Problem 5.2   Use the results in **Problem 5.1** to obtain Eq. (5.1.25).

5.1.1.2   *Quadrature eigenstates*

The phase-space average of $F_{\rm W}\,(x,p)$ is therefore equal to the quantum-mechanical expectation value of F. This result is useful for relating the measurement data with the $\rho_{\rm W}(x,p)$ of an unknown quantum state, where the operator F now refers to one of the measurement outcomes.

As an example, we investigate the expectation values of the quadrature eigenstates $F = |x_\vartheta\rangle\,\langle x_\vartheta|$. Its Wigner function is equal to

$$f_{\rm W}^{\rm quad}\,(x,p) = \int \mathrm{d}y\,\mathrm{e}^{\mathrm{i}py}\,\langle x-\tfrac{y}{2}\,|\,x_\vartheta\rangle\,\langle x_\vartheta\,|\,x+\tfrac{y}{2}\rangle, \tag{5.1.27}$$

where we take, from Eq. (4.2.129),

$$\langle x\,|\,x_\vartheta\rangle \propto \frac{\mathrm{e}^{-\frac{\mathrm{i}}{2}\cot\vartheta\left(x^2-\frac{2xx_\vartheta}{\cos\vartheta}\right)}}{\sqrt{2\pi\,\sin\vartheta}} \tag{5.1.28}$$

and neglect the irrelevant phase factors. The integrand becomes

$$\mathrm{e}^{\mathrm{i}py}\,\langle x-\tfrac{y}{2}\,|\,x_\vartheta\rangle\,\langle x_\vartheta\,|\,x+\tfrac{y}{2}\rangle$$

$$= \frac{1}{2\pi\,\sin\vartheta}\,\mathrm{e}^{\mathrm{i}py}\,\mathrm{e}^{-\frac{\mathrm{i}}{2}\cot\vartheta\left[\left(x-\frac{y}{2}\right)^2-\left(x+\frac{y}{2}\right)^2-\frac{2x-y}{\cos\vartheta}x_\vartheta+\frac{2x+y}{\cos\vartheta}x_\vartheta\right]}$$

$$= \frac{1}{2\pi\,\sin\vartheta}\,\mathrm{e}^{\mathrm{i}y\left(p+\frac{x\cos\vartheta-x_\vartheta}{\sin\vartheta}\right)}, \tag{5.1.29}$$

so that, after introducing the new variable $y' = y/\sin\vartheta$,

$$f_{\rm W}^{\rm quad}\,(x,p) = \delta\,(x\cos\vartheta + p\sin\vartheta - x_\vartheta). \tag{5.1.30}$$

This form of the Wigner function is another indication of the unphysical nature of the projector $|x_\vartheta\rangle\,\langle x_\vartheta|$. The probability distribution

$p(x_\vartheta, \vartheta) = \langle|x_\vartheta\rangle \langle x_\vartheta|\rangle$ that is expressed in terms of the phase-space average of $f_W^{\text{quad}}(x, p)$ can now be derived as follows:

$$p(x_\vartheta, \vartheta) = \int \frac{\mathrm{d}x\,\mathrm{d}p}{2\pi} \rho_W(x, p)\, \delta\left(x\cos\vartheta + p\sin\vartheta - x_\vartheta\right)$$

$$\left(\begin{array}{l} x' = x\cos\vartheta + p\sin\vartheta \\ p_\vartheta = -x\sin\vartheta + p\cos\vartheta \end{array}\right)$$

$$= \int \frac{\mathrm{d}x'\mathrm{d}p_\vartheta}{2\pi} \rho_W(x'\cos\vartheta - p_\vartheta\sin\vartheta, x'\sin\vartheta + p_\vartheta\cos\vartheta)\, \delta\left(x' - x_\vartheta\right)$$

$$= \int \frac{\mathrm{d}p_\vartheta}{2\pi} \rho_W(x_\vartheta\cos\vartheta - p_\vartheta\sin\vartheta, x_\vartheta\sin\vartheta + p_\vartheta\cos\vartheta). \qquad (5.1.31)$$

This expression relates the probability distribution $p(x_\vartheta, \theta)$ of the quadrature eigenstates to an integral of the Wigner function $\rho_W(x, p)$ that is evaluated over straight lines of various angles defined by the reference coherent state. This integral is known as the *Radon* transform*.

In principle, one may apply the inverse Radon transform to recover the Wigner function $\rho_W(x, p)$. To obtain this inverse transform, we start from the expression of in Eq. (5.1.31) and carry out a Fourier transform on the probability distribution $p(x_\vartheta, \vartheta)$,

$$\tilde{p}(k, \vartheta) = \int \mathrm{d}x_\vartheta\, e^{-ikx_\vartheta}\, p(x_\vartheta, \vartheta)$$

$$= \int \mathrm{d}x_\vartheta\, e^{-ikx_\vartheta} \int \frac{\mathrm{d}p_\vartheta}{2\pi} \rho_W(x_\vartheta\cos\vartheta - p_\vartheta\sin\vartheta, x_\vartheta\sin\vartheta + p_\vartheta\cos\vartheta)$$

$$\left(\begin{array}{l} x_\vartheta = x\cos\vartheta + p\sin\vartheta \\ p_\vartheta = -x\sin\vartheta + p\cos\vartheta \end{array}\right)$$

$$= \int \frac{\mathrm{d}x\,\mathrm{d}p}{2\pi} e^{-ik(x\cos\vartheta + p\sin\vartheta)} \rho_W(x, p). \qquad (5.1.32)$$

The above integral is in fact the two-variable Fourier transform of $\rho_W(x, p)$, with the new variables $k' = k\cos\vartheta$ and $l' = k\sin\vartheta$, that is

$$\tilde{p}(k', l') = \int \frac{\mathrm{d}x\,\mathrm{d}p}{2\pi} e^{-i(k'x + l'p)} \rho_W(x, p). \qquad (5.1.33)$$

---

*Johann Karl August Radon (1887–1956).

This means that the function $\rho_{\mathrm{W}}(x,p)$ can be easily recovered with an application of the inverse Fourier transform, thus arriving at the integral

$$\rho_{\mathrm{W}}(x,p) = \int \frac{\mathrm{d}k'\mathrm{d}l'}{2\pi} \, \mathrm{e}^{\mathrm{i}(k'x+l'p)} \, \widetilde{p}(k',l') \quad \begin{pmatrix} k' = k\cos\vartheta \\ l' = k\sin\vartheta \end{pmatrix}$$

$$= \int_0^{2\pi} \frac{\mathrm{d}\vartheta}{2\pi} \int_0^\infty \mathrm{d}k \, k \, \mathrm{e}^{\mathrm{i}k(x\cos\vartheta + p\sin\vartheta)} \, \widetilde{p}(k,\vartheta). \tag{5.1.34}$$

The integration variable k introduced here is nonnegative, following the usual parametrization in polar coordinates. We can make contact with the Fourier transform $\widetilde{p}(k,\vartheta)$ by allowing k to also take negative values. This can be done by breaking up the integral

$$\int_0^{2\pi} \frac{\mathrm{d}\vartheta}{2\pi} \int_0^\infty \mathrm{d}k \, k \, \mathrm{e}^{\mathrm{i}k(x\cos\vartheta + p\sin\vartheta)} \, \widetilde{p}(k,\vartheta)$$

$$= \int_0^\pi \frac{\mathrm{d}\vartheta}{2\pi} \int_0^\infty \mathrm{d}k \, k \, \mathrm{e}^{\mathrm{i}k(x\cos\vartheta + p\sin\vartheta)} \, \widetilde{p}(k,\vartheta)$$

$$+ \int_\pi^{2\pi} \frac{\mathrm{d}\vartheta}{2\pi} \int_0^\infty \mathrm{d}k \, k \, \mathrm{e}^{\mathrm{i}k(x\cos\vartheta + p\sin\vartheta)} \, \widetilde{p}(k,\vartheta) \tag{5.1.35}$$

into two separate integrals for different ranges of ϑ, with the second integral

$$\int_\pi^{2\pi} \frac{\mathrm{d}\vartheta}{2\pi} \int_0^\infty \mathrm{d}k \, k \, \mathrm{e}^{\mathrm{i}k(x\cos\vartheta + p\sin\vartheta)} \, \widetilde{p}(k,\vartheta) \quad (\vartheta'' = \vartheta - \pi)$$

$$= \int_0^\pi \frac{\mathrm{d}\vartheta''}{2\pi} \int_0^\infty \mathrm{d}k \, k \, \mathrm{e}^{-\mathrm{i}k(x\cos\vartheta'' + p\sin\vartheta'')} \, \widetilde{p}(k,\vartheta'' + \pi) \quad (k'' = -k)$$

$$= \int_0^\pi \frac{\mathrm{d}\vartheta''}{2\pi} \int_{-\infty}^0 \mathrm{d}k'' \, (-k'') \, \mathrm{e}^{\mathrm{i}k''(x\cos\vartheta'' + p\sin\vartheta'')} \, \underbrace{\widetilde{p}(-k'',\vartheta'' + \pi)}_{= \, \widetilde{p}(k'',\vartheta'')}.$$

$$\tag{5.1.36}$$

When all is said and done, the compact form for the inverse Radon transform is given by

$$\rho_{\rm W}(x,p) = \int_0^\pi \frac{{\rm d}\vartheta}{2\pi} \int {\rm d}k\, |k| \int {\rm d}x_\vartheta\, {\rm e}^{{\rm i}k(x\cos\vartheta + p\sin\vartheta - x_\vartheta)}\, p(x_\vartheta, \vartheta). \quad (5.1.37)$$

5.1.1.3 *Coherent states*

We have just calculated the Wigner function for the quadrature eigenstates, and in that exercise, we find that such unphysical eigenstates will lead to unbounded Wigner functions. Let us now investigate the Wigner functions for some physical quantum states. The first example we shall look at is the familiar set of coherent states $\{|\alpha'\rangle\langle\alpha'^*|\}$. The Wigner function for a coherent state (refer to Fig. 5.1) can be found rather effortless with the aid of the identity in Eq. (4.2.93) for the parity operator:

$$\begin{aligned}
\rho_{\rm W}^{\rm coh}(\alpha,\alpha^*;\alpha',\alpha'^*) &= 2\,{\rm tr}\Big\{|\alpha'\rangle\langle\alpha'^*|(-1)^{(A^\dagger-\alpha^*)(A-\alpha)}\Big\} \\
&= 2\,\langle\alpha'^*|\Big[{\rm e}^{-2(A^\dagger-\alpha^*)(A-\alpha)}\Big]_{\rm N}|\alpha'\rangle \\
&= 2\,{\rm e}^{-2|\alpha'-\alpha|^2}. \quad (5.1.38)
\end{aligned}$$

Therefore, the Wigner function for the coherent state $|\alpha'\rangle\langle\alpha'^*|$ is a Gaussian function centered at the value of α'. Such a quantum state that possesses a Gaussian Wigner function is known as a *Gaussian state*.

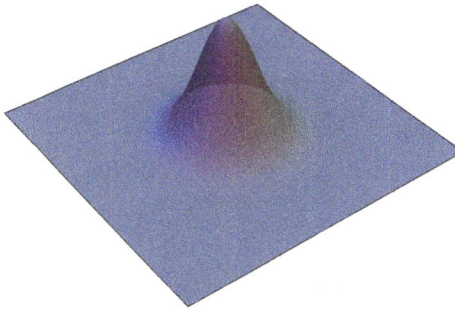

Fig. 5.1   Wigner function for a coherent state.

Problem 5.3 A slightly more general class of Gaussian states is the well-known set of *squeezed coherent states* $|\alpha'; z\rangle \langle\alpha'^*; z|$ that are characterized by the complex *squeeze* parameter $z = r\,e^{i\theta}$, with coherent-state wave functions of the form

$$\langle\alpha^*|\alpha'; z\rangle = \sqrt{\operatorname{sech} r}\, e^{-\frac{1}{2}|\alpha|^2-\frac{1}{2}|\alpha'|^2}$$

$$\times \exp\left(\frac{\alpha^*\alpha'}{\cosh r} + \frac{\tanh r}{2}\left(\alpha^{*2}\,e^{i\theta} - \alpha'^2\,e^{-i\theta}\right)\right). \quad (5.1.39)$$

Calculate the Wigner function for such a state. To simplify matters, you may consult Eq. (5.1.116) and Eq. (5.1.130) in advance.

5.1.1.4 *Odd and even coherent states*

For the next example, we turn to a more interesting class of states — the *odd* or *even* superpositions of two coherent states. By this we mean that the ket $|\alpha'\rangle_\pm$ of interest has the structure

$$|\alpha'_\pm\rangle = (|\alpha'\rangle \pm |-\alpha'\rangle)\mathcal{N}, \quad (5.1.40)$$

where the normalization constant $\mathcal{N}$ can be found using Eq. (4.2.40):

$$1 = \langle\alpha'^*_\pm|\alpha'_\pm\rangle$$
$$= \mathcal{N}^2\left((\langle\alpha'^*| \pm \langle-\alpha'^*|)(|\alpha'\rangle \pm |-\alpha'\rangle)\right)$$
$$= \mathcal{N}^2\left(2 \pm 2\,e^{-2|\alpha'|^2}\right)$$
$$\Rightarrow \quad \mathcal{N} = \frac{1}{\sqrt{2\left(1 \pm e^{-2|\alpha'|^2}\right)}}. \quad (5.1.41)$$

The Wigner function

$$\rho_W^{o/e}(\alpha, \alpha^*; \alpha', \alpha'^*) = 2\operatorname{tr}\left\{|\alpha'_\pm\rangle\langle\alpha'^*_\pm|(-1)^{(A^\dagger-\alpha^*)(A-\alpha)}\right\} \quad (5.1.42)$$

consists of four distinct terms.

There are two terms in the sum that are respectively proportional to exponential functions of α, namely $2\,e^{-2|\alpha'-\alpha|^2}$ and $2\,e^{-2|\alpha'+\alpha|^2}$, as though the system is prepared in a mixture of two coherent states $|\alpha'\rangle\langle\alpha'^*|$ and $|-\alpha'\rangle\langle-\alpha'^*|$. The other two terms are more interesting as they involve the cross terms that relate $\langle\alpha'^*|$ with $|-\alpha'\rangle$, as well as $\langle-\alpha'^*|$ and $|\alpha'\rangle$. Let us examine one of these cross terms again with the normal-ordered form of the parity operator:

$$2\,\mathrm{tr}\left\{|\alpha'\rangle\,\langle-\alpha'^*|\,(-1)^{\left(A^\dagger-\alpha^*\right)(A-\alpha)}\right\} = 2\,\langle-\alpha'^*|\left[e^{-2\left(A^\dagger-\alpha^*\right)(A-\alpha)}\right]_N|\alpha'\rangle$$

$$= 2\underbrace{\langle-\alpha'^*|\alpha'\rangle}_{\textstyle = \,e^{-2|\alpha'|^2}}e^{-2(-\alpha'^*-\alpha^*)(\alpha'-\alpha)}$$

$$= 2\,e^{-2|\alpha|^2}\,e^{2\left(\alpha^*\alpha'-\alpha\alpha'^*\right)}$$

$$= 2\,e^{-2|\alpha|^2}\,e^{4i\,\mathrm{Im}\{\alpha^*\alpha'\}}. \qquad (5.1.43)$$

The other cross term leads to an answer that is the complex conjugate of the first, so that

$$\rho_W^{o/e}(\alpha,\alpha^*;\alpha',\alpha'^*)$$
$$= 2\,\mathcal{N}^2\left[e^{-2|\alpha'-\alpha|^2}+e^{-2|\alpha'+\alpha|^2}\underbrace{\pm 2\,e^{-2|\alpha|^2}\,\cos\left(4\,\mathrm{Im}\,\{\alpha^*\alpha'\}\right)}_{\text{interference terms}}\right].$$

$$(5.1.44)$$

It is now clear that the two cross terms give rise to the oscillatory cosine function in the Wigner function $\rho_W^{o/e}(\alpha,\alpha^*;\alpha',\alpha'^*)$ (refer to Fig. 5.2, for instance). These terms are the *interference terms* that contribute to the additional interference effects originating from the quantum superposition. Because of this behavior, the function $\rho_W^{o/e}(\alpha,\alpha^*;\alpha',\alpha'^*)$ is negative at certain regions of interference. In general, a superposition of coherent states will have such features.

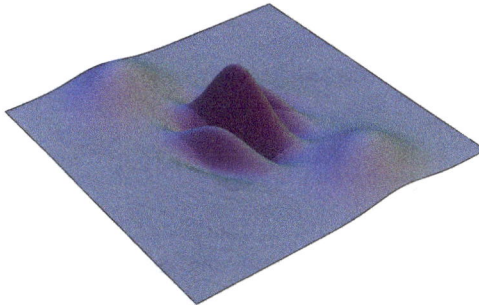

Fig. 5.2   Wigner function for an even superposition of coherent states.

5.1.1.5 *Fock states*

As the third example, let us derive the Wigner functions $\rho_{\mathrm{w}}^{\mathrm{Fock}}(\alpha, \alpha^*; n)$ of the Fock states. To be systematic about this, we shall establish a general formula for the Wigner function of a general quantum state ρ expressed in the Fock basis,

$$\rho = \sum_{m=0}^{\infty} \sum_{n=0}^{\infty} |m\rangle \underbrace{\langle m| \rho |n\rangle}_{= \, \rho_{mn}} \langle n| . \tag{5.1.45}$$

Starting from the definition in Eq. (5.1.1),

$$\rho_{\mathrm{w}}(x, p) = \int dy \, e^{ipy} \left\langle x - \tfrac{y}{2} \middle| \rho \middle| x + \tfrac{y}{2} \right\rangle$$

$$= \sum_{m=0}^{\infty} \sum_{n=0}^{\infty} \rho_{mn} \int dy \, e^{ipy} \left\langle x - \tfrac{y}{2} \middle| m \right\rangle \left\langle n \middle| x + \tfrac{y}{2} \right\rangle . \tag{5.1.46}$$

Using the identity for the wave function $\langle x | n \rangle$ in Eq. (4.2.117),

$$\rho_{\mathrm{w}}(x, p) = \sum_{m=0}^{\infty} \sum_{n=0}^{\infty} \frac{\rho_{mn}}{\sqrt{2^{m+n} \pi \, m! \, n!}}$$

$$\times \int dy \, e^{ipy} \, e^{-\frac{1}{2}\left(x - \frac{y}{2}\right)^2} \, e^{-\frac{1}{2}\left(x + \frac{y}{2}\right)^2} H_m\left(x - \frac{y}{2}\right) H_n\left(x + \frac{y}{2}\right)$$

$$= e^{-x^2} \sum_{m=0}^{\infty} \sum_{n=0}^{\infty} \frac{\rho_{mn}}{\sqrt{2^{m+n} \pi \, m! \, n!}}$$

$$\times \int dy \, e^{-\frac{y^2}{4} + ipy} H_m\left(x - \frac{y}{2}\right) H_n\left(x + \frac{y}{2}\right) . \tag{5.1.47}$$

After a change of variable, the task now is to evaluate the integral

$$I_{m,n} = \int dy \, e^{-y^2 + 2ipy} H_m(x - y) H_n(x + y) . \tag{5.1.48}$$

We can make use of the generating function for the Hermite polynomials in Eq. (4.2.125) to write down the obvious relation

$$H_n(x) = \left(\frac{\partial}{\partial t}\right)^n e^{2xt - t^2} \Bigg|_{t=0} \tag{5.1.49}$$

directly from the Taylor series of $H_n(x)$ in Eq. (4.2.125). With this relation, the integral turns into

$$
I_{m,n} = \left(\frac{\partial}{\partial t}\right)^m \left(\frac{\partial}{\partial t'}\right)^n \left[e^{2xt-t^2}\, e^{2xt'-t'^2} \int dy\, e^{-y^2+2(\mathrm{i}p-t+t')y}\right]\Bigg|_{t=t'=0}
$$

$$
= \sqrt{\pi}\, \left(\frac{\partial}{\partial t}\right)^m \left(\frac{\partial}{\partial t'}\right)^n \left[e^{2xt-t^2}\, e^{2xt'-t'^2}\, e^{(\mathrm{i}p-t+t')^2}\right]\Bigg|_{t=t'=0}, \qquad (5.1.50)
$$

where

$$
e^{2xt-t^2}\, e^{2xt'-t'^2}\, e^{(\mathrm{i}p-t+t')^2} = e^{-p^2}\, e^{2(x-\mathrm{i}p)t}\, e^{2(x+\mathrm{i}p)t'}\, e^{-2\,tt'}. \qquad (5.1.51)
$$

The alternative form

$$
I_{m,n} = \sqrt{\pi}\, e^{-p^2} \left(\frac{\partial}{\partial t}\right)^m \left(\frac{\partial}{\partial t'}\right)^n \left[e^{2(x-\mathrm{i}p)t}\, e^{2(x+\mathrm{i}p)t'}\, e^{-2\,tt'}\right]\Bigg|_{t=t'=0}, \qquad (5.1.52)
$$

expressed as a double derivative, allows for a much more convenient way to simplify the integral $I_{m,n}$.

To proceed, we shall consider separate cases for all values of m and n. When $m \geq n$, Eq. (5.1.52) can be evaluated by first taking the derivatives with respect to m,

$$
I_{m \geq n} = \sqrt{\pi}\, e^{-p^2} \left(\frac{\mathrm{d}}{\mathrm{d}t'}\right)^n \left\{[2\,(x-\mathrm{i}p-t')]^m\, e^{2(x+\mathrm{i}p)t'}\right\}\Bigg|_{t'=0}. \qquad (5.1.53)
$$

This expression takes a very familiar form and to elucidate matters, we introduce a new variable

$$
y = 2\,(x+\mathrm{i}p)\,(x-\mathrm{i}p-t'), \qquad (5.1.54)
$$

along with the relation

$$
\frac{\mathrm{d}}{\mathrm{d}t'} = -2\,(x+\mathrm{i}p)\,\frac{\mathrm{d}}{\mathrm{d}y}, \qquad (5.1.55)
$$

and turn Eq. (5.1.53) into

$$
I_{m \geq n} = (-2)^n \sqrt{\pi}\, e^{-p^2}\, e^{2(x^2+p^2)}\, (x+\mathrm{i}p)^{n-m} \left(\frac{\mathrm{d}}{\mathrm{d}y}\right)^n \left(y^m\, e^{-y}\right)\Bigg|_{y=2(x^2+p^2)}. \qquad (5.1.56)
$$

It is now clear that the nth-order derivative in Eq. (5.1.56) is a part of the Rodrigues formula

$$\left(\frac{d}{dy}\right)^n \left(y^{n+\nu}\,e^{-y}\right) = n!\,y^\nu\,e^{-y}\,L_n^{(\nu)}(y) \qquad (5.1.57)$$

for the *associated Laguerre** polynomial in y of degree n and index ν. In general, ν can be any complex number. For convenience we shall choose it to be a positive number. With this knowledge, we have

$$I_{m\geq n} = (-1)^n\,2^m\,n!\,\sqrt{\pi}\,e^{-p^2}\,(x-ip)^{m-n}\,L_n^{(m-n)}\!\left(2x^2+2p^2\right). \qquad (5.1.58)$$

For the other case $m < n$, we simply note the symmetry of Eq. (5.1.52) and obtain the answer

$$I_{m<n} = (-1)^m\,2^n\,m!\,\sqrt{\pi}\,e^{-p^2}\,(x+ip)^{n-m}\,L_m^{(n-m)}\!\left(2x^2+2p^2\right) \qquad (5.1.59)$$

by interchanging m with n and $x-ip$ with $x+ip$. Equations (5.1.58) and (5.1.59) can be compactly written into a single equation by defining the two new variables $n_> = \max\{m,n\}$ and $n_< = \min\{m,n\}$, and rewriting

$$x + ip = \sqrt{2}\,\alpha = \sqrt{2}\,|\alpha|\,e^{i\phi} \qquad (5.1.60)$$

in terms of the polar coordinates $(|\alpha|,\phi)$, leading to

$$I_{m,n} = (-1)^{n_<}\,2^{n_>}\,n_<!\,\sqrt{\pi}\,e^{-p^2}\,\left|\sqrt{2}\,\alpha\right|^{n_>-n_<}\,e^{-i(m-n)\phi}\,L_{n_<}^{(n_>-n_<)}\!\left(4\,|\alpha|^2\right). \qquad (5.1.61)$$

Problem 5.4   Arrive at Eq. (5.1.61) using the identities

$$H_n(x) = n!\oint_{\substack{\text{unit}\\\text{circle}}} \frac{dz}{2\pi i}\,\frac{e^{2xz-z^2}}{z^{n+1}}, \qquad (5.1.62)$$

$$e^{-xy+\alpha x+\beta y} = \sum_{j=0}^{\infty}\sum_{k=0}^{\infty}\frac{(-1)^k\,x^j y^k}{j!}\,\alpha^{j-k}\,L_k^{(j-k)}(2\,\alpha\beta). \qquad (5.1.63)$$

---

*Edmond Nicolas Laguerre (1834–1886).

Problem 5.5 Beginning with Eq. (5.1.57), establish that

$$\sum_{n=0}^{\infty} t^n \mathrm{L}_n^{(\nu)}(y) = \frac{e^{-\frac{yt}{1-t}}}{(1-t)^{\nu+1}} \quad \text{for } |t| \le 1, \tag{5.1.64}$$

thus revealing the generating function for the associated Laguerre polynomials $\mathrm{L}_n^{(\nu)}(y)$. With this, prove the orthogonality relation

$$\int_0^{\infty} dy\, y^{\nu}\, e^{-y}\, \mathrm{L}_m^{(\nu)}(y)\, \mathrm{L}_n^{(\nu)}(y) = \frac{(n+\nu)!}{n!}\,\delta_{m,n}. \tag{5.1.65}$$

The final expression for the Wigner function in terms of α is then given by

$$\rho_{\mathrm{W}}(\alpha,\alpha^*) = 2\,e^{-2|\alpha|^2} \sum_{m=0}^{\infty}\sum_{n=0}^{\infty} (-1)^{n_<}\, \rho_{mn}\, \frac{2^{n_>}}{2^{n_<}}\sqrt{\frac{n_<!}{n_>!}}$$
$$\times\, |\alpha|^{n_>-n_<}\, e^{-i(m-n)\phi}\, \mathrm{L}_{n_<}^{(n_>-n_<)}\big(4|\alpha|^2\big). \tag{5.1.66}$$

We now have a formula of $\rho_{\mathrm{W}}(\alpha,\alpha^*)$ for any ρ that is written in the Fock basis. This is extremely useful when one is interested in plotting the approximate Wigner function for a statistical operator that is represented by a square matrix.

In reaching the final expression, we have made a connection between Hermite polynomials and Laguerre polynomials, and it took us some time to get there. While we are at it, we shall demonstrate that it is possible to arrive at this expression very quickly with the normal-ordered form of the parity operator. All we need are the identities in Eq. (4.2.93) and Eq. (4.2.59).

With just a few lines of calculation, we have

$$\langle n| \left[e^{-2(A^\dagger - \alpha^*)(A-\alpha)} \right]_{\mathrm{N}} |m\rangle$$
$$= \frac{\sqrt{m!\,n!}}{4\pi^2} \oint_{\substack{\text{unit} \\ \text{circle}}} d\alpha' \oint_{\substack{\text{unit} \\ \text{circle}}} d\alpha'' \frac{\langle a'^* | a''\rangle}{\alpha'^{*\,n+1}\alpha''^{\,m+1}}\, e^{-2(\alpha'^* - \alpha^*)(\alpha'' - \alpha)}$$

$$= -\mathrm{e}^{-2|\alpha|^2} \frac{\sqrt{m!\,n!}}{4\pi^2} \oint_{\substack{\text{unit} \\ \text{circle}}} \mathrm{d}\alpha'^* \oint_{\substack{\text{unit} \\ \text{circle}}} \mathrm{d}\alpha'' \frac{\mathrm{e}^{-\alpha'^*\alpha''+2(\alpha'^*\alpha+\alpha^*\alpha'')}}{\alpha'^{*n+1}\alpha''^{m+1}},$$

$$(5.1.67)$$

from which we can readily extract the components in the sum of Eq. (5.1.66). Let us do this for the case where $m \geq n$, in which we first evaluate the integral in α'', then evaluate the integral in α'^*, both using the Cauchy residue theorem introduced in Chapter 4:

$$-\mathrm{e}^{-2|\alpha|^2} \frac{\sqrt{m!\,n!}}{4\pi^2} \oint_{\substack{\text{unit} \\ \text{circle}}} \mathrm{d}\alpha'^* \oint_{\substack{\text{unit} \\ \text{circle}}} \mathrm{d}\alpha'' \frac{\mathrm{e}^{-\alpha'^*\alpha''+2(\alpha'^*\alpha+\alpha^*\alpha'')}}{\alpha'^{*n+1}\alpha''^{m+1}}$$

$$= -\mathrm{i}\, \frac{\mathrm{e}^{-2|\alpha|^2}}{2\pi} \sqrt{\frac{n!}{m!}} \oint_{\substack{\text{unit} \\ \text{circle}}} \mathrm{d}\alpha'^* \frac{\mathrm{e}^{2\alpha'^*\alpha}}{\alpha'^{*n+1}} \left(\frac{\partial}{\partial\alpha''}\right)^m \mathrm{e}^{(-\alpha'^*+2\alpha^*)\alpha''} \Bigg|_{\alpha''=0}$$

$$= \frac{\mathrm{e}^{-2|\alpha|^2}}{2\pi\mathrm{i}} \sqrt{\frac{n!}{m!}} \oint_{\substack{\text{unit} \\ \text{circle}}} \mathrm{d}\alpha'^* \frac{(2\alpha^* - \alpha'^*)^m\, \mathrm{e}^{2\alpha'^*\alpha}}{\alpha'^{*n+1}}$$

$$= \frac{\mathrm{e}^{-2|\alpha|^2}}{\sqrt{m!\,n!}} \left(\frac{\partial}{\partial\alpha'^*}\right)^n \left[(2\alpha^* - \alpha'^*)^m\, \mathrm{e}^{2\alpha'^*\alpha}\right]\Bigg|_{\alpha'^*=0}. \qquad (5.1.68)$$

By a change of variable $y = 2\,(2\alpha^* - \alpha'^*)\,\alpha$, such that

$$\frac{\partial}{\partial\alpha'^*} = -2\alpha \frac{\mathrm{d}}{\mathrm{d}y}, \qquad (5.1.69)$$

and invoking the Rodrigues formula in Eq. (5.1.57) for the associated Laguerre polynomials, the last equation turns into

$$(-1)^n \frac{\mathrm{e}^{2|\alpha|^2}}{\sqrt{m!\,n!}} (2\alpha)^{n-m} \left(\frac{\mathrm{d}}{\mathrm{d}y}\right)^n (y^m\, \mathrm{e}^{-y})\Bigg|_{y=4|\alpha|^2}$$

$$= (-1)^n \mathrm{e}^{-2|\alpha|^2} 2^{m-n} \sqrt{\frac{n!}{m!}}\, \alpha^{*m-n} \mathrm{L}_n^{(m-n)}\left(4|\alpha|^2\right), \qquad (5.1.70)$$

which is, indeed, part of the components of the sum in Eq. (5.1.66).

From Eq. (5.1.66), it is straightforward to extract the Wigner functions for the Fock states. These are the summands of Eq. (5.1.66) with the

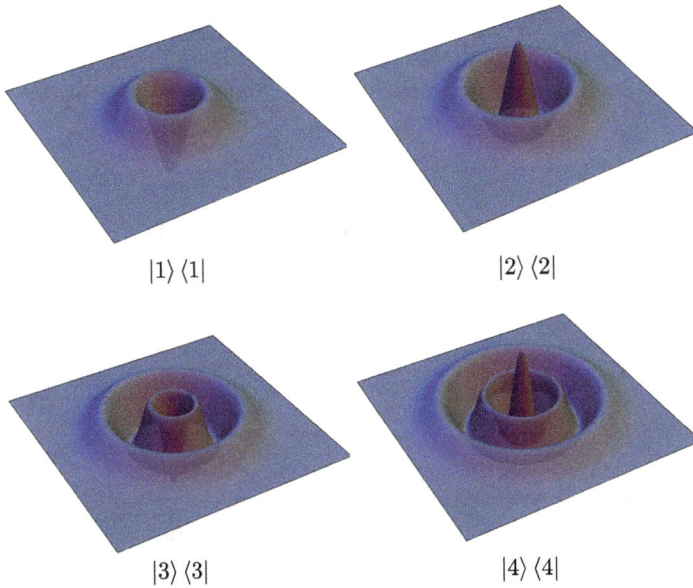

$|1\rangle\langle 1|$ $|2\rangle\langle 2|$

$|3\rangle\langle 3|$ $|4\rangle\langle 4|$

Fig. 5.3   Wigner functions for the Fock states.

summation indices $m = n$,

$$\rho_{\mathrm{W}}^{\mathrm{Fock}}(\alpha, \alpha^*; n) = 2\,(-1)^n\,\mathrm{e}^{-2\,|\alpha|^2}\,\mathrm{L}_n\big(4\,|\alpha|^2\big), \qquad (5.1.71)$$

where the Laguerre polynomials $\mathrm{L}_n\big(4\,|\alpha|^2\big)$ of degree n are defined as the associated Laguerre polynomials with zero index.

Problem 5.6   Write down the Wigner function for the displaced Fock states $\rho = D(\alpha')\,|n\rangle\langle n|\,D(-\alpha')$.

Since $\mathrm{L}_0(y) = 1$, the function $\rho_{\mathrm{W}}^{\mathrm{Fock}}(\alpha, \alpha^*; 0)$ for the vacuum state corresponds to the Wigner function of a Gaussian state, where the parameter α' of $\rho_{\mathrm{W}}^{\mathrm{coh}}(\alpha, \alpha^*; \alpha', \alpha'^*)$ in Eq. (5.1.38) is set to zero. The rest of the Fock states with $n > 0$ give nonpositive Wigner functions (refer to Fig. 5.3). Quantum states that possess nonpositive Wigner functions are classified under the category of *nonclassical states*.

Problem 5.7   After performing tomography on an unknown quantum state with the quadrature-eigenstate measurement introduced in Sec. 4.2.3.2, the resulting probability distribution is given by

$$p(x_\vartheta) = \frac{e^{-x_\vartheta^2}}{\sqrt{\pi}\,(1+\epsilon)}\left(1 + 2\epsilon x_\vartheta^2\right), \qquad (5.1.72)$$

where $0 < \epsilon < 1$. Infer the unknown quantum state directly from this distribution.

Problem 5.8   Prove that

$$\langle m|\, D(\alpha)\, |n\rangle$$

$$= |\alpha|^{n_> - n_<}\, e^{i(m-n)\phi - \frac{1}{2}|\alpha|^2} (-1)^{(n-m)\eta(n-m)} \sqrt{\frac{n_<!}{n_>!}}\, L_{n_<}^{(n_>-n_<)}\!\left(|\alpha|^2\right),$$

$$(5.1.73)$$

where $\eta(x)$ is the Heaviside* step function

$$\eta(x) = \begin{cases} 1 & \text{if } x \geq 0, \\ 0 & \text{if } x < 0. \end{cases} \qquad (5.1.74)$$

Problem 5.9   By directly calculating the Wigner function $\rho_{\mathrm{w}}(\alpha, \alpha^*)$ for the displaced Fock states $\rho = D(\alpha')\,|n\rangle\langle n|\, D(-\alpha')$ with Eq. (5.1.23) and the result in **Problem 5.8**, and later comparing the answer with that in **Problem 5.6**, establish the identity

$$\sum_{m=0}^{\infty} (-1)^{m-n}\frac{n_<!}{n_>!}\, k^{(n_>-n_<)}\left[L_{n_<}^{(n_>-n_<)}(k)\right]^2 = e^{-k}\, L_n(4\,k), \quad (5.1.75)$$

where $k = |\alpha - \alpha'|^2$ in this case.

---

*Oliver Heaviside (1850–1925).

Problem 5.10 Confirm the identity

$$(-1)^{A^\dagger A} = \int \frac{(\mathrm{d}\gamma)}{\pi} \, |\gamma\rangle \langle -\gamma^*| \qquad (5.1.76)$$

for the parity operator $(-1)^{A^\dagger A}$. Use this identity to rederive Eq. (5.1.66) by starting with the form of $\rho_{\mathrm{w}}(\alpha, \alpha^*)$ that is presented in Eq. (5.1.23).

Problem 5.11 Express $\rho_{\mathrm{w}}(\alpha, \alpha^*)$ in terms of the coherent-state basis, that is show that

$$\rho_{\mathrm{w}}(\alpha, \alpha^*)$$
$$= 2\,\mathrm{e}^{-2|\alpha|^2} \int \frac{(\mathrm{d}\alpha')}{\pi} \int \frac{(\mathrm{d}\alpha'')}{\pi} \, \rho\,(\alpha'^*, \alpha'') \, \mathrm{e}^{2\left(\alpha^*\alpha' + \alpha\alpha''^*\right)} \langle -\alpha''^* | \alpha' \rangle,$$
$$(5.1.77)$$

where $\rho\,(\alpha'^*, \alpha'') = \langle \alpha'^* | \rho | \alpha'' \rangle$.

5.1.2 *Glauber-Sudarshan P function*

5.1.2.1 *Definitions and properties*

The concept of classical states goes beyond the domains of Wigner functions. Coherent states may be regarded as the key ingredients for a generic classical state because in some respects, they correspond to statistical behavior similar to properties of classical electromagnetic field amplitudes and classical currents. This is because the expectation value of the electric-field operator in quantum electrodynamics are simply complex numbers converted directly from the photonic ladder operators ($A \to \alpha$, $A^\dagger \to \alpha^*$) when evaluated with coherent states. More explicitly, a classical state may be defined to be any statistical mixture of coherent states.

This discussion leads us to another quasi-probability distribution function that is intimately related to the notion of classical states. This distribution function is the *Glauber–Sudarshan* P function*, which is

---

*Roy Jay Glauber (1925–) and Ennackal Chandy George Sudarshan (1931–).

denoted by $\rho_P(\alpha, \alpha^*)$, and is defined in terms of the coherent states as

$$\rho = \int \frac{(d\alpha)}{\pi} \, |\alpha\rangle \, \rho_P(\alpha, \alpha^*) \, \langle\alpha^*|, \qquad (5.1.78)$$

where $\int(d\alpha)/\pi$ — an integration over all α — is just the complex notation for the phase-space integration measure that was first introduced in Eq. (5.1.6).

With this decomposition, the simplest classical pure state, that is the coherent state $|\alpha'\rangle \, \langle\alpha'^*|$, has a P function equal to

$$\rho_P^{\mathrm{coh}}(\alpha, \alpha^*; \alpha', \alpha'^*) = \pi \, \delta(\alpha - \alpha'), \qquad (5.1.79)$$

where we define the two-dimensional delta function as

$$\delta(\alpha) \equiv \delta(\mathrm{Re}\,\{\alpha\}) \, \delta(\mathrm{Im}\,\{\alpha\}). \qquad (5.1.80)$$

It follows that the set of classical states can then be defined as the set of states with $\rho_P(\alpha, \alpha^*)$ that are either positive functions, or objects which are *no more singular* than the delta function.

A word of caution. In using such adjectives as "classical" and "nonclassical" when describing states of affairs, one must always remember that these states are quantum-mechanical. The states that we are discussing describe sources with single-mode quantum degrees of freedom. As such, callous usage of words such as "macroscopic states", "cats", or even "kittens" should be avoided.

One may also define the P function for any operator F and, just as with Wigner functions, the expectation value of F can also be readily expressed in terms of P functions inasmuch as

$$\mathrm{tr}\{\rho F\} = \int \frac{(d\alpha')}{\pi} \int \frac{(d\alpha'')}{\pi} \, e^{-|\alpha' - \alpha''|^2} \, \rho_P(\alpha', \alpha'^*) \, F_P(\alpha'', \alpha''^*).$$

$$(5.1.81)$$

Similar to the Wigner function, we can express $\rho_P(\alpha, \alpha^*)$ as an integral function of ρ in order to facilitate the study of its properties. To obtain such an integral function, we first consider the complex scalar function

$$
\langle -u^* | \rho | u \rangle = \int \frac{(\mathrm{d}\alpha)}{\pi} \rho_P(\alpha, \alpha^*) \langle -u^* | \alpha \rangle \langle \alpha^* | u \rangle
$$

$$
= \int \frac{(\mathrm{d}\alpha)}{\pi} \rho_P(\alpha, \alpha^*) \left(e^{-\frac{1}{2}|u|^2 - \frac{1}{2}|\alpha|^2 - u^* \alpha} \right) \left(e^{-\frac{1}{2}|u|^2 - \frac{1}{2}|\alpha|^2 + u\alpha^*} \right)
$$

$$
= e^{-|u|^2} \int \frac{(\mathrm{d}\alpha)}{\pi} \rho_P(\alpha, \alpha^*) \, e^{-|\alpha|^2} \, e^{u\alpha^* - u^* \alpha}, \qquad (5.1.82)
$$

which involves an interesting exponential factor $e^{u\alpha^* - u^* \alpha}$ that we shall examine. An integral of this factor over the whole complex plane for α gives

$$
\int \frac{(\mathrm{d}\alpha)}{\pi^2} \, e^{u\alpha^* - u^* \alpha} = \int \frac{\mathrm{d}x \, \mathrm{d}y}{2\pi^2} \, e^{u \frac{x - iy}{\sqrt{2}} - u^* \frac{x + iy}{\sqrt{2}}} \qquad \left(\alpha = \frac{x + iy}{\sqrt{2}} \right)
$$

$$
= \int \frac{\mathrm{d}x \, \mathrm{d}y}{2\pi^2} \, e^{x \frac{u - u^*}{\sqrt{2}} - iy \frac{u + u^*}{\sqrt{2}}}
$$

$$
= \int \frac{\mathrm{d}x \, \mathrm{d}y}{2\pi^2} \, e^{ix \sqrt{2} \, \mathrm{Im}\{u\}} \, e^{-iy \sqrt{2} \, \mathrm{Re}\{u\}}
$$

$$
= 2 \, \delta \left(\sqrt{2} \, \mathrm{Im}\{u\} \right) \delta \left(\sqrt{2} \, \mathrm{Re}\{u\} \right)
$$

$$
= \delta \left(\mathrm{Im}\{u\} \right) \delta \left(\mathrm{Re}\{u\} \right) = \delta(u). \qquad (5.1.83)
$$

The above integral is thus the Fourier transform of $1/\pi$ in disguise.

From this we realize that the exponential $e^{u\alpha^* - u^* \alpha}$ is then the usual two-dimensional Fourier basis functions that are used to establish the Fourier transform $\widetilde{f}(u, u^*)$ of a complex function $f(\alpha, \alpha^*)$, that is

$$
\widetilde{f}(u, u^*) = \int \frac{(\mathrm{d}\alpha)}{\pi} \, e^{u\alpha^* - u^* \alpha} \, f(\alpha, \alpha^*), \qquad (5.1.84)
$$

and the inverse Fourier transform

$$
f(\alpha, \alpha^*) = \int \frac{(\mathrm{d}u)}{\pi} \, e^{\alpha u^* - \alpha^* u} \, \widetilde{f}(u, u^*). \qquad (5.1.85)
$$

This hints at a simple method of obtaining the integral function for $\rho_P(\alpha, \alpha^*)$: Perform the inverse Fourier transform of $e^{|u|^2} \langle -u^* | \rho | u \rangle$. This leads to

$$\int \frac{(du)}{\pi} e^{\alpha u^* - \alpha^* u} e^{|u|^2} \langle -u^* | \rho | u \rangle$$

$$= \int (d\alpha') \rho_P(\alpha', \alpha'^*) e^{-|\alpha'|^2} \underbrace{\int \frac{(du)}{\pi^2} e^{(\alpha - \alpha')u^* - (\alpha^* - \alpha'^*)u}}_{= \delta(\alpha - \alpha')}$$

$$= \rho_P(\alpha, \alpha^*) e^{-|\alpha|^2}, \tag{5.1.86}$$

or

$$\rho_P(\alpha, \alpha^*) = e^{|\alpha|^2} \int \frac{(du)}{\pi} e^{\alpha u^* - \alpha^* u} e^{|u|^2} \langle -u^* | \rho | u \rangle. \tag{5.1.87}$$

So, the P function is also an expectation value of a Hermitian operator, the *P operator*, that we shall denote by P_α — $\rho_P(\alpha, \alpha^*) = \text{tr}\{\rho P_\alpha\}$ —, and in terms of the ladder operators A and $A^\dagger$, this operator is given by

$$P_\alpha = e^{|\alpha|^2} \int \frac{(du)}{\pi} |u\rangle e^{\alpha u^* - \alpha^* u} e^{|u|^2} \langle -u^* |$$

$$= e^{|\alpha|^2} \sum_{k=0}^{\infty} \frac{1}{k!} \int \frac{(du)}{\pi} |u\rangle e^{\alpha u^* - \alpha^* u} |u|^{2k} \langle -u^* |$$

$$= e^{|\alpha|^2} \sum_{k=0}^{\infty} \frac{1}{k!} \left(\frac{\partial}{\partial \alpha} \right)^k \left(-\frac{\partial}{\partial \alpha^*} \right)^k \int \frac{(du)}{\pi} |u\rangle e^{\alpha u^* - \alpha^* u} \langle -u^* |$$

$$= e^{|\alpha|^2} e^{-\left| \frac{\partial}{\partial \alpha} \right|^2} \int \frac{(du)}{\pi} |u\rangle e^{\alpha u^* - \alpha^* u} \langle -u^* |$$

$$= e^{|\alpha|^2} e^{-\left| \frac{\partial}{\partial \alpha} \right|^2} \left[e^{-\alpha^* A} (-1)^{A^\dagger A} e^{-\alpha A^\dagger} \right], \tag{5.1.88}$$

where the last equality is a consequence of the identity in **Problem 5.10**.

It is easy to rewrite the P operator such that it is normally ordered. One way to do this is to invoke Eq. (4.2.76) and the BCH relation to obtain the results

$$e^{-\alpha^* A} F(A^\dagger) = F(A^\dagger - \alpha^*) e^{-\alpha^* A},$$

(5.1.89)

$$F(A) e^{-\alpha A^\dagger} = e^{-\alpha A^\dagger} F(A - \alpha),$$

after which straightforward manipulations would, hence, lead to

$$e^{-\alpha^* A} (-1)^{A^\dagger A} e^{-\alpha A^\dagger} = e^{-\alpha^* A} \left[e^{-2A^\dagger A} \right]_N e^{-\alpha A^\dagger}$$

$$= \left[e^{-2(A^\dagger - \alpha^*)A} \right]_N \underbrace{e^{-\alpha^* A} e^{-\alpha A^\dagger}}_{= e^{|\alpha|^2} e^{-\alpha A^\dagger} e^{-\alpha^* A} \ \ (\text{BCH})}$$

$$= e^{|\alpha|^2} e^{-\alpha A^\dagger} (-1)^{(A^\dagger - \alpha^*)(A - \alpha)} e^{-\alpha^* A}. \qquad (5.1.90)$$

Therefore, in terms of the P operator,

$$\rho_P(\alpha, \alpha^*) = \text{tr}\{\rho \, P_\alpha\},$$

$$P_\alpha = e^{|\alpha|^2} e^{-\left|\frac{\partial}{\partial \alpha}\right|^2} \left[e^{|\alpha|^2} e^{-\alpha A^\dagger} (-1)^{(A^\dagger - \alpha^*)(A - \alpha)} e^{-\alpha^* A} \right]. \qquad (5.1.91)$$

5.1.2.2 *Coherent states*

Evidently, the Hermitian operator P_α involves an infinite series of derivatives and we shall see that this is the origin of the typically highly-singular nature of $\rho_P(\alpha, \alpha^*)$. To practice working with the P operator P_α, let us derive the P function for the coherent state $|\alpha'\rangle \langle \alpha'^*|$ using Eq. (5.1.91):

$$\rho_P^{\text{coh}}(\alpha, \alpha^*) = \langle \alpha'^* | P_\alpha | \alpha' \rangle$$

$$= e^{|\alpha|^2} e^{-\left|\frac{\partial}{\partial \alpha}\right|^2} \left[e^{|\alpha|^2} e^{-\alpha \alpha'^*} e^{-2(\alpha'^* - \alpha^*)(\alpha' - \alpha)} e^{-\alpha^* \alpha'} \right]$$

$$= e^{|\alpha|^2 - |\alpha'|^2} e^{-\left|\frac{\partial}{\partial \alpha}\right|^2} e^{-|\alpha - \alpha'|^2}, \qquad (5.1.92)$$

where the singular contribution

$$e^{-\left|\frac{\partial}{\partial \alpha}\right|^2} e^{-|\alpha - \alpha'|^2} = \sum_{k=0}^{\infty} \frac{1}{k!} \left(\frac{d}{dy} \right)^k (y^k e^{-y}) \Bigg|_{y = |\alpha - \alpha'|^2}$$

$$= e^{-|\alpha - \alpha'|^2} \sum_{k=0}^{\infty} L_k(|\alpha - \alpha'|^2). \qquad (5.1.93)$$

As we know, from the generating function in Eq. (5.1.64) of **Problem 5.5** for the associated Laguerre polynomials,

$$\sum_{k=0}^{\infty} L_k\left(|\alpha - \alpha'|^2\right) = \left.\frac{e^{-\frac{|\alpha-\alpha'|^2}{1-t}}}{1-t}\right|_{1\geq t\to 1}$$

$$= \pi\,\delta(\alpha - \alpha'). \tag{5.1.94}$$

This implies that

$$e^{-\left|\frac{\partial}{\partial\alpha}\right|^2} e^{-|\alpha-\alpha'|^2} = \pi\,\delta(\alpha - \alpha'), \tag{5.1.95}$$

and we thus obtain the correct P function for the coherent state.

5.1.2.3 *Odd and even coherent states*

For nonclassical states, the action of the exponential operator $\exp\left(-|\partial/\partial\alpha|^2\right)$ of the P_α operator results in highly singular features. It is interesting to reveal these features of $\rho_P(\alpha, \alpha^*)$ when this happens. For this, we first calculate the P function for the odd/even superposition of coherent states that are introduced in Eq. (5.1.40).

There are, altogether, four terms that constitute the P function

$$\rho_P^{o/e}(\alpha, \alpha^*; \alpha', \alpha'^*) = \langle\alpha'^*_\pm| P_\alpha |\alpha'_\pm\rangle$$

$$= \mathcal{N}^2 \left(\langle\alpha'^*| \pm \langle-\alpha'^*|\right) P_\alpha \left(|\alpha'\rangle \pm |-\alpha'\rangle\right). \tag{5.1.96}$$

The two expectation values $\mathcal{N}^2 \langle\alpha'^*| P_\alpha |\alpha'\rangle$ and $\mathcal{N}^2 \langle-\alpha'^*| P_\alpha |-\alpha'\rangle$ are respectively proportional to the delta functions $\delta(\alpha - \alpha')$ and $\delta(\alpha + \alpha')$. In analyzing one of the other two interference terms, we find

$$\langle\alpha'^*| P_\alpha |-\alpha'\rangle = e^{|\alpha|^2-2|\alpha'|^2} e^{-\left|\frac{\partial}{\partial\alpha}\right|^2} \left[e^{|\alpha|^2} e^{-\alpha\alpha'^*} e^{2(\alpha'^*-\alpha^*)(\alpha'+\alpha)} e^{\alpha^*\alpha'}\right]$$

$$= e^{|\alpha|^2-|\alpha'|^2} e^{-\left|\frac{\partial}{\partial\alpha}\right|^2} e^{-(\alpha^*-\alpha'^*)(\alpha+\alpha')}$$

$$= e^{|\alpha|^2-|\alpha'|^2} e^{\alpha'\frac{\partial}{\partial\alpha}-\alpha'^*\frac{\partial}{\partial\alpha^*}} e^{-\left|\frac{\partial}{\partial\alpha}\right|^2} e^{-|\alpha|^2}$$

$$= \pi\,e^{|\alpha|^2-|\alpha'|^2} e^{\alpha'\frac{\partial}{\partial\alpha}-\alpha'^*\frac{\partial}{\partial\alpha^*}} \delta(\alpha), \tag{5.1.97}$$

where the actions of the exponential operators are essentially the answers to the following problem.

Problem 5.12 Show that

$$e^{\alpha' \frac{\partial}{\partial \alpha} + \alpha''^* \frac{\partial}{\partial \alpha^*}} f(\alpha, \alpha^*) = f(\alpha + \alpha', \alpha^* + \alpha''^*) \tag{5.1.98}$$

for a complex function $f(\alpha, \alpha^*)$ by first switching to the real and imaginary parts of α and α'. You may use the relation

$$e^{z \frac{\partial}{\partial x}} f(x) = f(x + z). \tag{5.1.99}$$

The complete expression for the P function is thus given by

$$\rho_P^{o/e}(\alpha, \alpha^*; \alpha', \alpha'^*)$$
$$= \mathcal{N}^2 \pi \, e^{|\alpha|^2 - |\alpha'|^2}$$
$$\times \left[\delta(\alpha - \alpha') + \delta(\alpha + \alpha') \pm \underbrace{\left(e^{\alpha' \frac{\partial}{\partial \alpha} - \alpha'^* \frac{\partial}{\partial \alpha^*}} + e^{-\alpha' \frac{\partial}{\partial \alpha} + \alpha''^* \frac{\partial}{\partial \alpha^*}} \right) \delta(\alpha)}_{\text{interference terms}} \right].$$

$$\tag{5.1.100}$$

To better understand the interference terms, we convert one of those terms back into an integral using the integral representation of the delta function in Eq. (5.1.83), taking for example

$$\pi \, e^{\alpha' \frac{\partial}{\partial \alpha} - \alpha'^* \frac{\partial}{\partial \alpha^*}} \delta(\alpha)$$

$$= \int \frac{(\mathrm{d}u)}{\pi} \, e^{(\alpha + \alpha')u^* - (\alpha^* - \alpha'^*)u} \qquad \left(u = \frac{x'' + \mathrm{i}p''}{\sqrt{2}} \right)$$

$$= \int \frac{\mathrm{d}x'' \mathrm{d}p''}{2\pi} \, e^{\frac{1}{2}[x + x' + \mathrm{i}(p + p')](x'' - \mathrm{i}p'') - \frac{1}{2}[x - x' - \mathrm{i}(p - p')](x'' + \mathrm{i}p'')}$$

$$= \int \frac{\mathrm{d}x''}{2\pi} \, e^{(x' + \mathrm{i}p)x''} \int \mathrm{d}p'' \, e^{(-\mathrm{i}x + p')p''}. \tag{5.1.101}$$

The two integrals of x'' and p'' are, by themselves, strictly divergent and are, thus, much more singular than delta functions.

Problem 5.13   Despite the divergent nature of the quantity

$$\delta_{ix}(k) = \int \frac{dx'}{2\pi} \, e^{(x+ik)x'}, \tag{5.1.102}$$

show that $\delta_{ix}(k)$ behaves just like a delta function that replaces k with ix by verifying that the integral

$$\int dk \, f(k) \, \delta_{ix}(k) = f(ix) \tag{5.1.103}$$

for the class of Gaussian functions

$$f(k) = \sqrt{\frac{a}{\pi}} \, e^{-ak^2+bk} \tag{5.1.104}$$

with $\mathrm{Re}\,\{a\} \geq 0$. One may take Eq. (5.1.103) to be valid for any $f(k)$, provided that one understands the integral as a limit of another well-defined function.

Therefore, the P function for the odd/even coherent state consists of two parts. There is one part that corresponds to the contributions that come from a mixture of two coherent states, giving rise to two separate delta functions. There is the other part that comes from the interference terms. The latter part is responsible for the highly singular nature of $\rho_P^{o/e}(\alpha,\alpha^*;\alpha',\alpha'^*)$ in Eq. (5.1.100) that arises from the exponential operator $\exp\left(-|\partial/\partial\alpha|^2\right)$. The mathematical object

$$\delta(\alpha+\alpha',\alpha^*-\alpha'^*) \equiv e^{\alpha'\frac{\partial}{\partial\alpha}-\alpha'^*\frac{\partial}{\partial\alpha^*}} \, \delta(\alpha), \tag{5.1.105}$$

for instance, is a product of two infinite sums of high-order derivatives of the delta function $\delta(\alpha)$ that is highly singular. Hence, these superposition states are nonclassical, and this is consistent with the observation made in Sec. 5.1.1. Such mathematical objects, which include the delta functions, make up the class of *improper functions*.

By virtue of **Problem 5.13**, it is worth noting that the improper function $\delta(\alpha-\beta,\alpha^*-\beta'^*)$, which we will encounter again later, behaves very much like a delta function in the sense that

$$\int (d\alpha)\,\delta(\alpha - \beta, \alpha^* - \beta'^*) = 1$$

and $\quad \displaystyle\int (d\alpha)\, f(\alpha, \alpha^*)\, \delta(\alpha - \beta, \alpha^* - \beta'^*) = f(\beta, \beta'^*) \qquad (5.1.106)$

for any well-behaved function $f(\alpha, \alpha^*)$.

At this point, one may still doubt the validity of (5.1.106). As a reassurance, the reader is reminded that improper functions like these can only appear in the intermediate steps during the evaluations of well-defined integrals, and that the manipulations of these highly singular objects are permissible as long as they are logically consistent to give correct final answers. This is justified when the integrals involving these highly-singular improper functions are understood as proper limits of other well-defined integrals.

Problem 5.14 Is the squeezed coherent state (introduced in **Problem 5.3**) classical? Answer this question by inspecting its P function.

5.1.2.4 Fock states

We may also express the function $\rho_P(\alpha, \alpha^*)$ with the help of the Fock basis and this is easily done by substituting the Fock representation of ρ, introduced in Eq. (5.1.45), into Eq. (5.1.87),

$$\rho_P(\alpha, \alpha^*) = e^{|\alpha|^2} \sum_{m=0}^{\infty} \sum_{n=0}^{\infty} \rho_{mn} \int \frac{(du)}{\pi}\, e^{\alpha u^* - \alpha^* u}\, e^{|u|^2} \langle -u^* | m \rangle \langle n | u \rangle$$

$$= e^{|\alpha|^2} \sum_{m=0}^{\infty} \sum_{n=0}^{\infty} \frac{\rho_{mn}}{\sqrt{m!\,n!}} \int \frac{(du)}{\pi}\, e^{\alpha u^* - \alpha^* u}\, (-u^*)^m\, u^n$$

$$= e^{|\alpha|^2} \sum_{m=0}^{\infty} \sum_{n=0}^{\infty} \frac{\rho_{mn}}{\sqrt{m!\,n!}} \left(-\frac{\partial}{\partial \alpha}\right)^m \left(-\frac{\partial}{\partial \alpha^*}\right)^n \int \frac{(du)}{\pi}\, e^{\alpha u^* - \alpha^* u}$$

$$= \pi\, e^{|\alpha|^2} \sum_{m=0}^{\infty} \sum_{n=0}^{\infty} \frac{\rho_{mn}}{\sqrt{m!\,n!}} \left(-\frac{\partial}{\partial \alpha}\right)^m \left(-\frac{\partial}{\partial \alpha^*}\right)^n \delta(\alpha). \qquad (5.1.107)$$

So, in terms of complex derivatives, the Fock representation of the P function is given by

$$\rho_{\mathrm{P}}(\alpha, \alpha^*) = \pi\, e^{|\alpha|^2} \sum_{m=0}^{\infty} \sum_{n=0}^{\infty} \frac{\rho_{mn}}{\sqrt{m!\, n!}} \left(-\frac{\partial}{\partial\alpha} \right)^m \left(-\frac{\partial}{\partial\alpha^*} \right)^n \delta(\alpha). \quad (5.1.108)$$

The high-order derivatives of the delta function are a reflection of the precarious nature that is inherent in the P functions of typical quantum states.

By definition, the double sums in Eq. (5.1.108) converge to a nonnegative function for all classical states. Let us, at least, check this for the coherent state $\rho = |\alpha'\rangle \langle \alpha'^*|$. The matrix elements ρ_{mn} in the Fock basis are

$$\rho_{mn} = \langle m|\alpha'\rangle \langle \alpha'^*|n\rangle$$

$$= e^{-|\alpha'|^2} \frac{\alpha'^m \alpha'^{*n}}{\sqrt{m!\, n!}}. \quad (5.1.109)$$

With these elements, we have

$$\rho_{\mathrm{P}}^{\mathrm{coh}}(\alpha; \alpha') = \pi\, e^{|\alpha|^2 - |\alpha'|^2} \sum_{m=0}^{\infty} \sum_{n=0}^{\infty} \frac{\alpha'^m \alpha'^{*n}}{m!\, n!} \left(-\frac{\partial}{\partial\alpha} \right)^m \left(-\frac{\partial}{\partial\alpha^*} \right)^n \delta(\alpha)$$

$$= \pi\, e^{|\alpha|^2 - |\alpha'|^2}\, e^{-\alpha' \frac{\partial}{\partial\alpha} - \alpha'^* \frac{\partial}{\partial\alpha^*}}\, \delta(\alpha). \quad (5.1.110)$$

It then follows, from **Problem 5.12**, that $\rho_{\mathrm{P}}^{\mathrm{coh}}(\alpha; \alpha') = \pi\, \delta(\alpha - \alpha')$, as it should be.

Problem 5.15   Use Eq. (5.1.108) to derive Eq. (5.1.100).

There is actually a simple relation one can use to infer the Wigner function $\rho_{\mathrm{W}}(\alpha, \alpha^*)$ from the P function $\rho_{\mathrm{P}}(\alpha, \alpha^*)$. To derive it, we insert Eq. (5.1.78) into Eq. (5.1.2), after which one obtains

$$\rho_{\mathrm{W}}(\alpha, \alpha^*) = \operatorname{tr}\{\rho\, W_\alpha\}$$

$$= \int \frac{(d\alpha')}{\pi}\, \rho_{\mathrm{P}}(\alpha', \alpha'^*)\, \operatorname{tr}\{|\alpha'\rangle \langle \alpha'^*|\, W_\alpha\}, \quad (5.1.111)$$

and upon referring to Eq. (5.1.38) for the Wigner function for coherent states, we have the expression

$$\rho_{\mathrm{W}}(\alpha, \alpha^*) = 2 \int \frac{(\mathrm{d}\alpha')}{\pi} \rho_{\mathrm{P}}(\alpha', \alpha'^*)\, \mathrm{e}^{-2|\alpha-\alpha'|^2}. \qquad (5.1.112)$$

As $\mathrm{e}^{-2|\alpha-\alpha'|^2} > 0$, any negativity of $\rho_{\mathrm{W}}(\alpha, \alpha^*)$ must originate from $\rho_{\mathrm{P}}(\alpha, \alpha^*)$. From Eq. (5.1.112), we can therefore conclude that the state ρ is nonclassical if $\rho_{\mathrm{W}}(\alpha, \alpha^*) \not\geq 0$. A nonnegative $\rho_{\mathrm{W}}(\alpha, \alpha^*)$, however, is *not* necessarily derived from a classical state, for there exist nonclassical states, with nonpositive or highly singular P functions, that can give positive Wigner functions, a message that is hopefully brought across by **Problems 5.3** and **5.14**.

To give a demonstration of Eq. (5.1.112), we shall consider the Fock states $|n\rangle\langle n|$. From Eq. (5.1.108), its P function is simply

$$\rho_{\mathrm{P}}^{\mathrm{Fock}}(\alpha, \alpha^*; n) = \frac{\pi\, \mathrm{e}^{|\alpha|^2}}{n!} \left|\frac{\partial}{\partial\alpha}\right|^{2n} \delta(\alpha). \qquad (5.1.113)$$

Making use of the Fourier representation of $\delta(\alpha)$ in Eq. (5.1.83), we have

$$\begin{aligned}
\rho_{\mathrm{P}}^{\mathrm{Fock}}(\alpha, \alpha^*; n) &= \frac{\pi\, \mathrm{e}^{|\alpha|^2}}{n!} \left|\frac{\partial}{\partial\alpha}\right|^{2n} \int \frac{(\mathrm{d}u)}{\pi^2}\, \mathrm{e}^{\alpha u^* - \alpha^* u} \\
&= \frac{\mathrm{e}^{|\alpha|^2}}{n!} \int \frac{(\mathrm{d}u)}{\pi} \left(-|u|^2\right)^n \mathrm{e}^{\alpha u^* - \alpha^* u}. \qquad (5.1.114)
\end{aligned}$$

Equation (5.1.112) then implies that

$$\begin{aligned}
&\rho_{\mathrm{W}}^{\mathrm{Fock}}(\alpha, \alpha^*; n) \\
&= \frac{2}{n!} \int \frac{(\mathrm{d}\alpha')}{\pi}\, \mathrm{e}^{-2|\alpha-\alpha'|^2}\, \mathrm{e}^{|\alpha'|^2} \int \frac{(\mathrm{d}u)}{\pi} \left(-|u|^2\right)^n \mathrm{e}^{\alpha' u^* - \alpha'^* u} \\
&= \frac{2\, \mathrm{e}^{-2|\alpha|^2}}{n!} \int \frac{(\mathrm{d}u)}{\pi} \left(-|u|^2\right)^n \int \frac{(\mathrm{d}\alpha')}{\pi}\, \mathrm{e}^{-|\alpha'|^2 + (2\alpha^* + u^*)\alpha' + (2\alpha - u)\alpha'^*}.
\end{aligned}$$
$$(5.1.115)$$

At this point, we note that there is a two-dimensional Gaussian integral which we need to calculate. As this is a rather common type of integral, we will spend some time to evaluate the more general form

$$I_{\text{gauss}}(a, b_1, b_2, c_1, c_2) = \int \frac{(\mathrm{d}\alpha')}{\pi} \, \mathrm{e}^{-a|\alpha'|^2 + b_1\alpha' + b_2\alpha'^* + c_1\alpha'^2 + c_2\alpha'^{*2}}$$

(5.1.116)

with respect to the complex parameters a, b_1, b_2, c_1 and c_2.

This integral will converge under a certain condition for a, c_1 and c_2 and to look for such a condition, it is convenient to collect the terms in the exponent and form linear mathematical objects inasmuch as

$$I_{\text{gauss}}(a, b_1, b_2, c_1, c_2)$$

$$= \int \frac{(\mathrm{d}\alpha')}{\pi} \, \exp\left(-\begin{pmatrix} \alpha'^* & \alpha' \end{pmatrix} \mathbf{M}' \begin{pmatrix} \alpha' \\ \alpha'^* \end{pmatrix} + \boldsymbol{b}'^{\mathrm{T}} \begin{pmatrix} \alpha' \\ \alpha'^* \end{pmatrix}\right),$$

(5.1.117)

where the 2×2 matrix

$$\mathbf{M}' = \begin{pmatrix} \frac{a}{2} & -c_2 \\ -c_1 & \frac{a}{2} \end{pmatrix},$$

(5.1.118)

and the complex column

$$\boldsymbol{b}' = \begin{pmatrix} b_1 \\ b_2 \end{pmatrix}.$$

(5.1.119)

Switching to the real parameters x' and p' using the transformation

$$\begin{pmatrix} \alpha' \\ \alpha'^* \end{pmatrix} = \frac{1}{\sqrt{2}} \begin{pmatrix} 1 & \mathrm{i} \\ 1 & -\mathrm{i} \end{pmatrix} \begin{pmatrix} x' \\ p' \end{pmatrix},$$

(5.1.120)

we get

$$I_{\text{gauss}}(a, b_1, b_2, c_1, c_2) = \int \frac{\mathrm{d}x'\mathrm{d}p'}{2\pi} \, \exp\left(-\begin{pmatrix} x' & p' \end{pmatrix} \widetilde{\mathbf{M}} \begin{pmatrix} x' \\ p' \end{pmatrix} + \boldsymbol{b}^{\mathrm{T}} \begin{pmatrix} x' \\ p' \end{pmatrix}\right),$$

(5.1.121)

where

$$\widetilde{\mathbf{M}} = \frac{1}{\sqrt{2}} \begin{pmatrix} 1 & 1 \\ -\mathrm{i} & \mathrm{i} \end{pmatrix} \begin{pmatrix} \frac{a}{2} & -c_2 \\ -c_1 & \frac{a}{2} \end{pmatrix} \frac{1}{\sqrt{2}} \begin{pmatrix} 1 & \mathrm{i} \\ 1 & -\mathrm{i} \end{pmatrix}$$

$$= \frac{1}{2} \begin{pmatrix} a - c_1 - c_2 & \mathrm{i}(c_2 - c_1) \\ \mathrm{i}(c_2 - c_1) & a + c_1 + c_2 \end{pmatrix}$$

(5.1.122)

and

$$b = \frac{1}{\sqrt{2}} \begin{pmatrix} 1 & 1 \\ i & -i \end{pmatrix} \begin{pmatrix} b_1 \\ b_2 \end{pmatrix}$$

$$= \frac{1}{\sqrt{2}} \begin{pmatrix} b_1 + b_2 \\ ib_1 - ib_2 \end{pmatrix}. \qquad (5.1.123)$$

The integral in Eq. (5.1.121) converges for a complex symmetric matrix $\widetilde{\mathbf{M}}$ when $\text{Re}\left\{\widetilde{\mathbf{M}}\right\} \geq 0$, which is a generalization of the convergence condition for Eq. (4.1.32). Since this matrix $\widetilde{\mathbf{M}}$ is already symmetric, the condition for a, c_1 and c_2 can thus be derived by noting that

$$0 \leq \text{Det}\left\{\text{Re}\left\{\widetilde{\mathbf{M}}\right\}\right\}$$

$$= \frac{1}{4}\left[(\text{Re}\{a\})^2 - (\text{Re}\{c_1 + c_2\})^2 - (\text{Im}\{c_1 - c_2\})^2\right], \qquad (5.1.124)$$

which leads to the inequality criterion

$$\text{Re}\{a\} \geq |c_1 + c_2^*|. \qquad (5.1.125)$$

Applying Eq. (4.1.31), which is also valid for the complex symmetric matrix $\widetilde{\mathbf{M}}$ possessing a positive real part, we have

$$I_{\text{gauss}}(a, b_1, b_2, c_1, c_2) = \frac{1}{2\sqrt{\text{Det}\left\{\widetilde{\mathbf{M}}\right\}}} e^{\frac{1}{4} b^{\mathsf{T}} \widetilde{\mathbf{M}}^{-1} b}. \qquad (5.1.126)$$

As the determinant of $\widetilde{\mathbf{M}}$ is given by

$$\text{Det}\left\{\widetilde{\mathbf{M}}\right\} = \det\{\mathbf{M}'\} = \frac{1}{4}\left(a^2 - 4c_1 c_2\right), \qquad (5.1.127)$$

the inverse of this 2×2 matrix is simply

$$\widetilde{\mathbf{M}}^{-1} = \frac{2}{a^2 - 4c_1 c_2} \begin{pmatrix} a + c_1 + c_2 & -i(c_2 - c_1) \\ -i(c_2 - c_1) & a - c_1 - c_2 \end{pmatrix}, \qquad (5.1.128)$$

so that

$$\frac{1}{4} b^{\mathsf{T}} \widetilde{\mathbf{M}}^{-1} b$$

$$= \frac{1}{4(a^2 - 4c_1 c_2)} \begin{pmatrix} b_1 + b_2 \\ ib_1 - ib_2 \end{pmatrix} \cdot \begin{pmatrix} a + c_1 + c_2 & -i(c_2 - c_1) \\ -i(c_2 - c_1) & a - c_1 - c_2 \end{pmatrix} \cdot \begin{pmatrix} b_1 + b_2 \\ ib_1 - ib_2 \end{pmatrix}$$

$$= \frac{ab_1 b_2 + c_1 b_2^2 + c_2 b_1^2}{a^2 - 4c_1 c_2}. \qquad (5.1.129)$$

After this simplification, one obtains

$$I_{\text{gauss}}(a, b_1, b_2, c_1, c_2) = \frac{1}{\sqrt{a^2 - 4c_1c_2}} \exp\left(\frac{ab_1b_2 + c_1b_2^2 + c_2b_1^2}{a^2 - 4c_1c_2}\right),$$

$$\text{Re}\{a\} \geq |c_1 + c_2^*|. \tag{5.1.130}$$

Problem 5.16 Beginning with

$$\langle x_\vartheta | \alpha \rangle = \langle x | e^{-i\vartheta A^\dagger A} | \alpha \rangle \Big|_{x=x_\vartheta}, \tag{5.1.131}$$

verify that

$$\langle x_\vartheta | \alpha \rangle = \frac{1}{\pi^{\frac{1}{4}}} e^{-\frac{1}{2}|\alpha|^2 - \frac{1}{2}(\alpha e^{-i\vartheta})^2 - \frac{1}{2}x^2 + \sqrt{2}\alpha e^{-i\vartheta}x} \tag{5.1.132}$$

in two ways: first by using Eq. (4.2.20), and second, by using Eq. (4.2.61).

For the special case where $c_1 = c_2 = 0$, the general result in Eq. (5.1.130) reduces to

$$I_{\text{gauss}}(a, b_1, b_2, 0, 0) = \frac{1}{a} e^{\frac{b_1b_2}{a}}, \tag{5.1.133}$$

an expression that is worth committing to memory.

Taking $a = 1$, $b_1 = 2\alpha^* + u^*$ and $b_2 = 2\alpha - u$, the double integral in Eq. (5.1.115) can be further simplified as follows:

$$\begin{aligned}
\rho_{\text{W}}^{\text{Fock}}(\alpha, \alpha^*; n) &= \frac{2\,e^{-2|\alpha|^2}}{n!} \int \frac{(du)}{\pi} \left(-|u|^2\right)^n e^{(2\alpha^* + u^*)(2\alpha - u)} \\
&= \frac{2\,e^{2|\alpha|^2}}{n!} \int \frac{(du)}{\pi} \left(-|u|^2\right)^n e^{-|u|^2 - 2\alpha^* u + 2\alpha u^*} \\
&= \frac{2\,e^{2|\alpha|^2}}{n!} \left|\frac{1}{2}\frac{\partial}{\partial\alpha}\right|^{2n} \int \frac{(du)}{\pi} e^{-|u|^2 - 2\alpha^* u + 2\alpha u^*} \\
&= \frac{2\,e^{2|\alpha|^2}}{n!} \left|\frac{1}{2}\frac{\partial}{\partial\alpha}\right|^{2n} e^{-4|\alpha|^2} \\
&= \frac{2\,(-1)^n e^{2|\alpha|^2}}{n!} \left(\frac{\partial}{\partial\alpha}\right)^n \left(\alpha^n e^{-4|\alpha|^2}\right), \tag{5.1.134}
\end{aligned}$$

where Eq. (5.1.133) is made use of the second time. From here, we can introduce a new variable $t = 4|\alpha|^2$, turning the partial derivative with respect to α into

$$\frac{\partial}{\partial\alpha} = 4\alpha^* \frac{d}{dt}, \qquad (5.1.135)$$

such that Eq. (5.1.134) is cast into the form

$$\frac{2(-1)^n e^{2|\alpha|^2}}{n!} \left(\frac{\partial}{\partial\alpha}\right)^n \left(\alpha^n e^{-4|\alpha|^2}\right)$$
$$= \frac{2(-1)^n e^{2|\alpha|^2}}{n!} \left(\frac{d}{dt}\right)^n \left(t^n e^{-t}\right)\Big|_{t=4|\alpha|^2}$$
$$= 2(-1)^n e^{-2|\alpha|^2} L_n\left(4|\alpha|^2\right), \qquad (5.1.136)$$

which is indeed the result that was initially obtained as Eq. (5.1.71).

Problem 5.17 Take a different route starting from Eq. (5.1.113) and use the binomial expansion

$$\left[\left(\frac{\partial}{\partial x}\right)^2 + \left(\frac{\partial}{\partial p}\right)^2\right]^n = \sum_{k=0}^n \binom{n}{k} \left(\frac{\partial}{\partial x}\right)^{2k} \left(\frac{\partial}{\partial p}\right)^{2n-2k} \qquad (5.1.137)$$

to express $\rho_W^{\text{Fock}}(\alpha, \alpha^*; n)$ in Eq. (5.1.112) as a sum involving Hermite polynomials. Thereafter, by comparing with Eq. (5.1.71), show that

$$\left(-\frac{1}{4}\right)^n \frac{1}{n!} \sum_{k=0}^n \binom{n}{k} H_{2k}(x) H_{2n-2k}(y) = L_n\left(x^2 + y^2\right) \qquad (5.1.138)$$

for any x and y.

5.1.3 *Husimi Q function*

There is a third quasi-probability distribution function that more closely resembles a proper probability distribution. This function is known as the *Husimi* Q function* defined as

$$\rho_Q(\alpha, \alpha^*) = \langle\alpha^*|\rho|\alpha\rangle, \qquad (5.1.139)$$

---
*Kôdi Husimi (1909–2008).

where $|\alpha\rangle \langle \alpha^*|$ may be regarded as the corresponding Q *operator*. This is clearly a positive function of α that is bounded from above by one for all statistical operators ρ.

A relation between the Q function and the P function, which is of the same type as Eq. (5.1.112), follows immediately from the definition of $\rho_P(\alpha, \alpha^*)$ in Eq. (5.1.78),

$$\rho_Q(\alpha, \alpha^*) = \langle \alpha^* | \rho | \alpha \rangle$$

$$= \int \frac{(\mathrm{d}\alpha')}{\pi} \rho_P(\alpha', \alpha'^*) |\langle \alpha^* | \alpha' \rangle|^2, \qquad (5.1.140)$$

which, after an application of Eq. (4.2.40), gives

$$\rho_Q(\alpha, \alpha^*) = \int \frac{(\mathrm{d}\alpha')}{\pi} \rho_P(\alpha', \alpha'^*) \, e^{-|\alpha - \alpha'|^2}. \qquad (5.1.141)$$

From Eqs. (5.1.112) and (5.1.141), it is certainly natural to think that the Q function should also be related to the Wigner function by an integral of this type, which it is. This third relation can in fact be found by using the property of the Wigner function stated as Eq. (5.1.25), implying that

$$\rho_Q(\alpha, \alpha^*) = \langle \alpha^* | \rho | \alpha \rangle$$

$$= \int \frac{(\mathrm{d}\alpha')}{\pi} \rho_W(\alpha', \alpha'^*) \, \rho_W^{\mathrm{coh}}(\alpha, \alpha^*, \alpha', \alpha'^*), \qquad (5.1.142)$$

so that

$$\rho_Q(\alpha, \alpha^*) = 2 \int \frac{(\mathrm{d}\alpha')}{\pi} \rho_W(\alpha', \alpha'^*) \, e^{-2|\alpha - \alpha'|^2}. \qquad (5.1.143)$$

5.2 Connections among quasi-probability distribution functions

5.2.1 *Formalism*

We have investigated the three distribution functions $\rho_W(\alpha, \alpha^*)$, $\rho_P(\alpha, \alpha^*)$ and $\rho_Q(\alpha, \alpha^*)$, and found that they are related in terms of *convolution integrals* with Gaussian functions. An integral of this type is known as a

Gauss–Weierstrass* transform and is more generally defined by

$$f_{\mathrm{GW}}(\beta, \beta^*) = k \int \frac{(\mathrm{d}\alpha)}{\pi} f(\alpha, \alpha^*) e^{-k|\beta - \alpha|^2} \qquad (5.2.1)$$

for an arbitrary real function $f(\alpha, \alpha^*)$. By assigning the notation

$$f_{\mathrm{GW}}(\alpha, \alpha^*) \xleftarrow{k} f(\alpha, \alpha^*) \qquad (5.2.2)$$

to represent the above transform, there is a simple pictorial representation (refer to Fig. 5.4) that summarizes the relationships among the three distribution functions.

One way to understand these notations is to think of the transform defined in Eq. (5.2.2) as a *"smoothing operation"* on $f(\alpha, \alpha^*)$ with a Gaussian weight. The reciprocal of the parameter $k > 0$ characterizes the "intensity" of the "smoothing operation". Given a highly singular P function $\rho_{\mathrm{P}}(\alpha, \alpha^*)$, one can obtain the Wigner function $\rho_{\mathrm{W}}(\alpha, \alpha^*)$ by an application of the "smoothing operation" with $k = 2$. The resulting function $\rho_{\mathrm{W}}(\alpha, \alpha^*)$, which is regular as compared to $\rho_{\mathrm{P}}(\alpha, \alpha^*)$, may still contain negativity that can be completely removed by another application of the "smoothing operation" with $k = 2$, yielding the nonnegative Q function $\rho_{\mathrm{Q}}(\alpha, \alpha^*)$. The function $\rho_{\mathrm{P}}(\alpha, \alpha^*)$ can also be directly transformed into $\rho_{\mathrm{Q}}(\alpha, \alpha^*)$ under a more intensive "smoothing operation" with $k = 1$.

It may seem, at first sight, that once a Gauss–Weierstrass transform is carried out on $\rho_{\mathrm{W}}(\alpha, \alpha^*)$, say, to give the nonnegative $\rho_{\mathrm{Q}}(\alpha, \alpha^*)$, any information that is contained in the negative regions of $\rho_{\mathrm{W}}(\alpha, \alpha^*)$ is lost in the process. This "loss" of information is, however, only apparent, for this transform is, in fact, invertible by an "anti-smoothing operation". One can

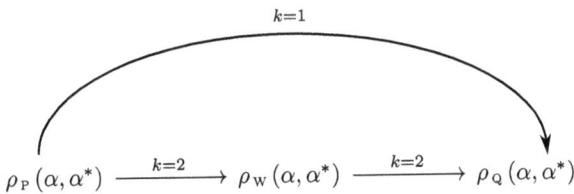

Fig. 5.4 Relationships among the Wigner, P and Q functions.

---

*Johann Carl Friedrich Gauss (1777–1855) and Karl Theodor Wilhelm Weierstrass (1815–1897).

show this by first noting that

$$f_{\mathrm{GW}}(\beta, \beta^*) = k \int \frac{(\mathrm{d}\alpha)}{\pi} f(\alpha) e^{-k|\beta - \alpha|^2} \qquad (\alpha' = \beta - \alpha)$$

$$= k \int \frac{(\mathrm{d}\alpha')}{\pi} f(\beta - \alpha', \beta^* - \alpha'^*) e^{-k|\alpha'|^2}$$

$$= k \int \frac{(\mathrm{d}\alpha')}{\pi} e^{-k|\alpha'|^2} e^{-\alpha' \frac{\partial}{\partial \beta} - \alpha'^* \frac{\partial}{\partial \beta^*}} f(\beta, \beta^*)$$

$$= e^{\frac{1}{k} \left| \frac{\partial}{\partial \beta} \right|^2} f(\beta, \beta^*), \tag{5.2.3}$$

where we have recognized the action of the translation operator that is presented in Eq. (5.1.98) and made use of the result in Eq. (5.1.133). We have thus expressed the Gauss–Weierstrass integral transform as an exponential operator acting on a complex function.

Thus, the *inverse* Gauss–Weierstrass transform is consequently the inverse of Eq. (5.2.3), or

$$f(\alpha, \alpha^*) = e^{-\frac{1}{k} \left| \frac{\partial}{\partial \alpha} \right|^2} f_{\mathrm{GW}}(\alpha, \alpha^*). \tag{5.2.4}$$

Expanding Eq. (5.2.4) into a deconvolution integral transform requires nothing more than another straightforward application of Eq. (5.1.133),

$$e^{-\frac{1}{k} \left| \frac{\partial}{\partial \alpha} \right|^2} f_{\mathrm{GW}}(\alpha, \alpha^*) = k \int \frac{(\mathrm{d}\beta')}{\pi} e^{-k|\beta'|^2} e^{\mathrm{i}\beta' \frac{\partial}{\partial \alpha} + \mathrm{i}\beta'^* \frac{\partial}{\partial \alpha^*}} f_{\mathrm{GW}}(\alpha, \alpha^*),$$

$$\tag{5.2.5}$$

where we again have

$$e^{\mathrm{i}\beta' \frac{\partial}{\partial \alpha} + \mathrm{i}\beta'^* \frac{\partial}{\partial \alpha^*}} f_{\mathrm{GW}}(\alpha, \alpha^*) = f_{\mathrm{GW}}(\alpha + \mathrm{i}\beta', \alpha^* + \mathrm{i}\beta'^*), \tag{5.2.6}$$

from Eq. (5.1.98), or

$$f(\alpha, \alpha^*) = k \int \frac{(\mathrm{d}\beta')}{\pi} e^{-k|\beta'|^2} f_{\mathrm{GW}}(\alpha + \mathrm{i}\beta', \alpha^* + \mathrm{i}\beta'^*). \tag{5.2.7}$$

In terms of the real variables x' and p' that constitute $\beta' = (x' + \mathrm{i}p')/\sqrt{2}$ along with the real variables x and p for α, the inverse transform in Eq. (5.2.5) turns into

$$k \int \frac{\mathrm{d}x'\mathrm{d}p'}{2\pi}\, \mathrm{e}^{-\frac{k}{2}\left(x'^2+p'^2\right)} f_{\mathrm{GW}}(x+\mathrm{i}x', p+\mathrm{i}p'). \qquad (5.2.8)$$

We can write (5.2.8) as contour integrals by a further introduction of a pair of complex variables $s = x + \mathrm{i}x'$ and $t = p + \mathrm{i}p'$, and the final form for the inverse Gauss–Weierstrass transform is given by

$$f(\alpha,\alpha^*) = -\frac{k}{2\pi} \int\limits_{x-\mathrm{i}\infty}^{x+\mathrm{i}\infty} \mathrm{d}s \int\limits_{p-\mathrm{i}\infty}^{p+\mathrm{i}\infty} \mathrm{d}t\, \mathrm{e}^{\frac{k}{2}\left[(s-x)^2+(t-p)^2\right]} f_{\mathrm{GW}}(s,t). \qquad (5.2.9)$$

The form of the integral in Eq. (5.2.9) is analogous to that of the two-dimensional inverse Laplace* transform of $f_{\mathrm{GW}}(s,t)$, with the exponential functions replaced by Gaussian functions. The contours for this integral are infinitely long lines parallel to the imaginary axes of the respective complex planes for s and t that are positioned at some fixed values $\mathrm{Re}\,\{s\} = x$ and $\mathrm{Re}\,\{t\} = p$.

More generally, all positions for the vertical lines are equally good for the contour integration as long as $f_{\mathrm{GW}}(s,t)$ is a well-behaved function, owing to which we are free to substitute x and p in the integration limits with arbitrary real variables x_0 and p_0 as we see fit. In the end, we obtain the very general inverse Gauss–Weierstrass transform

$$f(\alpha,\alpha^*) = -\frac{k}{2\pi} \int\limits_{x_0-\mathrm{i}\infty}^{x_0+\mathrm{i}\infty} \mathrm{d}s \int\limits_{p_0-\mathrm{i}\infty}^{p_0+\mathrm{i}\infty} \mathrm{d}t\, \mathrm{e}^{\frac{k}{2}\left[(s-x)^2+(t-p)^2\right]} f_{\mathrm{GW}}(s,t), \qquad (5.2.10)$$

and this transform ultimately does not depend on the pre-chosen x_0 and p_0.

Problem 5.18 Show that $f(\alpha,\alpha^*)$ in Eq. (5.2.10) is independent of the arbitrary variables x_0 and p_0 whenever $f_{\mathrm{GW}}(s,t)$ is well-defined by calculating the derivatives $\partial f(\alpha,\alpha^*)/\partial x_0$ and $\partial f(\alpha,\alpha^*)/\partial p_0$.

Problem 5.19 By a direct insertion of Eq. (5.2.1) into the right-hand side of Eq. (5.2.10), convince yourself that the latter equation does indeed give the correct inverse Gauss–Weierstrass transform.

---

*Pierre-Simon, Marquis de Laplace (1749–1827).

For our purposes, we shall assume that $f_{\mathrm{GW}}(s,t)$ is well-behaved for all $\mathrm{Re}\{s\}$ and $\mathrm{Re}\{t\}$, so that we may set $x_0 = p_0 = 0$. After the variable change $s \to is$ and $t \to it$, Eq. (5.2.10) will then take the simpler form

$$f(\alpha,\alpha^*) = k \int \frac{ds\,dt}{2\pi}\, \mathrm{e}^{-\frac{k}{2}\left[(s+ix)^2+(t+ip)^2\right]} f_{\mathrm{GW}}(is,it), \qquad (5.2.11)$$

where s and t are now real. Remember that $f_{\mathrm{GW}}(is,it)$ means $f_{\mathrm{GW}}(is,it) = f_{\mathrm{GW}}(x,p)\big|_{x=is,p=it}$. In short, the expressions in Eqs. (5.2.7) and (5.2.11) are equivalent.

5.2.2 *Applications*

Let us demonstrate that Eq. (5.2.11) indeed gives the correct inverse transform by using it to obtain the Wigner function of the coherent state $|\alpha'\rangle\,\langle\alpha'^*|$ starting with its Q function

$$\rho_{\mathrm{Q}}^{\mathrm{coh}}(\alpha,\alpha^*;\alpha',\alpha'^*) = |\langle\alpha^*|\alpha'\rangle|^2 = \mathrm{e}^{-|\alpha-\alpha'|^2}. \qquad (5.2.12)$$

Taking $k = 2$ and

$$\begin{aligned}
f_{\mathrm{GW}}^{\mathrm{coh}}(is,it) &= \rho_{\mathrm{Q}}^{\mathrm{coh}}(x,p;x',p')\big|_{x=is,p=it} \\
&= \mathrm{e}^{-\frac{1}{2}(x-x')^2-\frac{1}{2}(p-p')^2}\bigg|_{x=is,p=it} \\
&= \mathrm{e}^{\frac{1}{2}\left[(s+ix')^2+(t+ip')^2\right]},
\end{aligned} \qquad (5.2.13)$$

the inverse Gauss–Weierstrass transform of $f_{\mathrm{GW}}^{\mathrm{coh}}(is,it)$ is

$$\begin{aligned}
f^{\mathrm{coh}}(\alpha,\alpha^*) &= \int \frac{ds\,dt}{\pi}\, \mathrm{e}^{-\left[(s+ix)^2+(t+ip)^2\right]} \mathrm{e}^{\frac{1}{2}\left[(s+ix')^2+(t+ip')^2\right]} \\
&= \mathrm{e}^{x^2+p^2}\,\mathrm{e}^{-\frac{1}{2}(x'^2+p'^2)} \int \frac{ds\,dt}{\pi}\, \mathrm{e}^{-\frac{1}{2}s^2+is(x'-2x)}\,\mathrm{e}^{-\frac{1}{2}t^2+it(p'-2p)} \\
&= 2\,\mathrm{e}^{x^2+p^2}\,\mathrm{e}^{-\frac{1}{2}(x'^2+p'^2)}\,\mathrm{e}^{-\frac{1}{2}(x'-2x)^2}\,\mathrm{e}^{-\frac{1}{2}(p'-2p)^2} \\
&= 2\,\mathrm{e}^{-(x-x')^2-(p-p')^2} \\
&= 2\,\mathrm{e}^{-2|\alpha-\alpha'|^2} = \rho_{\mathrm{W}}(\alpha,\alpha^*),
\end{aligned} \qquad (5.2.14)$$

as we would expect. The P function $\rho_{\mathrm{P}}^{\mathrm{coh}}(\alpha, \alpha^*)$ can also be recovered by setting $k = 1$, giving us

$$
f^{\mathrm{coh}}(\alpha, \alpha^*) = \int \frac{ds\,dt}{2\pi}\, e^{-\frac{1}{2}\left[(s+ix)^2+(t+ip)^2\right]}\, e^{\frac{1}{2}\left[(s+ix')^2+(t+ip')^2\right]}
$$

$$
= e^{\frac{1}{2}\left(x^2+p^2\right)}\, e^{-\frac{1}{2}\left(x'^2+p'^2\right)} \int \frac{ds\,dt}{2\pi}\, e^{is(x'-x)}\, e^{it(p'-p)}
$$

$$
= 2\pi\,\delta(x-x')\,\delta(p-p')
$$

$$
= \pi\,\boldsymbol{\delta}(\alpha - \alpha') = \rho_{\mathrm{P}}^{\mathrm{coh}}(\alpha, \alpha^*; \alpha', \alpha'^*). \tag{5.2.15}
$$

It is also straightforward to obtain the right expressions with Eq. (5.2.7) and

$$
f_{\mathrm{GW}}^{\mathrm{coh}}(\alpha + i\beta', \alpha^* + i\beta'^*) = \rho_{\mathrm{Q}}^{\mathrm{coh}}(\alpha + i\beta', \alpha^* + i\beta'^*)
$$

$$
= e^{-(\alpha^*+i\beta'^*-\alpha'^*)(\alpha+i\beta'-\alpha')}
$$

$$
= e^{-|\alpha-\alpha'|^2}\, e^{|\beta'|^2 - i(\alpha^*-\alpha'^*)\beta' - i(\alpha-\alpha')\beta'^*}. \tag{5.2.16}
$$

When $k = 2$, we have

$$
f^{\mathrm{coh}}(\alpha, \alpha^*) = 2\,e^{-|\alpha-\alpha'|^2} \int \frac{(d\beta')}{\pi}\, e^{-|\beta'|^2 - i(\alpha^*-\alpha'^*)\beta' - i(\alpha-\alpha')\beta'^*}
$$

$$
= 2\,e^{-2|\alpha-\alpha'|^2}. \tag{5.2.17}
$$

On the other hand, $k = 1$ readily gives

$$
f^{\mathrm{coh}}(\alpha, \alpha^*) = e^{-|\alpha-\alpha'|^2} \int \frac{(d\beta')}{\pi}\, e^{-i(\alpha^*-\alpha'^*)\beta' - i(\alpha-\alpha')\beta'^*}
$$

$$
= \pi\,\boldsymbol{\delta}(\alpha - \alpha'). \tag{5.2.18}
$$

Now that we have all the necessary tools, we can show that the apparent loss of information after a Gauss–Weierstrass transform of a distribution function is completely recoverable, and any nonpositive features that were initially present in that distribution function before the transform can be fully restored. For this, we consider, say, the odd/even coherent states

$|\alpha'_{\pm}\rangle \langle \alpha'^{*}_{\pm}|$. We start with its Q function

$$\rho_Q^{o/e}(\alpha, \alpha^*; \alpha', \alpha'^*)$$

$$= |\langle \alpha^* | \alpha'_{\pm} \rangle|^2$$

$$= \mathcal{N}^2 \left| e^{-\frac{1}{2}|\alpha|^2 - \frac{1}{2}|\alpha'|^2 + \alpha^* \alpha'} \pm e^{-\frac{1}{2}|\alpha|^2 - \frac{1}{2}|\alpha'|^2 - \alpha^* \alpha'} \right|^2$$

$$= \mathcal{N}^2 e^{-|\alpha|^2 - |\alpha'|^2} \left| e^{\alpha^* \alpha'} \pm e^{-\alpha^* \alpha'} \right|^2$$

$$= \mathcal{N}^2 \left(e^{-|\alpha - \alpha'|^2} + e^{-|\alpha + \alpha'|^2} \pm 2 e^{-|\alpha|^2 - |\alpha'|^2} \mathrm{Re} \left\{ e^{2i \, \mathrm{Im}\{\alpha^* \alpha'\}} \right\} \right).$$

$$(5.2.19)$$

For this state, it is clearly expedient to work with Eq. (5.2.7).

An inverse Gauss–Weierstrass transform with $k = 2$ on the Q function would, therefore, result in four terms. The first two contributions from the Q function give the usual terms $2\mathcal{N}^2 e^{-2|\alpha - \alpha'|^2}$ and $2\mathcal{N}^2 e^{-2|\alpha + \alpha'|^2}$. One of the interference terms leads to

$$f_{\mathrm{GW}}^{\mathrm{interference}}(\alpha + i\beta', \alpha^* + i\beta'^*)$$

$$= e^{-(\alpha^* + i\beta'^*)(\alpha + i\beta') - |\alpha'|^2} e^{(\alpha^* + i\beta'^*)\alpha' - (\alpha + i\beta')\alpha'^*}$$

$$= e^{-|\alpha|^2 - |\alpha'|^2} e^{|\beta'|^2 - i(\alpha^* + \alpha'^*)\beta' - i(\alpha - \alpha')\beta'^*} e^{2i \, \mathrm{Im}\{\alpha^* \alpha'\}}, \quad (5.2.20)$$

so that an integration of this term in phase space gives

$$2 \int \frac{(\mathrm{d}\beta')}{\pi} e^{-2|\beta'|^2} f_{\mathrm{GW}}^{\mathrm{interference}}(\alpha + i\beta', \alpha^* + i\beta'^*)$$

$$= 2 e^{-|\alpha|^2 - |\alpha'|^2} e^{2i \, \mathrm{Im}\{\alpha^* \alpha'\}} \int \frac{(\mathrm{d}\beta')}{\pi} e^{-|\beta'|^2 - i(\alpha^* + \alpha'^*)\beta' - i(\alpha - \alpha')\beta'^*}$$

$$= 2 e^{-2|\alpha|^2} e^{4i \, \mathrm{Im}\{\alpha^* \alpha'\}}.$$

$$(5.2.21)$$

The other interference term is just the complex conjugate of this one. Collecting all four terms together,

$$f^{o/e}(\alpha, \alpha^*)$$

$$= 2\mathcal{N}^2 \left[e^{-2|\alpha - \alpha'|^2} + e^{-2|\alpha + \alpha'|^2} \pm e^{-2|\alpha|^2} \left(e^{4i \, \mathrm{Im}\{\alpha^* \alpha'\}} + e^{-4i \, \mathrm{Im}\{\alpha^* \alpha'\}} \right) \right]$$

$$= 2\mathcal{N}^2 \left[e^{-2|\alpha - \alpha'|^2} + e^{-2|\alpha + \alpha'|^2} \pm 2 e^{-2|\alpha|^2} \cos\left(4 \, \mathrm{Im}\{\alpha^* \alpha'\}\right) \right]$$

$$= \rho_W^{o/e}(\alpha, \alpha^*; \alpha', \alpha'^*).$$

$$(5.2.22)$$

We can play the same game and set $k = 1$ to obtain the highly singular P function directly from the Q function. Again, the first two terms will give us the familiar P functions $\pi\,\delta(\alpha - \alpha')$ and $\pi\,\delta(\alpha + \alpha')$ of the two coherent states $|\alpha'\rangle\langle\alpha'^*|$ and $|-\alpha'\rangle\langle-\alpha'^*|$. Singularity arises from the interference terms, where one of them supplies

$$\int \frac{(\mathrm{d}\beta')}{\pi}\,\mathrm{e}^{-|\beta'|^2}\,f_{\mathrm{GW}}^{\mathrm{interference}}(\alpha + \mathrm{i}\beta', \alpha^* + \mathrm{i}\beta'^*)$$

$$= \mathrm{e}^{-|\alpha|^2 - |\alpha'|^2}\,\mathrm{e}^{2\mathrm{i}\,\mathrm{Im}\{\alpha^*\alpha'\}}\underbrace{\int \frac{(\mathrm{d}\beta')}{\pi}\,\mathrm{e}^{-\mathrm{i}(\alpha^* + \alpha'^*)\beta' - \mathrm{i}(\alpha - \alpha')\beta'^*}}_{= \pi\,\delta(\alpha - \alpha', \alpha^* + \alpha'^*)}$$

$$= \pi\,\mathrm{e}^{-|\alpha|^2 - \alpha(-\alpha^*)}\,\mathrm{e}^{\alpha^*\alpha - \alpha(-\alpha^*)}\delta(\alpha - \alpha', \alpha^* + \alpha'^*)$$

$$= \pi\,\mathrm{e}^{2|\alpha|^2}\,\delta(\alpha - \alpha', \alpha^* + \alpha'^*). \tag{5.2.23}$$

The second-last equality is a consequence of the properties listed in Eq. (5.1.106).

Altogether, we find that

$$f^{\mathrm{o/e}}(\alpha, \alpha^*) = \pi\mathcal{N}^2\Big\{\delta(\alpha - \alpha') + \delta(\alpha + \alpha')$$

$$\pm\,\mathrm{e}^{2|\alpha|^2}\,[\delta(\alpha - \alpha', \alpha^* + \alpha'^*) + \delta(\alpha + \alpha', \alpha^* - \alpha'^*)]\Big\} \tag{5.2.24}$$

is indeed $\rho_{\mathrm{P}}^{\mathrm{o/e}}(\alpha, \alpha^*; \alpha', \alpha'^*)$ since

$$\mathrm{e}^{|\alpha|^2 - |\alpha'|^2}\,\delta(\alpha - \alpha', \alpha^* + \alpha'^*) = \mathrm{e}^{|\alpha|^2 - \alpha(-\alpha^*)}\,\delta(\alpha - \alpha', \alpha^* + \alpha'^*)$$

$$= \mathrm{e}^{2|\alpha|^2}\,\delta(\alpha - \alpha', \alpha^* + \alpha'^*). \tag{5.2.25}$$

One can also set $k = 2$ and obtain the P function $\rho_{\mathrm{P}}^{\mathrm{o/e}}(\alpha, \alpha^*; \alpha', \alpha'^*)$ from the Wigner function $\rho_{\mathrm{W}}^{\mathrm{o/e}}(\alpha, \alpha^*; \alpha', \alpha'^*)$. The relevant Gauss–Weierstrass transform for one of the interference terms is given by

$$f_{\mathrm{GW}}^{\mathrm{interference}}(\alpha + \mathrm{i}\beta', \alpha^* + \mathrm{i}\beta'^*)$$

$$= 2\,\mathrm{e}^{-2(\alpha^* + \mathrm{i}\beta'^*)(\alpha + \mathrm{i}\beta')}\,\mathrm{e}^{2(\alpha^* + \mathrm{i}\beta'^*)\alpha' - 2(\alpha + \mathrm{i}\beta')\alpha'^*}$$

$$= 2\,\mathrm{e}^{-2|\alpha|^2 + 4\mathrm{i}\,\mathrm{Im}\{\alpha^*\alpha'\}}\,\mathrm{e}^{2|\beta'|^2 - 2\mathrm{i}(\alpha^* + \alpha'^*)\beta' - 2\mathrm{i}(\alpha - \alpha')\beta'^*}. \tag{5.2.26}$$

Its inverse transform reads

$$2 \int \frac{(\mathrm{d}\beta')}{\pi} \, \mathrm{e}^{-2|\beta'|^2} \, f_{\mathrm{GW}}^{\mathrm{interference}}(\alpha + \mathrm{i}\beta', \alpha^* + \mathrm{i}\beta'^*)$$

$$= 4\,\mathrm{e}^{-2|\alpha|^2 + 4\mathrm{i}\,\mathrm{Im}\{\alpha^*\alpha'\}} \underbrace{\int \frac{(\mathrm{d}\beta')}{\pi} \, \mathrm{e}^{-2\mathrm{i}(\alpha^* + \alpha'^*)\beta' - 2\mathrm{i}(\alpha - \alpha')\beta'^*}}_{= \frac{\pi}{4}\,\delta(\alpha - \alpha', \alpha^* + \alpha'^*)}$$

$$= \pi\,\mathrm{e}^{2|\alpha|^2}\,\delta(\alpha - \alpha', \alpha^* + \alpha'^*), \tag{5.2.27}$$

which is consistent with Eqs. (5.2.24) and (5.1.100).

The manipulation of the improper functions in Eqs. (5.2.25) and (5.2.27) is very similar to that of delta functions. More generally, any two improper functions are equal when their products with an infinitely differentiable (or smooth) *test function* $f_{\mathrm{t}}(\alpha, \alpha^*)$, which has rapidly decreasing derivatives of all orders with $|\alpha|^2$, give the same integral over the whole complex plane for *all* such test functions. In other words, $\rho_{\mathrm{P}}(\alpha, \alpha^*)$ and $\rho'_{\mathrm{P}}(\alpha, \alpha^*)$ are equal if

$$\int \frac{(\mathrm{d}\alpha)}{\pi} \, \rho_{\mathrm{P}}(\alpha, \alpha^*)\, f_{\mathrm{t}}(\alpha, \alpha^*) = \int \frac{(\mathrm{d}\alpha)}{\pi} \, \rho'_{\mathrm{P}}(\alpha, \alpha^*)\, f_{\mathrm{t}}(\alpha, \alpha^*) \tag{5.2.28}$$

for any $f_{\mathrm{t}}(\alpha, \alpha^*)$, and only then.

We now convince the reader that the manipulation in Eqs. (5.2.25) and (5.2.27) is valid by showing explicitly that Eqs. (5.2.24) and (5.1.100) are equal in the sense that both distribution functions represent the *same* quantum state $|\alpha'_{\pm}\rangle\langle\alpha'^*_{\pm}|$. It is enough to demonstrate this for the interference terms. From the first interference term in Eq. (5.1.100), neglecting all irrelevant prefactors for the moment, one may inspect

$$F(\alpha', \alpha'^*) = \int \frac{(\mathrm{d}\alpha)}{\pi} \, |\alpha\rangle \left[\pi\, \mathrm{e}^{|\alpha|^2 - |\alpha'|^2}\, \delta(\alpha - \alpha', \alpha^* + \alpha'^*)\right] \langle\alpha^*|$$

$$= \pi\,\mathrm{e}^{-|\alpha'|^2} \int \frac{(\mathrm{d}\alpha)}{\pi} \, |\alpha\rangle\, \mathrm{e}^{|\alpha|^2} \left[\int \frac{(\mathrm{d}u)}{\pi^2} \, \mathrm{e}^{(\alpha - \alpha')u^* - (\alpha^* + \alpha'^*)u}\right] \langle\alpha^*|$$

$$= \pi\,\mathrm{e}^{-|\alpha'|^2} \int \frac{(\mathrm{d}u)}{\pi^2} \, \mathrm{e}^{-\alpha' u^* - \alpha'^* u} \int \frac{(\mathrm{d}\alpha)}{\pi} \, |\alpha\rangle\, \mathrm{e}^{|\alpha|^2 + \alpha u^* - \alpha^* u} \, \langle\alpha^*|.$$

$$\tag{5.2.29}$$

It is more convenient to obtain the matrix elements of $F(\alpha', \alpha'^*)$ in the coherent-state basis $\{|\beta\rangle \langle \beta^*|\}$,

$$
\langle \beta^* | F(\alpha', \alpha'^*) | \beta' \rangle
$$

$$
= \pi \, e^{-\frac{1}{2}|\beta|^2 - \frac{1}{2}|\beta'|^2 - |\alpha'|^2} \int \frac{(du)}{\pi^2} \, e^{-\alpha'u^* - \alpha'^* u} \underbrace{\int \frac{(d\alpha)}{\pi} \, e^{(\beta^* + u^*)\alpha + (\beta' - u)\alpha^*}}_{= \pi \delta(u - \beta', u^* + \beta^*)}
$$

$$
= e^{-\frac{1}{2}|\beta|^2 - \frac{1}{2}|\beta'|^2 - |\alpha'|^2} \, e^{\alpha'\beta^* - \alpha'^*\beta'}, \tag{5.2.30}
$$

thus confirming the expectation that

$$
\langle \beta^* | F(\alpha', \alpha'^*) | \beta' \rangle = \langle \beta^* | \alpha' \rangle \langle -\alpha'^* | \beta' \rangle, \tag{5.2.31}
$$

of course.

Beginning with the first interference term in Eq. (5.2.24) this time, one may look at

$$
F'(\alpha', \alpha'^*) = \int \frac{(d\alpha)}{\pi} \, |\alpha\rangle \left[\pi \, e^{2|\alpha|^2} \, \delta(\alpha - \alpha', \alpha^* + \alpha'^*) \right] \langle \alpha^* |
$$

$$
= \pi \int \frac{(du)}{\pi^2} \, e^{-\alpha'u^* - \alpha'^* u} \int \frac{(d\alpha)}{\pi} \, |\alpha\rangle \, e^{2|\alpha|^2 + \alpha u^* - \alpha^* u} \langle \alpha^* |. \tag{5.2.32}
$$

with its matrix elements defined as

$$
\langle \beta^* | F'(\alpha', \alpha'^*) | \beta' \rangle
$$

$$
= \pi \, e^{-\frac{1}{2}|\beta|^2 - \frac{1}{2}|\beta'|^2} \int \frac{(du)}{\pi^2} \, e^{-\alpha'u^* - \alpha'^* u} \int \frac{(d\alpha)}{\pi} \, e^{|\alpha|^2 + (\beta^* + u^*)\alpha + (\beta' - u)\alpha^*}. \tag{5.2.33}
$$

We point out that although the Gaussian integral with respect to α diverges in Eq. (5.2.33), the function $\langle \beta^* | F(\alpha', \alpha'^*) | \beta' \rangle$ is still well-defined. To illustrate this, we express the divergent Gaussian integral

$$
\int \frac{(d\alpha)}{\pi} \, e^{|\alpha|^2 + (\beta^* + u^*)\alpha + (\beta' - u)\alpha^*}
$$

$$
= \sum_{k=0}^{\infty} \frac{1}{k!} \int \frac{(d\alpha)}{\pi} \, |\alpha|^{2k} \, e^{(\beta^* + u^*)\alpha + (\beta' - u)\alpha^*}
$$

$$= \sum_{k=0}^{\infty} \frac{1}{k!} \left(\frac{\partial}{\partial \beta^*}\right)^k \left(\frac{\partial}{\partial \beta'}\right)^k \int \frac{(\mathrm{d}\alpha)}{\pi} \, \mathrm{e}^{(\beta^* + u^*)\alpha + (\beta' - u)\alpha^*}$$

$$= \pi \, \mathrm{e}^{\frac{\partial}{\partial \beta^*} \frac{\partial}{\partial \beta'}} \, \delta(u - \beta', u^* + \beta^*) \tag{5.2.34}$$

in the form of an infinite series of derivatives of improper functions that would directly lead to

$$\langle \beta^* | F'(\alpha', \alpha'^*) | \beta' \rangle = \mathrm{e}^{-\frac{1}{2}|\beta|^2 - \frac{1}{2}|\beta'|^2} \underbrace{\mathrm{e}^{\frac{\partial}{\partial \beta^*} \frac{\partial}{\partial \beta'}} \, \mathrm{e}^{\alpha' \beta^* - \alpha'^* \beta'}}_{= \, \mathrm{e}^{-|\alpha'|^2 + \alpha' \beta^* - \alpha'^* \beta'}}$$

$$= \langle \beta^* | \alpha' \rangle \langle -\alpha'^* | \beta' \rangle. \tag{5.2.35}$$

Therefore, $F(\alpha', \alpha'^*) = F'(\alpha', \alpha'^*)$. One may carry out the same set of calculations for the second interference term in each of the two distribution functions to find that Eq. (5.1.100) and Eq. (5.2.24) indeed represent the same quantum state.

   This little expedition has taught us that all the three distribution functions $\rho_{\mathrm{P}}(\alpha, \alpha^*)$, $\rho_{\mathrm{W}}(\alpha, \alpha^*)$ and $\rho_{\mathrm{Q}}(\alpha, \alpha^*)$ contain exactly the *same* information about the source. They are simply different representations of the same measurement data. Given one distribution function, a different distribution function can be constructed with either a Gauss–Weierstrass transform or its inverse. The roles of these functions are very similar to those of the position wave function $\psi(x)$ and the momentum wave function $\psi(p)$ that describe a physical system, with $\psi(x)$ and $\psi(p)$ related to each other by a Fourier transform and each wave function containing no more information than the other. This important observation suggests the existence of an underlying framework that ties all these distribution functions together in a coherent manner, which we shall analyze in the next section.

Problem 5.20   From the Q function for the Fock state $|n\rangle \langle n|$, obtain Eqs. (5.1.71) and (5.1.113) using the inverse Gauss–Weierstrass transform in Eq. (5.2.11).

Problem 5.21   Derive the P function

$$\rho_{\mathrm{P}}^{\mathrm{Fock}}(\alpha, \alpha^*; n) = (-1)^n \, \pi \, e^{2|\alpha|^2} \, \mathrm{L}_n \left(\left| \frac{\partial}{\partial \alpha} \right|^2 \right) \delta(\alpha) \qquad (5.2.36)$$

using the inverse Gauss–Weierstrass transform in Eq. (5.2.11) on the Wigner function and show that this distribution function is equal to that in Eq. (5.1.113), in the sense that both functions represent the state $|n\rangle \langle n|$.

5.3   An equivalence theorem and generalized quasi-probability distributions

To begin, let us recall a few common ways to write an operator function $F(A, A^\dagger)$ of A and $A^\dagger$. We have already come across a particularly convenient way of ordering the ladder operators in $F(A, A^\dagger)$ in Sec. 4.2, that is putting all $A^\dagger$s on the *left* of all As. This is the normal ordering procedure, where

$$F(A, A^\dagger) = F_{\mathrm{N}}(A, A^\dagger) = \sum_{mn} f_{mn}^{(\mathrm{N})} \left[A^{\dagger m} A^n \right]_{\mathrm{N}}. \qquad (5.3.1)$$

A normally-ordered operator function can always be readily turned into complex numbers when it is sandwiched with a coherent bra $\langle \alpha^* |$ and a coherent ket $|\alpha'\rangle$. Hence, $\langle \alpha^* | F_{\mathrm{N}}(A, A^\dagger) | \alpha' \rangle$ is well-behaved as long as the operator function $F_{\mathrm{N}}(A, A^\dagger)$ is itself an entire function of A and $A^\dagger$. Therefore, under these circumstances, $F_{\mathrm{N}}(A, A^\dagger)$ always exists. There is another way to express the normal-ordered form $\left[A^{\dagger m} A^n \right]_{\mathrm{N}}$ in terms of the displacement operator $D(\alpha)$, that is

$$\left[A^{\dagger m} A^n \right]_{\mathrm{N}} = \left(\frac{\partial}{\partial \alpha} \right)^m \left(-\frac{\partial}{\partial \alpha^*} \right)^n e^{\alpha A^\dagger} e^{-\alpha^* A} \bigg|_{\alpha=0}$$

$$= \left(\frac{\partial}{\partial \alpha} \right)^m \left(-\frac{\partial}{\partial \alpha^*} \right)^n [D(\alpha)]_{\mathrm{N}} \bigg|_{\alpha=0}. \qquad (5.3.2)$$

The same function $F(A, A^\dagger)$ can also be written so that all $A^\dagger$s are now on the *right* of all As. This procedure is called *antinormal ordering* and gives

$$F(A, A^\dagger) = F_A(A, A^\dagger) = \sum_{mn} f_{mn}^{(A)} \left[A^{\dagger m} A^n \right]_A, \qquad (5.3.3)$$

where the antinormal-ordered form

$$\left[A^{\dagger m} A^n \right]_A = A^n A^{\dagger m}. \qquad (5.3.4)$$

Unlike $F_N(A, A^\dagger)$, not all operator functions $F(A, A^\dagger)$ have valid forms of $F_A(A, A^\dagger)$ when antinormally ordered, since $\langle \alpha^* | F_A(A, A^\dagger) | \alpha' \rangle$ is not always well-defined because of the derivatives that arise from ladder operators meeting the eigenkets/eigenbras of their adjoints.

Problem 5.22 What are the operators $F_N(A, A^\dagger)$ and $F_A(A, A^\dagger)$ for the operator function

$$F(A, A^\dagger) = \int \frac{(d\alpha')}{\pi} D(\alpha'), \qquad (5.3.5)$$

where $D(\alpha)$ is the displacement operator.

With the displacement operator, we can again express

$$\left[A^{\dagger m} A^n \right]_A = \left(\frac{\partial}{\partial \alpha} \right)^m \left(-\frac{\partial}{\partial \alpha^*} \right)^n e^{-\alpha^* A} e^{\alpha A^\dagger} \Big|_{\alpha=0}$$

$$= \left(\frac{\partial}{\partial \alpha} \right)^m \left(-\frac{\partial}{\partial \alpha^*} \right)^n [D(\alpha)]_A \Big|_{\alpha=0}. \qquad (5.3.6)$$

The third way of writing the operator function $F(A, A^\dagger)$ is to symmetrically order products of A and $A^\dagger$, that is, to perform Weyl ordering,

$$F(A, A^\dagger) = F_{WS}(A, A^\dagger) = \sum_{mn} f_{mn}^{(WS)} \left[A^{\dagger m} A^n \right]_{WS}. \qquad (5.3.7)$$

The Weyl-ordered form $\left[A^{\dagger m} A^n \right]_{WS}$ is inherent in the displacement operator $D(\alpha)$ and to see this, we may investigate the terms in the Taylor series

$$D(\alpha) = \sum_{l=0}^{\infty} \frac{(\alpha A^\dagger - \alpha^* A)^l}{l!}. \qquad (5.3.8)$$

Following the arguments in Sec. 4.2.3.2, we have

$$
(\alpha A^\dagger - \alpha^* A)^l = \sum_{k=0}^{l} \binom{l}{k} \alpha^{l-k} (-\alpha^*)^k \left[A^{\dagger l-k} A^k \right]_{\text{WS}},
\qquad (5.3.9)
$$

which implies that

$$
D(\alpha) = \sum_{l=0}^{\infty} \frac{1}{l!} \sum_{k=0}^{l} \binom{l}{k} \alpha^{l-k} (-\alpha^*)^k \left[A^{\dagger l-k} A^k \right]_{\text{WS}}
$$

$$
= \underbrace{\sum_{l=0}^{\infty} \sum_{k=0}^{l}} \frac{\alpha^{l-k} (-\alpha^*)^k}{(l-k)!\, k!} \left[A^{\dagger l-k} A^k \right]_{\text{WS}}
\qquad (j = l - k)
$$

$$
= \sum_{k=0}^{\infty} \sum_{l=k}^{\infty}
$$

$$
= \sum_{j=0}^{\infty} \sum_{k=0}^{\infty} \frac{\alpha^j (-\alpha^*)^k}{j!\, k!} \left[A^{\dagger j} A^k \right]_{\text{WS}}.
\qquad (5.3.10)
$$

Hence,

$$
\left(\frac{\partial}{\partial \alpha} \right)^m \left(-\frac{\partial}{\partial \alpha^*} \right)^n D(\alpha) \bigg|_{\alpha=0}
$$

$$
= \sum_{j=0}^{\infty} \sum_{k=0}^{\infty} \frac{\left[A^{\dagger j} A^k \right]_{\text{WS}}}{j!\, k!} \left[\left(\frac{\partial}{\partial \alpha} \right)^m \alpha^j \right] \left[\left(-\frac{\partial}{\partial \alpha^*} \right)^n (-\alpha^*)^k \right] \bigg|_{\alpha=0}
$$

$$
= \sum_{j=0}^{\infty} \sum_{k=0}^{\infty} \frac{(-1)^{k-n}}{j!\, k!} \left[A^{\dagger j} A^k \right]_{\text{WS}} \left[\frac{j!}{(j-m)!} \alpha^{j-m} \right] \left[\frac{k!}{(k-n)!} \alpha^{*k-n} \right] \bigg|_{\alpha=0}
$$

$$
= \sum_{j=0}^{\infty} \sum_{k=0}^{\infty} \frac{(-1)^{k-n}}{(j-m)!\,(k-n)!} \left[A^{\dagger j} A^k \right]_{\text{WS}} \delta_{j,m}\, \delta_{k,n}
$$

$$
= \left[A^{\dagger m} A^n \right]_{\text{WS}}.
\qquad (5.3.11)
$$

We are now ready to show that the three distribution functions $\rho_{\text{P}}(\alpha, \alpha^*)$, $\rho_{\text{Q}}(\alpha, \alpha^*)$ and $\rho_{\text{W}}(\alpha, \alpha^*)$ are weight functions that appear in the expectation value $\langle F(A, A^\dagger) \rangle$ of the operator function $F(A, A^\dagger)$ when it is ordered differently. Suppose $F(A, A^\dagger)$ is normally ordered. Then, the expectation value of its components $\left[A^{\dagger m} A^n\right]_N$ is given by

$$\left\langle \left[A^{\dagger m} A^n\right]_N \right\rangle = \int \frac{(d\alpha')}{\pi} \rho_P(\alpha', \alpha'^*) \langle \alpha'^* | \left[A^{\dagger m} A^n\right]_N |\alpha'\rangle$$

$$= \int \frac{(d\alpha')}{\pi} \rho_P(\alpha', \alpha'^*) \alpha'^{*m} \alpha'^n \qquad (5.3.12)$$

or

$$\langle F_N(A, A^\dagger) \rangle = \int \frac{(d\alpha')}{\pi} \rho_P(\alpha', \alpha'^*) \, F_N(\alpha', \alpha'^*), \qquad (5.3.13)$$

which states that the expectation value of the normally ordered version of $F(A, A^\dagger)$, namely $F_N(A, A^\dagger)$, is equivalent to the phase-space average of $F_N(\alpha', \alpha'^*)$, with the operators *directly replaced* by the corresponding complex numbers, weighted with the P function.

Similarly, for the antinormally ordered version, we have

$$\left\langle \left[A^{\dagger m} A^n\right]_A \right\rangle = \int \frac{(d\alpha')}{\pi} \mathrm{tr}\left\{ \rho \, A^n |\alpha'\rangle \langle \alpha'^* | A^{\dagger m} \right\}$$

$$= \int \frac{(d\alpha')}{\pi} \rho_Q(\alpha', \alpha'^*) \alpha'^{*m} \alpha'^n, \qquad (5.3.14)$$

which leads to

$$\langle F_A(A, A^\dagger) \rangle = \int \frac{(d\alpha')}{\pi} \rho_Q(\alpha', \alpha'^*) \, F_A(\alpha', \alpha'^*). \qquad (5.3.15)$$

To obtain the third equivalence statement, note that in working out **Problem 5.22**, the reader would have encountered the useful relation

$$2(-1)^{A^\dagger A} = \int \frac{(d\alpha')}{\pi} D(\alpha') \qquad (5.3.16)$$

between the parity operator and the displacement operator. Referring to Eqs. (4.2.75) and (5.1.23), we have

$$\rho_W(\alpha, \alpha^*) = \int \frac{(d\alpha')}{\pi} e^{\alpha \alpha'^* - \alpha^* \alpha'} \, \mathrm{tr}\{\rho \, D(\alpha')\}, \qquad (5.3.17)$$

which implies that $\rho_{\rm W}(\alpha, \alpha^*)$ and $\langle D(\alpha')\rangle = \text{tr}\{\rho\, D(\alpha')\}$ are Fourier transforms of each other. Equivalently, the relation

$$D(\alpha') = \int \frac{(d\alpha)}{\pi}\, e^{\alpha'\alpha^* - \alpha'^*\alpha}\, W_\alpha \qquad (5.3.18)$$

states that the displacement operator is the Fourier transform of the Wigner operator.

With Eq. (5.3.17), the third equivalence statement is established with the Weyl-ordered version of $F(A, A^\dagger)$ inasmuch as

$$\left\langle \left[A^{\dagger m} A^n \right]_{\rm ws} \right\rangle = \left(\frac{\partial}{\partial\alpha} \right)^m \left(-\frac{\partial}{\partial\alpha^*} \right)^n \text{tr}\{\rho\, D(\alpha)\}\bigg|_{\alpha=0}$$

$$= \int \frac{(d\alpha')}{\pi}\, \rho_{\rm W}(\alpha', \alpha'^*) \left(\frac{\partial}{\partial\alpha} \right)^m \left(-\frac{\partial}{\partial\alpha^*} \right)^n e^{\alpha\alpha'^* - \alpha^*\alpha'}\bigg|_{\alpha=0}$$

$$= \int \frac{(d\alpha')}{\pi}\, \rho_{\rm W}(\alpha', \alpha'^*)\, \alpha'^{*m}\alpha'^n, \qquad (5.3.19)$$

giving us

$$\left\langle F_{\rm ws}\left(A, A^\dagger\right) \right\rangle = \int \frac{(d\alpha')}{\pi}\, \rho_{\rm W}(\alpha', \alpha'^*)\, F_{\rm ws}(\alpha', \alpha'^*). \qquad (5.3.20)$$

These equivalence statements directly relate the quantum-mechanical expectation values to their respective phase-space averages, revealing the important roles of the distribution functions in these statements. These statements make up the *optical equivalence theorem* and are a further indication that these distribution functions contain the same amount of information about a given quantum system, for

$$\left\langle F(A, A^\dagger) \right\rangle = \int \frac{(d\alpha')}{\pi}\, \rho_{\rm P}(\alpha', \alpha'^*)\, F_{\rm N}(\alpha', \alpha'^*)$$

$$= \int \frac{(d\alpha')}{\pi}\, \rho_{\rm Q}(\alpha', \alpha'^*)\, F_{\rm A}(\alpha', \alpha'^*)$$

$$= \int \frac{(d\alpha')}{\pi}\, \rho_{\rm W}(\alpha', \alpha'^*)\, F_{\rm ws}(\alpha', \alpha'^*), \qquad (5.3.21)$$

where we emphasize that the different numerical functions $F_{\rm N}(\alpha', \alpha'^*)$, $F_{\rm A}(\alpha', \alpha'^*)$ and $F_{\rm ws}(\alpha', \alpha'^*)$ are derived from the one and the same operator function $F(A, A^\dagger)$.

This motivates us to establish a *generalized* quasi-probability distribution function that encompasses all the three distribution functions that we have discussed. In the same structure as Eq. (5.3.17) for the Wigner function, we may also express $\langle D(u) \rangle$ in terms of the P function. This can be done with ease by referring to the first equation of Eq. (5.3.21) and taking $F(A, A^\dagger) = D(\alpha) = e^{\alpha A^\dagger - \alpha^* A}$. Then,

$$F_{\mathrm{N}}(\alpha', \alpha'^*) = e^{-\frac{1}{2}|\alpha|^2} \, e^{\alpha A^\dagger - \alpha^* A} \Bigg|_{\substack{A \longrightarrow \alpha' \\ A^\dagger \longrightarrow \alpha'^*}}$$

$$= e^{-\frac{1}{2}|\alpha|^2} \, e^{\alpha \alpha'^* - \alpha^* \alpha'}, \tag{5.3.22}$$

which leads to

$$\left\langle e^{\frac{1}{2}|\alpha|^2} D(\alpha) \right\rangle = \int \frac{(\mathrm{d}\alpha')}{\pi} \, e^{\alpha \alpha'^* - \alpha^* \alpha'} \, \rho_{\mathrm{P}}(\alpha', \alpha'^*). \tag{5.3.23}$$

An inverse Fourier transform produces

$$\rho_{\mathrm{P}}(\alpha, \alpha^*) = \int \frac{(\mathrm{d}u)}{\pi} \, e^{\alpha u^* - \alpha^* u} \left\langle e^{\frac{1}{2}|u|^2} D(u) \right\rangle. \tag{5.3.24}$$

Thus, the P function is the inverse Fourier transform of $e^{\frac{1}{2}|u|^2} \langle D(u) \rangle$. Lastly, from the second equation of Eq. (5.3.21), it is straightforward to verify that the Q function is the inverse Fourier transform of $e^{-\frac{1}{2}|u|^2} \langle D(u) \rangle$,

$$\rho_{\mathrm{Q}}(\alpha, \alpha^*) = \int \frac{(\mathrm{d}u)}{\pi} \, e^{\alpha u^* - \alpha^* u} \left\langle e^{-\frac{1}{2}|u|^2} D(u) \right\rangle. \tag{5.3.25}$$

Problem 5.23 Show that, for $\alpha' = |\alpha'| e^{i\phi}$,

$$\left\langle \left[A^{\dagger m} A^n \right] \right\rangle_{\mathrm{N}}$$

$$= \left(-\frac{1}{2} \right)^{n_<} n_<! \int \frac{(\mathrm{d}\alpha')}{\pi} \Bigg\{ |\alpha'|^{n_> - n_<} e^{-i(m-n)\phi}$$

$$\times L_{n_<}^{(n_> - n_<)} \left(2|\alpha'|^2 \right) \rho_{\mathrm{W}}(\alpha', \alpha'^*) \Bigg\}. \tag{5.3.26}$$

We can therefore define the generalized distribution function

$$\rho_s(\alpha) = \int \frac{(du)}{\pi} e^{\alpha u^* - \alpha^* u} \left\langle e^{\frac{s}{2}|u|^2} D(u) \right\rangle \qquad (5.3.27)$$

that incorporates the functions $\rho_Q(\alpha)$ ($s = -1$), $\rho_W(\alpha)$ ($s = 0$) and $\rho_P(\alpha)$ ($s = 1$). The variable $-1 \leq s \leq 1$ characterizes the different types of ordering for the operator function $F(A, A^\dagger)$, and this relationship is quite obvious since

$$\int \frac{(d\alpha')}{\pi} \rho_s(\alpha') \alpha'^{*m} \alpha'^n = \left(\frac{\partial}{\partial \alpha} \right)^m \left(-\frac{\partial}{\partial \alpha^*} \right)^n \langle [D(\alpha)]_s \rangle \bigg|_{\alpha=0}, \qquad (5.3.28)$$

where we have introduced the general s-*ordered form* of the displacement operator $D(\alpha)$ as

$$[D(\alpha)]_s = e^{\frac{s}{2}|\alpha|^2} D(\alpha) \bigg|_{s \text{ ordering}}. \qquad (5.3.29)$$

Problem 5.24 Demonstrate the identity

$$\left\langle \left[A^{\dagger m} e^{-A^\dagger A} A^n \right]_N \right\rangle = \left(\frac{\partial}{\partial \alpha} \right)^m \left(\frac{\partial}{\partial \alpha^*} \right)^n \left[e^{|\alpha|^2} \rho_Q(\alpha) \right] \bigg|_{\alpha=0}, \qquad (5.3.30)$$

and show explicitly that the right-hand side is indeed a phase-space average involving the corresponding P function.

The concepts of quasi-probability distributions as alternative representations of the measurement data, introduced thus far, apply to single-mode quantum states of light. These concepts can be readily generalized to multi-mode quantum states of light and data obtained from such sources may be represented with these functions.

Further reading

(1) K. E. Cahill and R. J. Glauber, Density operators and quasi-probability distributions, *Phys. Rev.* **177**, 1882 (1969).
   [The article that started the equivalence theorem of quasi-probability distributions.]
(2) B.-G. Englert, On the operator bases underlying Wigner's, Kirkwood's and Glauber's phase space, *J. Phys. A: Math. Gen.* **22**, 625 (1989).
   [A proposal to use operator ordering to formulate phase-space functions and greatly facilitate calculations.]

Hints to All Problems

Chapter 1

Problem 1.1: To ensure that $\widehat{\rho}$ always has unit trace, one may consider the paramterization $\widehat{\rho} = \frac{\widehat{\rho_0}}{\mathrm{tr}\{\widehat{\rho_0}\}}$. Thereafter, simply perform a variation on the relevant average cost function with respect to ρ_0.

Problem 1.2: No additional hints necessary.

Problem 1.3: Write the operators with kets and bras and manipulate these linear objects.

Problem 1.4: Write the operators with kets and bras and manipulate these linear objects.

Problem 1.5: Like vectors, any Hermitian operator can be written as a linear combination of trace-orthonormal basis operators.

Problem 1.6: No additional hints necessary.

Problem 1.7: Will there be any contradiction if this were really the case?

Problem 1.8: In the course of the calculation, a form for $\mathbf{Q}\mathbf{Q}^-$ may be necessary. Since we know that any plausible $\mathbf{Q}^-$ will lead to the same set of dual operators for a minimally complete POM, one convenient choice may be the symmetric form

$$\mathbf{Q}\mathbf{Q}^- \cong \mathbf{1}_{D^2} - \frac{1}{D^2} \begin{pmatrix} 1 & \cdots & 1 \\ \vdots & \ddots & \vdots \\ 1 & \cdots & 1 \end{pmatrix},$$

which is a projector that has $D^2 - 1$ nonzero eigenvalues that clearly satisfies the relation $\mathbf{Q}\mathbf{Q}^-\mathbf{Q} = \mathbf{Q}$. Indeed, this choice is always available for any informationally complete POM and to show this, **Problem 1.10** and the fact that $\mathbf{Q}$ has a one-dimensional cokernel may prove useful.

Problem 1.9: It is enough to look at the relations in Eq. (1.2.91) and start with two MP pseudo-inverses for $\mathbf{Q}$.

Problem 1.10: Remember that $\mathbf{Q}$ has $D^2 - 1$ linearly independent columns, and one can construct a full-rank square matrix out of $\mathbf{Q}$. In the end, one should find that

$$\mathbf{Q}^- = \left(\mathbf{Q}^{\mathrm{T}}\mathbf{Q}\right)^{-1}\mathbf{Q}^{\mathrm{T}}.$$

Why is this the MP pseudo-inverse?

Problem 1.11: No additional hints necessary.

Problem 1.12: No additional hints necessary.

Problem 1.13: Work out $\mathcal{F}\,|\Theta_j\rangle\!\rangle$ by writing $\mathcal{F}$ with the set of basis operators involving $\frac{1}{\sqrt{D}}$ and the traceless Ω_ks, recalling the completeness of the basis. The results in **Problem 1.10** and the fact that $\sum_j \mathbf{Q}_{jk} = 0$ should be of use.

Problem 1.14: No additional hints necessary.

Problem 1.15: For the first part of the problem, one may deduce that $y \cdot (\mathbf{\Pi} - \mathbf{G}) \cdot y$ is positive for any y just from the definitions of $\mathbf{\Pi}$ and $\mathbf{G}$. Next, by using a set of basis operators containing $D^2 - 1$ traceless operators, one should arrive at

$$\mathbf{G} = \frac{1}{D}\mathbf{\Pi}\boldsymbol{1}\boldsymbol{1}^{\mathrm{T}}\mathbf{\Pi} + \mathbf{Q}\mathbf{Q}^{\mathrm{T}},$$

where $\mathbf{1}$ is an M-dimensional column of ones. The eigenvector for the kernel of $\mathbf{\Pi} - \mathbf{G}$ should follow immediately.

Problem 1.16: Answers are in the book.

Problem 1.17: Remember that

$$\int (\mathrm{d}\mathbb{D})\, \mathcal{L}\,(\mathbb{D}; \rho_{\text{true}}) = 1.$$

Problem 1.18: One may make use of the relation

$$\overline{n_j n_k} = (N^2 - N)p_j p_k + N\delta_{j,k}p_k$$

for the number of occurrences n_j, which shall be revisited in Chapter 4. Then, the definition of $\mathbf{Q}$ should immediately lead to

$$\mathbf{F}(\rho) = N\mathbf{Q}^{\mathsf{T}}\mathbf{P}^{-1}\mathbf{Q},$$

where $\mathbf{P}$ is represented by the matrix of elements $\mathbf{P}_{jk} = \delta_{j,k}p_j$.

Problem 1.19: From the answers to **Problem 1.18**, it is probably easier to show that the quantity $\mathrm{Sp}\left\{\left(\mathbf{Q}^{\mathsf{T}}\mathbf{P}^{-1}\mathbf{Q}\right)^{-1}\right\}$ is concave in $\mathbf{P}$. For this, one can first write $\mathbf{P} = \lambda\mathbf{P}_1 + (1 - \lambda)\mathbf{P}_2$, where $0 \leq \lambda \leq 1$ and $\mathbf{P}_1$ and $\mathbf{P}_2$ are the dyadics evaluated respectively with ρ_1 and ρ_2. Next, confirm that the second-order derivative of this quantity with respect to λ is always nonpositive. The inequality

$$\mathrm{Sp}\,\{\mathbf{AB}\} \leq \|\mathbf{A}\|_2\, \mathrm{Sp}\,\{\mathbf{B}\} \text{ for any } \mathbf{A} \geq 0 \text{ and } \mathbf{B} \geq 0$$

and Eq. (3.1.28) for dyadics are helpful for this task.

Problem 1.20: In establishing the inequality in (1.3.18), one can use the inequality

$$\mathrm{tr}\{\mathcal{AB}\} \leq \mathrm{tr}\{\mathcal{A}\}\mathrm{tr}\{\mathcal{B}\} \text{ for any } \mathcal{A} \geq 0 \text{ and } \mathcal{B} \geq 0.$$

If this inequality is unfamiliar to you, derive it as an exercise. To derive the inequality in (1.3.19), first, with a clever use of the simple result $\dfrac{x^2 + y^2}{2xy} \geq 1$ for any $x > 0$ and $y > 0$, show that

$$\mathrm{Sp}\,\{\mathcal{A}\}\, \mathrm{Sp}\,\{\mathcal{A}^{-1}\} \geq \dim\{\mathcal{A}\}^2 \text{ for any } \mathcal{A} \geq 0,$$

where the equality is achieved for any $\mathcal{A}$ that is a multiple of the identity. Then, an application of (1.3.18) should do the job.

Chapter 2

Problem 2.1: Perhaps, calculate the corresponding Gram matrix.

Problem 2.2: The concepts of singular-value decompositions, introduced in Chapter 1, and eigenvalue equations are very useful.

Problem 2.3: No additional hints necessary.

Problem 2.4: Use the relations

$$a \cdot \sigma \, b \cdot \sigma = a \cdot b + \mathrm{i} a \times b \cdot \sigma$$

and

$$a \times (b \times c) = b \, a \cdot c - c \, a \cdot b.$$

If they are not familiar to you, derive them as an exercise.

Problem 2.5: Either compare the two linear constraints that go with these two measurement outcomes with Eq. (2.1.22) to directly obtain the linear transformation, or be more systematic about things and first look for the unitary transformation that rotates the standard von Neumann outcomes to these two, and then the orthogonal dyadic **O**. If the second method is preferred, recall that the unitary dyadic

$$\mathsf{U} = \sum_j b_j a_j^{\mathrm{T}}$$

transforms the orthonormal basis columns a_j into the orthonormal basis columns b_j.

Problem 2.6: This unitary matrix is $2D^2 \times 2D^2$. So, there are $2D^2$ eigenvalues with $2D^2$ orthonormal eigenvectors. As there is a clear block structure for this matrix, it is instructive to look at the 2×2 matrices

$$\mathsf{U}_1 = \frac{1}{\sqrt{2}} \begin{pmatrix} 1 & \mathrm{i} \\ 1 & -\mathrm{i} \end{pmatrix} \quad \text{and} \quad \mathsf{U}_2 = \frac{1}{\sqrt{2}} \begin{pmatrix} 1 & \mathrm{i} \\ -1 & \mathrm{i} \end{pmatrix}.$$

These matrices are related to the eigenvalues of the transposition superoperator $\mathcal{T}$. Thus, finding the eigenvalues and eigenvectors of $\mathcal{T}$ is key to solving this problem, and to do this, recall that any matrix can be decomposed into two parts, each of which are spanned by eigenvectors of T.

Problem 2.7:   Use the factorial function

$$x! = \int_0^\infty dt\, t^x\, e^{-t}.$$

Problem 2.8:   Separately consider adding and subtracting N_t by one.

Problem 2.9:   Note that

$$\log s = \int_0^\infty \frac{dt}{t} \left(e^{-t} - e^{-st} \right)$$

for positive s.

Problem 2.10: No additional hints necessary.

Problem 2.11: No additional hints necessary.

Chapter 3

Problem 3.1:   Write a positive matrix in the computational basis $\{e_j\}$ and consider its spectral decomposition to prove this statement.

Problem 3.2:   For the first part of this problem, just calculate the purity of $\rho + H$ for a pure ρ. For the second part, the two necessary conditions may come from the constraint on $\rho + H$ and its purity.

Problem 3.3:   One obvious operator to start with is the identity.

Problem 3.4:   One may follow the hints for solving **Problem 1.19** and consider a convex sum of operators. One may start with Eq. (3.1.33) for any positive operator $\mathcal{A}$.

Problem 3.5:   Recall the property of $R(\widehat{\rho}_{\mathrm{ML}})$.

Problem 3.6:   To obtain Eq. (3.1.33), confirm that

$$\log a = \int_0^\infty dt \left(\frac{1}{t+1} - \frac{1}{t+a} \right) \quad \text{for } a \geq 0$$

and make use of Eq. (3.1.28). To obtain Eq. (3.1.34), first prove the integral identity

$$\sqrt{a} = \int_0^\infty \frac{dt}{\pi\sqrt{t}}\frac{a}{a+t} \quad \text{for } a \geq 0$$

using a simple substitution and, again, apply Eq. (3.1.28). As a generalization of Eq. (3.1.34), we have the operator variation

$$\delta \mathcal{A}^r = \frac{\sin(r\pi)}{\pi}\int_0^\infty dt\, t^r \frac{1}{t+\mathcal{A}}\delta\mathcal{A}\frac{1}{t+\mathcal{A}} \quad \text{for } \mathcal{A} \geq 0 \text{ and } 0 < r < 1.$$

Problem 3.7: No additional hints necessary.

Problem 3.8: Consider the spectral decomposition of ρ for single-qubits.

Problem 3.9: Either remember the strategies for solving **Problem 2.5** and apply them to this problem, or simply minimize the purity of the general state estimator. Either way, one gets

$$s_{\text{MLME}} \,\widehat{=}\, \begin{pmatrix} \dfrac{2}{\sqrt{3}}\,(\widehat{p}_2 - \widehat{p}_3) \\[2mm] -\,(2\widehat{p}_1 - 1)\sin\vartheta \\[2mm] \dfrac{2\widehat{p}_1 - 1 + s_2\sin\vartheta}{\cos\vartheta} \end{pmatrix}.$$

Problem 3.10: There is a degree of freedom for these Lagrange multipliers.

Problem 3.11: Write a simple numerical program that performs the steepest-ascent ML algorithm for imperfect detections with the already established iterative equations. One should find that repeated runs with different starting states for the iterative algorithm always give the ML estimator

$$\widehat{\rho}_{\text{ML}} \,\widehat{=}\, \begin{pmatrix} 0.985483 & 0.0725539 + 0.0950889i \\ 0.0725539 - 0.0950889i & 0.0145167 \end{pmatrix}$$

and ML probabilities $\widehat{p}_1 = 0.0860887$, $\widehat{p}_2 = 0.492674$ and $\widehat{p}_3 = 0.421237$.

Problem 3.12: This problem describes one feasible way to carry out constrained function optimizations. One possibility is to design a barrier function that resembles a likelihood function with a plateau structure. This barrier function may involve the Heaviside step function $\eta(x)$, in which case it is useful to recall the limit

$$\frac{1}{2}\left[1 + \tanh(\kappa x)\right]\Big|_{\kappa \to \infty} = \eta(x).$$

You are advised to check numerically that the steepest-ascent algorithm correctly gives the minimum value of $f(x, y)$ within the region. The output should yield a minimum function value of about -1.066 at $x = 0.99$ and $y = 0.86$.

Chapter 4

Problem 4.1: Simply make use of the property of SIC POMs stated in Eq. (1.2.53), the results from doing **Problem 1.12**, and Eq. (1.2.94) for the expression of the LS dual operators. One would notice that both the six-outcome POM and the SIC POM offer the same tomographic accuracy for LIN estimators using these dual operators.

Problem 4.2: Recall the orthonormality relation in Eq. (1.2.49) that is valid for minimally complete POMs and write all dual superkets using a complete set of basis superkets $|\Gamma_j\rangle\rangle$, which are in turn spanned by all the POM-outcome superkets $|\Pi_j\rangle\rangle$.

Problem 4.3: Just compare the expressions in Eq. (4.1.9) and Eq. (4.1.22).

Problem 4.4: Consider the spectral decomposition of the positive dyadic **A** and write the multi-dimensional Gaussian integral in terms of multiple independent one-dimensional Gaussian integrals.

Problem 4.5: The result in Eq. (4.1.28), as well as Eq. (3.1.28) for dyadics should help. All that is left is to remember that $\bar{t} = \mu$, as well as the product rule for differentiation.

Problem 4.6: For Eq. (4.1.42), start with $\mathbf{AA}^{-1} = \mathbf{1}$. For Eq. (4.1.43), remember that a column is also a matrix. Note that in applying the chain rule and product rule for two real dyadic functions, the $m \times n$ **F(X)** and

$n \times p$ $\mathbf{G}(\mathbf{X})$, of $\mathbf{X}$, the following orders of dyadic multiplication,

$$\frac{\delta}{\delta \mathbf{X}} \mathbf{F}(\mathbf{G}(\mathbf{X})) = \mathbf{F}'(\mathbf{G}(\mathbf{X})) \frac{\delta}{\delta \mathbf{X}} \mathbf{G}(\mathbf{X}),$$

$$\frac{\delta}{\delta \mathbf{X}} [\mathbf{F}(\mathbf{X})\mathbf{G}(\mathbf{X})] = [\mathbf{1}_m \otimes \mathbf{G}(\mathbf{X})^{\mathrm{T}}] \frac{\delta}{\delta \mathbf{X}} \mathbf{F}(\mathbf{X}) + [\mathbf{F}(\mathbf{X}) \otimes \mathbf{1}_p] \frac{\delta}{\delta \mathbf{X}} \mathbf{G}(\mathbf{X})$$

are naturally respected, where the latter rule can also be deduced from Eq. (1.2.35).

Problem 4.7:   No additional hints necessary.

Problem 4.8:   No additional hints necessary.

Problem 4.9:   Remember that $A = \frac{X + \mathrm{i}P}{\sqrt{2}}$ is related to the position X and momentum P operators, and use the Gaussian integral identity in (4.1.32).

Problem 4.10: Calculate the relevant wave functions by constructing eigenvalue equations, and show that these wave functions are not normalizable.

Problem 4.11: Expand the kets with the Fock basis and look at the analyticity of the corresponding Fock-state wave functions, for which the binomial formula

$$(x + y)^n = \sum_{k=0}^{n} \binom{n}{k} x^k y^{n-k}$$

may be of use.

Problem 4.12: No additional hints necessary.

Problem 4.13: For the first part of this problem, consider the polar coordinate system for convenience. For the second part, it is probably convenient to work with the position basis and use the wave-function expression in **Problem 4.9**.

Problem 4.14: Begin with the definition

$$[F(A^\dagger A)]_N = \sum_{m=0}^{\infty} f_m A^{\dagger m} A^m$$

for the normal-ordered form of the operator function $F(A^\dagger A)$. From hereon, the Fock basis is a convenient choice to write $\left[F(A^\dagger A)\right]_N$, and the basic actions of A and $A^\dagger$ on the Fock states are the key ingredients to complete the calculations.

Problem 4.15: As always, the Fock basis is most compatible with the ladder operators A and $A^\dagger$ for such problems.

Problem 4.16: Study the characteristic function $\left\langle e^{ikI'_{\text{diff}}} \right\rangle$ by following the reasonings starting from Eq. (4.2.81) that led all the way to Eq. (4.2.97).

Problem 4.17: No additional hints necessary.

Problem 4.18: Insert the resolution of the identity

$$\int dx' \, |x'\rangle \, \langle x'| = 1$$

of position kets into $\langle p' | x_\vartheta \rangle$ and perform the integration.

Problem 4.19: No additional hints necessary.

Problem 4.20: Make use of the resolution of the identity

$$\int dx' \, |x'\rangle \, \langle x'| = 1$$

of position kets, and carefully simplify all trigonometric functions to arrive at the result for the first part of this problem. Otherwise, no additional hints are necessary for the rest of the problem.

Problem 4.21: One may start from the relation between the Hermite polynomials and their generating function in Eq. (4.2.125), and calculate the Fourier transform of the product of the generating function and $e^{-\frac{1}{2}x^2}$.

Chapter 5

Problem 5.1: The calculation is simply a straightforward manipulation of position eigenkets, and the consequences of this trace-orthogonality statement can be revealed by using the superket/superbra notation.

Problem 5.2: No additional hints necessary.

Problem 5.3:   For this problem, calculations may be greatly simplified by working in the coherent-state basis and using the relations in
Eq. (5.1.116) and Eq. (5.1.130). One should get the Gaussian function

$$\rho_W^{sq}(\alpha, \alpha'; z) = 2e^{-2\left\{\cosh(2r)|\alpha|^2 + |\alpha'|^2\right\}}$$

$$\times\, e^{4\,\mathrm{Re}\left\{\alpha\left(\alpha'^* \cosh r - \alpha' e^{-i\theta} \sinh r\right)\right\} + 2\sinh(2r)\mathrm{Re}\left\{\alpha^2 e^{-i\theta}\right\}},$$

which can be verified to be a properly normalized function by using
the relations Eq. (5.1.116) and Eq. (5.1.130), and reduces to the Wigner
function for coherent states when $z = 0$.

Problem 5.4:   No additional hints needed.

Problem 5.5:   Recall the useful identity

$$\frac{1}{(1-t)^{k+1}} = \sum_{n=0}^{\infty}\binom{k+n}{n}t^n$$

for any nonnegative k and $|t| \le 1$.

Problem 5.6:   Equation (5.1.23) should ring a bell.

Problem 5.7:   The concept behind solving this problem is the inverse
Radon transform. A clever usage of Eq. (5.1.37) should do it. The simple
expressions for $L_0(y)$ and $L_1(y)$ for the Laguerre polynomials is helpful.

Problem 5.8:   Make use of the identity in Eq. (4.2.60) and the Rodrigues
formula in Eq. (5.1.57) for the associated Laguerre polynomials.

Problem 5.9:   No additional hints necessary.

Problem 5.10:   Fock basis is convenient for the first part of this problem.
For the second part, a combined usage of Eq. (5.1.23) and the Rodrigues
formula in Eq. (5.1.57) for the associated Laguerre polynomials should do
the job.

Problem 5.11:   One can start with two applications of the overcompleteness relation for coherent states in Eq. (4.2.61). To simplify the subsequent
phase-space integrals, perhaps think of Eq. (4.2.93).

Problem 5.12: No additional hints necessary.

Problem 5.13: This problem shows that it is possible to manipulate improper functions with caution.

Problem 5.14: As it turns out, one obtains an improper function that is more singular than delta functions for $z \neq 0$.

Problem 5.15: No additional hints necessary.

Problem 5.16: The normal-ordered expression in Eq. (4.2.93) for exponential operator functions as well as the relations Eq. (5.1.116) and Eq. (5.1.130) are useful here.

Problem 5.17: One route to proceed is to consider the Fourier representation for one-dimensional delta functions,

$$\delta(x) = \int \frac{dk}{2\pi} e^{ikx}.$$

Problem 5.18: One should recall the fundamental theorem of calculus:

$$\frac{d}{da} \int_{g_1(a)}^{g_2(a)} dx\, f(x) = \left[f(g_2(a))\frac{d}{da}g_2(a) - f(g_1(a))\frac{d}{da}g_1(a) \right].$$

Problem 5.19: A change of integration variables and the Fourier representation for one-dimensional delta functions,

$$\delta(x) = \int \frac{dk}{2\pi} e^{ikx},$$

should help.

Problem 5.20: The calculations for this problem are facilitated by the Fourier representation for one-dimensional delta functions,

$$\delta(x) = \int \frac{dk}{2\pi} e^{ikx}.$$

Problem 5.21: One may note that

$$L_n(z) = \sum_{k=0}^{n} \binom{n}{k} \frac{(-z)^k}{k!}$$

and

$$\frac{z^n}{n!} = \sum_{k=0}^{n} \binom{n}{k} (-1)^k \, \mathrm{L}_k(z)$$

for any complex variable z. If these identities are unfamiliar to you, derive them as an exercise. These, together with the Rodrigues formula in Eq. (5.1.57) for the associated Laguerre polynomials and the relation in Eq. (4.2.57) should help.

Problem 5.22: Eventually, you would find that the normal-ordered form is just the parity operator, whereas the anitnormal-ordered form is an improper operator function.

Problem 5.23: By first writing the normal-ordered form using Eq. (5.3.2) and next invoking the results of **Problem 5.22** and the Rodrigues formula in Eq. (5.1.57) for the associated Laguerre polynomials, it should be straightforward to arrive at the answer.

Problem 5.24: A viable strategy would be to apply the overcompleteness resolution of the identity in Eq. (4.2.61) appropriately and, next, introduce the Gauss–Weierstrass transform.

Sample Solutions to All Problems

Sample solution to Problem 1.1

Varying the cost function $\mathbb{C}(\rho, \widehat{\rho})$ yields

$$\delta\mathbb{C}(\rho, \widehat{\rho}) = \mathrm{tr}\{\delta\widehat{\rho}(\log\widehat{\rho} - \log\rho)\}. \qquad \text{(SS 1)}$$

By using the parametrization

$$\widehat{\rho} = \frac{\widehat{\rho}_0}{\mathrm{tr}\{\widehat{\rho}_0\}} \qquad \text{(SS 2)}$$

to ensure that $\mathrm{tr}\{\widehat{\rho}\} = 1$, the total differential is simply

$$\delta\widehat{\rho} = \frac{\delta\widehat{\rho}_0}{\mathrm{tr}\{\widehat{\rho}_0\}} - \widehat{\rho}\,\frac{\mathrm{tr}\{\delta\widehat{\rho}_0\}}{\mathrm{tr}\{\widehat{\rho}_0\}}. \qquad \text{(SS 3)}$$

By defining the operator

$$\rho_0 = \exp\left\{\frac{\displaystyle\int (\mathrm{d}\rho)\,\mathcal{L}(\mathbb{D}; \rho)\log\rho}{\displaystyle\int (\mathrm{d}\rho')\,\mathcal{L}(\mathbb{D}; \rho')}\right\}, \qquad \text{(SS 4)}$$

the variation of the average cost function is then given by

$$\delta\overline{\mathbb{C}}(\widehat{\rho}) = \mathrm{tr}\left\{\int (\mathrm{d}\mathbb{D})\,\delta\widehat{\rho}\,(\log\widehat{\rho} - \log\rho_0)\right\}$$

$$= \mathrm{tr}\left\{\int (\mathrm{d}\mathbb{D})\,\frac{\delta\widehat{\rho}_0}{\mathrm{tr}\{\widehat{\rho}_0\}}\,(\log\widehat{\rho} - \log\rho_0 - \mathrm{tr}\{\widehat{\rho}\log\widehat{\rho}\} + \mathrm{tr}\{\widehat{\rho}\log\rho_0\})\right\}. \qquad \text{(SS 5)}$$

Setting $\delta\overline{\mathbb{C}}(\widehat{\rho}) = 0$, one gets the extremal equation

$$\log\widehat{\rho} - \mathrm{tr}\{\widehat{\rho}\log\widehat{\rho}\} = \log\rho_0 - \mathrm{tr}\{\widehat{\rho}\log\rho_0\}, \qquad \text{(SS 6)}$$

which is solved by the estimator

$$\widehat{\rho} = \frac{\rho_0}{\mathrm{tr}\{\rho_0\}}. \qquad \text{(SS 7)}$$

Since ρ_0 is the exponential of a Hermitian operator, the estimator $\widehat{\rho}$ is a positive estimator.

Sample solution to Problem 1.2

Under this representation, the left-hand side of Eq. (1.2.30) gives

$$\mathrm{tr}\{\Psi^\dagger\Phi\} = \mathrm{Tr}\left\{\begin{pmatrix}\psi_1^\dagger\\\psi_2^\dagger\\\vdots\\\psi_D^\dagger\end{pmatrix}\begin{pmatrix}\phi_1 & \phi_2 & \cdots & \phi_D\end{pmatrix}\right\}$$

$$= \sum_{j=1}^{D}\psi_j^\dagger\phi_j, \qquad \text{(SS 8)}$$

while the right-hand side of Eq. (1.2.30) gives

$$\langle\!\langle\Psi^\dagger|\Phi\rangle\!\rangle = \begin{pmatrix}\psi_1^\dagger & \psi_2^\dagger & \cdots & \psi_D^\dagger\end{pmatrix}\begin{pmatrix}\phi_1\\\phi_2\\\vdots\\\phi_D\end{pmatrix}$$

$$= \sum_{j=1}^{D}\psi_j^\dagger\phi_j, \qquad \text{(SS 9)}$$

and so the two sides are equal, indeed.

Since the mapping is such that

$$\Psi = \sum_{j=1}^{D}\sum_{k=1}^{D}|e_j\rangle\,\psi_{jk}\,\langle e_k| \longrightarrow |\Psi\rangle\!\rangle \cong \begin{pmatrix}\psi_1\\\psi_2\\\vdots\\\psi_D\end{pmatrix} = \sum_{j=1}^{D}\sum_{k=1}^{D}|e_k\rangle\,|e_j\rangle\,\psi_{jk},$$

$$\text{(SS 10)}$$

the two representations are related by a swap operation.

Sample solution to Problem 1.3

Making use of the computational basis, this result is obtained rather straightforwardly:

$$
|ACB\rangle\rangle = \sum_{j,k} |j\rangle\, a_{jk}\, \langle k| \sum_{j',k'} |j'\rangle\, c_{j'k'}\, \langle k'| \sum_{j'',k''} |j''\rangle\, b_{j''k''}\, |k''\rangle
$$

$$
= \sum_{j,k}\sum_{j',k'}\sum_{j'',k''} |j\rangle\, |k''\rangle\, a_{jk} c_{j'k'} b_{j''k''}\, \langle k|j'\rangle\, \underbrace{\langle k'|j''\rangle}_{=\,\langle j''|k'\rangle}
$$

$$
= \left(\sum_{j,k}\sum_{j'',k''} |j\rangle\, a_{jk}\, \langle k| \otimes |k''\rangle\, b_{j''k''}\, \langle j''| \right) |C\rangle\rangle
$$

$$
= A \otimes B^{\mathrm{T}}\, |C\rangle\rangle . \tag{SS 11}
$$

For the other representation, the result is given by

$$
|ACB\rangle\rangle = B^{\mathrm{T}} \otimes A\, |C\rangle\rangle , \tag{SS 12}
$$

since the column-stacking operation requires a swap in the positions of the tensor product of two kets, as compared to the first mapping.

Sample solution to Problem 1.4

If the meeting of the bras with the kets is understood to take place in the right order of ket spaces, then this result is obtained almost readily.

The object

$$
B^{\mathrm{T}} \otimes A = \left(\sum_{j'',k''} |j''\rangle\, b_{j''k''}\, \langle k''| \right)^{\mathrm{T}} \otimes \left(\sum_{j',k'} |j'\rangle\, a_{j'k'}\, \langle k'| \right)
$$

$$
\Rightarrow |B^{\mathrm{T}} \otimes A\rangle\rangle = \sum_{j',k',j'',k''} |k''\rangle\, |j'\rangle\, |j''\rangle\, |k'\rangle\, a_{j'k'} b_{j''k''} \tag{SS 13}
$$

involves the sets of kets $\{|j'\rangle\}$ and $\{|k''\rangle\}$ in the ket space $\mathcal{V}$, the set $\{|k'\rangle\}$ in $\mathcal{V}_1$, and the set $\{|j''\rangle\}$ in $\mathcal{V}_2$.

Thus, the tensorial object

$$\langle\!\langle 1_\nu| \otimes 1_{\nu_2} \otimes 1_{\nu_1} = \sum_{j,l,m} |l\rangle\,|m\rangle\,\langle j|\,\langle j|\,\langle l|\,\langle m| \qquad \text{(SS 14)}$$

acts on $|B^{\mathrm{T}} \otimes A\rangle\!\rangle$ to give

$$(\langle\!\langle 1_\nu| \otimes 1_{\nu_2} \otimes 1_{\nu_1})\,|B^{\mathrm{T}} \otimes A\rangle\!\rangle = \sum_{l,m} |l\rangle\,|m\rangle \sum_{j} b_{lj}a_{jm} = |BA\rangle\!\rangle. \qquad \text{(SS 15)}$$

Sample solution to Problem 1.5

Since for any Hermitian operator A of dimension D, the relations

$$A = \sum_{j=1}^{D^2} \mathrm{tr}\{A\Gamma_j\}\Gamma_j \qquad \text{(SS 16)}$$

is always true for a complete set of D^2 trace-orthonormal Hermitian basis operators Γ_j, the corresponding relation

$$|A\rangle\!\rangle = \sum_{j=1}^{D^2} |\Gamma_j\rangle\!\rangle\,\langle\!\langle\Gamma_j|A\rangle\!\rangle \qquad \text{(SS 17)}$$

for the superkets tells us that Eq. (1.2.71) must be true.

To explicitly demonstrate this completeness relation for the set of basis operators $\{1/\sqrt{2}, \sigma_x/\sqrt{2}, \sigma_y/\sqrt{2}, \sigma_z/\sqrt{2}\}$, we may first represent these operators with four-column vectors of dimension $D = 4$ using the mapping in Eq. (1.2.28) inasmuch as

$$|1\rangle\!\rangle \frac{1}{\sqrt{2}} \,\hat{=}\, \frac{1}{\sqrt{2}} \begin{pmatrix} 1 \\ 0 \\ 0 \\ 1 \end{pmatrix},$$

$$|\sigma_x\rangle\!\rangle \frac{1}{\sqrt{2}} \,\hat{=}\, \frac{1}{\sqrt{2}} \begin{pmatrix} 0 \\ 1 \\ 1 \\ 0 \end{pmatrix},$$

$$|\sigma_y\rangle\!\rangle \frac{1}{\sqrt{2}} \,\hat{=}\, \frac{1}{\sqrt{2}} \begin{pmatrix} 0 \\ -i \\ i \\ 0 \end{pmatrix},$$

$$|\sigma_z\rangle\!\rangle \frac{1}{\sqrt{2}} \,\hat{=}\, \frac{1}{\sqrt{2}} \begin{pmatrix} 1 \\ 0 \\ 0 \\ -1 \end{pmatrix}. \tag{SS 18}$$

Therefore, one has

$$|1\rangle\!\rangle \frac{1}{2}\langle\!\langle 1| + |\sigma_x\rangle\!\rangle \frac{1}{2}\langle\!\langle \sigma_x| + |\sigma_y\rangle\!\rangle \frac{1}{2}\langle\!\langle \sigma_y| + |\sigma_z\rangle\!\rangle \frac{1}{2}\langle\!\langle \sigma_z|$$

$$\hat{=} \frac{1}{2}\left[\begin{pmatrix} 1 \\ 0 \\ 0 \\ 1 \end{pmatrix} (1\ 0\ 0\ 1) + \begin{pmatrix} 0 \\ 1 \\ 1 \\ 0 \end{pmatrix} (0\ 1\ 1\ 0) \right.$$

$$\left. + \begin{pmatrix} 0 \\ -i \\ i \\ 0 \end{pmatrix} (0\ i\ -i\ 0) + \begin{pmatrix} 1 \\ 0 \\ 0 \\ -1 \end{pmatrix} (1\ 0\ 0\ -1) \right]$$

$$= \begin{pmatrix} 1 & 0 & 0 & 0 \\ 0 & 1 & 0 & 0 \\ 0 & 0 & 1 & 0 \\ 0 & 0 & 0 & 1 \end{pmatrix}. \tag{SS 19}$$

Sample solution to Problem 1.6

It is a trivial exercise to show that the operators are mutually trace-orthonormal, which implies that they form a complete basis for the D^2-dimensional space of Hermitian operators. Next, turning the operators into superkets using the mapping in Eq. (1.2.28), we have

$$|j\rangle\langle j| \longrightarrow |j\rangle\,|j\rangle\,,$$

$$\frac{|j\rangle\langle k| + |k\rangle\langle j|}{\sqrt{2}} \longrightarrow (|j\rangle\,|k\rangle + |k\rangle\,|j\rangle)\frac{1}{\sqrt{2}},$$

$$\frac{|j\rangle\langle k| - |k\rangle\langle j|}{i\sqrt{2}} \longrightarrow (|j\rangle\,|k\rangle - |k\rangle\,|j\rangle)\frac{1}{i\sqrt{2}}. \tag{SS 20}$$

The sum of all these superoperators then gives

$$
\sum_{j=1}^{D} |j\rangle |j\rangle \langle j| \langle j| + \sum_{k>j}^{D} \sum_{j=1}^{D} (|j\rangle |k\rangle + |k\rangle |j\rangle) \frac{1}{2} (\langle j| \langle k| + \langle k| \langle j|)
$$

$$
+ \sum_{k>j}^{D} \sum_{j=1}^{D} (|j\rangle |k\rangle - |k\rangle |j\rangle) \frac{1}{2} (\langle j| \langle k| - \langle k| \langle j|)
$$

$$
= \sum_{j=1}^{D} |j\rangle |j\rangle \langle j| \langle j| + \sum_{k>j}^{D} \sum_{j=1}^{D} (|j\rangle |k\rangle \langle j| \langle k| + |k\rangle |j\rangle \langle k| \langle j|)
$$

$$
= \sum_{k=1}^{D} \sum_{j=1}^{D} |j\rangle |k\rangle \langle j| \langle k| = \mathcal{I}. \tag{SS 21}
$$

Sample solution to Problem 1.7

Suppose that it is possible to have a set of D^2 traceless trace-orthonormal Hermitian operators. Then this operator basis can be used to span the entire Hermitian operator space, which implies that all Hermitian operators are traceless. Thus, such a situation is not possible. Indeed, only sets of up to $D^2 - 1$ such traceless operators are allowed.

Sample solution to Problem 1.8

A straightforward calculation should lead to the equation

$$
\mathrm{tr}\{\Pi_{j'}\Theta_j\} = \frac{\mathrm{tr}\{\Pi_{j'}\}}{D} + (\mathbf{Q}\mathbf{Q}^-)_{j'j} - \frac{1}{D} \sum_{l=1}^{D^2} \mathrm{tr}\{\Pi_l\} (\mathbf{Q}\mathbf{Q}^-)_{j'l}, \tag{SS 22}
$$

after which one is required to compute the matrix elements of $\mathbf{Q}\mathbf{Q}^-$. At this point, since any $\mathbf{Q}^-$ results in the same set of dual operators, it is a good strategy to choose one such that $\mathbf{Q}\mathbf{Q}^-$ takes a simple form and, at the same time, possesses $D^2 - 1$ nonzero eigenvalues.

From **Problem 1.10**, we can always choose $\mathbf{Q}\mathbf{Q}^-$ to be the symmetric form

$$
\mathbf{Q}\mathbf{Q}^- = \mathbf{Q} \left(\mathbf{Q}^{\mathsf{T}}\mathbf{Q}\right)^{-1} \mathbf{Q}^{\mathsf{T}} \tag{SS 23}
$$

that has exactly one zero eigenvalue for any $\mathbf{Q}$ of a minimally complete POM,

$$\mathbf{QQ}^{-}\begin{pmatrix}1\\\vdots\\1\end{pmatrix} = \mathbf{Q}\left(\mathbf{Q}^{\mathrm{T}}\mathbf{Q}\right)^{-1}\mathbf{Q}^{\mathrm{T}}\begin{pmatrix}1\\\vdots\\1\end{pmatrix} = \mathbf{0}. \qquad \text{(SS 24)}$$

Since $\mathbf{QQ}^{-}$ is a rank-$(D^2 - 1)$ projector, the only unique spectral decomposition for $\mathbf{QQ}^{-}$ of this form is

$$\mathbf{QQ}^{-} \mathrel{\widehat{=}} \mathbf{1}_{D^2} - \frac{1}{D^2}\begin{pmatrix}1 & \cdots & 1\\\vdots & \ddots & \vdots\\1 & \cdots & 1\end{pmatrix}. \qquad \text{(SS 25)}$$

With that, the final result to this calculation is immediate.

Sample solution to Problem 1.9

We assume that there are two M–P pseudo-inverses $\mathbf{Q}_{\mathrm{M-P}}^{-}$ and $\mathbf{Q}_{\mathrm{M-P}}'^{-}$ for the matrix $\mathbf{Q}$. Then,

$$\begin{aligned}\mathbf{QQ}_{\mathrm{M-P}}^{-} &= \left(\mathbf{QQ}_{\mathrm{M-P}}^{-}\right)^{\mathrm{T}}\\ &= \mathbf{Q}_{\mathrm{M-P}}^{-\ \mathrm{T}}\mathbf{Q}^{\mathrm{T}}\\ &= \mathbf{Q}_{\mathrm{M-P}}^{-\ \mathrm{T}}\mathbf{Q}^{\mathrm{T}}\mathbf{Q}_{\mathrm{M-P}}'^{-\ \mathrm{T}}\mathbf{Q}^{\mathrm{T}}\\ &= \mathbf{QQ}_{\mathrm{M-P}}^{-}\mathbf{QQ}_{\mathrm{M-P}}'^{-}\\ &= \mathbf{QQ}_{\mathrm{M-P}}'^{-}. \qquad \text{(SS 26)}\end{aligned}$$

Also, we have

$$\mathbf{Q}_{\mathrm{M-P}}^{-}\mathbf{Q} = \mathbf{Q}_{\mathrm{M-P}}'^{-}\mathbf{Q}. \qquad \text{(SS 27)}$$

Therefore

$$\begin{aligned}\mathbf{Q}_{\mathrm{M-P}}^{-} &= \mathbf{Q}_{\mathrm{M-P}}^{-}\mathbf{QQ}_{\mathrm{M-P}}^{-}\\ &= \mathbf{Q}_{\mathrm{M-P}}^{-}\mathbf{QQ}_{\mathrm{M-P}}'^{-}\\ &= \mathbf{Q}_{\mathrm{M-P}}'^{-}\mathbf{QQ}_{\mathrm{M-P}}'^{-} = \mathbf{Q}_{\mathrm{M-P}}'^{-}. \qquad \text{(SS 28)}\end{aligned}$$

Sample solution to Problem 1.10

If the measurement is informationally complete, then $\mathbf{Q}^{\mathsf{T}}\mathbf{Q}$ is a full-rank positive matrix. If a solution exists, the linear system of equation in Eq. (1.2.76) then implies that

$$t = \left(\mathbf{Q}^{\mathsf{T}}\mathbf{Q}\right)^{-1}\mathbf{Q}^{\mathsf{T}}f', \qquad \text{(SS 29)}$$

where the one-inverse

$$\mathbf{Q}^- = \left(\mathbf{Q}^{\mathsf{T}}\mathbf{Q}\right)^{-1}\mathbf{Q}^{\mathsf{T}} \qquad \text{(SS 30)}$$

is one of the many one-inverses that gives the same t. It turns out that this is also the M–P pseudo-inverse of $\mathbf{Q}$ since it solves Eq. (1.2.87). Alternatively, one can verify that

$$\left(\mathbf{Q}^-\mathbf{Q}\right)^{\mathsf{T}} = \left[\left(\mathbf{Q}^{\mathsf{T}}\mathbf{Q}\right)^{-1}\mathbf{Q}^{\mathsf{T}}\mathbf{Q}\right]^{\mathsf{T}}$$
$$= 1 = \mathbf{Q}^-\mathbf{Q} \qquad \text{(SS 31)}$$

and

$$\left(\mathbf{Q}\mathbf{Q}^-\right)^{\mathsf{T}} = \left[\mathbf{Q}\left(\mathbf{Q}^{\mathsf{T}}\mathbf{Q}\right)^{-1}\mathbf{Q}^{\mathsf{T}}\right]^{\mathsf{T}}$$
$$= \mathbf{Q}\left(\mathbf{Q}^{\mathsf{T}}\mathbf{Q}\right)^{-1}\mathbf{Q}^{\mathsf{T}}$$
$$= \mathbf{Q}\mathbf{Q}^-. \qquad \text{(SS 32)}$$

Sample solution to Problem 1.11

Using the traceless operator basis $\{\sigma_x/\sqrt{2}, \sigma_y/\sqrt{2}, \sigma_z/\sqrt{2}\}$, the coefficients of Π_j are collected in the matrix

$$\mathbf{Q} = \frac{1}{3\sqrt{2}}\begin{pmatrix} 1 & 0 & 0 \\ -1 & 0 & 0 \\ 0 & 1 & 0 \\ 0 & -1 & 0 \\ 0 & 0 & 1 \\ 0 & 0 & -1 \end{pmatrix}. \qquad \text{(SS 33)}$$

According to the solution to **Problem 1.10**, multiplying the inverse of the corresponding matrix

$$\mathbf{Q}^{\mathrm{T}}\mathbf{Q} = \frac{1}{9}\mathbb{1}_3 \qquad \text{(SS 34)}$$

to the left of $\mathbf{Q}^{\mathrm{T}}$ then gives the M–P pseudo-inverse

$$\mathbf{Q}^{-}_{\mathrm{M-P}} = \frac{3}{\sqrt{2}}\begin{pmatrix} 1 & -1 & 0 & 0 & 0 & 0 \\ 0 & 0 & 1 & -1 & 0 & 0 \\ 0 & 0 & 0 & 0 & 1 & -1 \end{pmatrix}. \qquad \text{(SS 35)}$$

Sample solution to Problem 1.12

From **Problem 1.11**, we have obtained $\mathbf{Q}^{-}_{\mathrm{M-P}}$ for the six-outcome POM. The task hereafter involves just routine calculations. For instance, using the traceless Pauli operators, we have

$$\begin{aligned}\Theta_1 &= \frac{1}{2} + \frac{1}{6}\sum_{l=1}^{6}\sum_{k=1}^{3}\Omega_k\left[\left(\mathbf{Q}^{-}_{\mathrm{M-P}}\right)_{k1} - \left(\mathbf{Q}^{-}_{\mathrm{M-P}}\right)_{kl}\right] \\ &= \frac{1}{2} + \frac{1}{6\sqrt{2}}\left[\sigma_x\left(3\sqrt{2} + \frac{12}{\sqrt{2}}\right)\right] \\ &= \frac{1}{2} + \frac{3}{2}\sigma_x, \qquad \text{(SS 36)}\end{aligned}$$

and

$$\begin{aligned}\Theta_2 &= \frac{1}{2} + \frac{1}{6}\sum_{l=1}^{6}\sum_{k=1}^{3}\Omega_k\left[\left(\mathbf{Q}^{-}_{\mathrm{M-P}}\right)_{k2} - \left(\mathbf{Q}^{-}_{\mathrm{M-P}}\right)_{kl}\right] \\ &= \frac{1}{2} - \frac{1}{6\sqrt{2}}\left[\sigma_x\left(3\sqrt{2} + \frac{12}{\sqrt{2}}\right)\right] \\ &= \frac{1}{2} - \frac{3}{2}\sigma_x. \qquad \text{(SS 37)}\end{aligned}$$

Similarly,

$$\Theta_{3,4} = \frac{1}{2} \pm \frac{3}{2}\sigma_y,$$

$$\Theta_{5,6} = \frac{1}{2} \pm \frac{3}{2}\sigma_z. \qquad \text{(SS 38)}$$

Sample solution to Problem 1.13

If $\operatorname{tr}\{\Pi_j\} = \dfrac{D}{M}$, then using the set of basis operators $\{1/\sqrt{D}, \Omega_k\}$,

$$
\begin{aligned}
\mathcal{F} &= \frac{M}{D} \sum_{j=1}^{M} |\Pi_j\rangle\rangle \langle\langle\Pi_j| \\
&= \frac{M}{D^3} |1\rangle\rangle \langle\langle 1| \left(\sum_{j=1}^{M} |\Pi_j\rangle\rangle \langle\langle\Pi_j| \right) |1\rangle\rangle \langle\langle 1| \\
&\quad + \frac{M}{D^2} |1\rangle\rangle \langle\langle 1| \left(\sum_{j=1}^{M} |\Pi_j\rangle\rangle \langle\langle\Pi_j| \right) \sum_{k=1}^{D^2-1} |\Omega_k\rangle\rangle \langle\langle\Omega_k| \\
&\quad + \frac{M}{D^2} \sum_{k=1}^{D^2-1} |\Omega_k\rangle\rangle \langle\langle\Omega_k| \left(\sum_{j=1}^{M} |\Pi_j\rangle\rangle \langle\langle\Pi_j| \right) |1\rangle\rangle \langle\langle 1| \\
&\quad + \frac{M}{D} \sum_{k=1}^{D^2-1} |\Omega_k\rangle\rangle \langle\langle\Omega_k| \left(\sum_{j=1}^{M} |\Pi_j\rangle\rangle \langle\langle\Pi_j| \right) \sum_{k=1}^{D^2-1} |\Omega_k\rangle\rangle \langle\langle\Omega_k| . \quad \text{(SS 39)}
\end{aligned}
$$

Since $\mathbf{Q}_{jk} = \operatorname{tr}\{\Pi_j \Omega_k\}$ and $\sum_j \mathbf{Q}_{jk} = 0$, two out of the above four terms, namely the second and the third, are zero and we get

$$
\mathcal{F} = |1\rangle\rangle \frac{1}{D} \langle\langle 1| + \frac{M}{D} \sum_{k=1}^{D^2-1} \sum_{k'=1}^{D^2-1} |\Omega_k\rangle\rangle \left(\mathbf{Q}^{\mathsf{T}}\mathbf{Q}\right)_{kk'} \langle\langle\Omega_{k'}| . \quad \text{(SS 40)}
$$

We know, from **Problem 1.10**, that $\mathbf{Q}_{\text{M-P}}^{-} = \left(\mathbf{Q}^{\mathsf{T}}\mathbf{Q}\right)^{-1} \mathbf{Q}^{\mathsf{T}}$ for an over-complete POM. Thus,

$$
\mathcal{F} |\Theta_j\rangle\rangle = \frac{|1\rangle\rangle}{D} + \frac{1}{D} \sum_{l=1}^{M} \sum_{k=1}^{D^2-1} |\Omega_k\rangle\rangle \left[\left(\mathbf{Q}^{\mathsf{T}}\right)_{kj} - \left(\mathbf{Q}^{\mathsf{T}}\right)_{kl}\right], \quad \text{(SS 41)}
$$

where the double sum can be further simplified to

$$\frac{1}{D} \sum_{l=1}^{M} \sum_{k=1}^{D^2-1} |\Omega_k\rangle\rangle \left[(\mathbf{Q}^{\mathrm{T}})_{kj} - (\mathbf{Q}^{\mathrm{T}})_{kl} \right] \qquad \left(\begin{array}{l} \text{no contribution from the} \\ \text{second term of the } l \text{ sum} \end{array} \right)$$

$$= \frac{M}{D} \sum_{k=1}^{D^2-1} |\Omega_k\rangle\rangle \langle\langle \Omega_k | \Pi_j \rangle\rangle$$

$$= \frac{M}{D} \left(\mathcal{I} - |1\rangle\rangle \frac{1}{D} \langle\langle 1| \right) |\Pi_j\rangle\rangle = |\Pi_j\rangle\rangle \frac{M}{D} - \frac{|1\rangle\rangle}{D}. \qquad \text{(SS 42)}$$

So,

$$|\Theta_j\rangle\rangle = \mathcal{F}^{-1} |\Pi_j\rangle\rangle \frac{M}{D}. \qquad \text{(SS 43)}$$

Sample solution to Problem 1.14

When $N = 5$, the column of frequencies is given by

$$\boldsymbol{f} = \frac{1}{5} \begin{pmatrix} 1 \\ 0 \\ 1 \\ 1 \\ 1 \\ 1 \end{pmatrix}, \quad \text{with } \boldsymbol{f}' = \frac{1}{30} \begin{pmatrix} 1 \\ -5 \\ 1 \\ 1 \\ 1 \\ 1 \end{pmatrix}, \qquad \text{(SS 44)}$$

which corresponds to

$$\boldsymbol{t} = \frac{3}{\sqrt{2}} \begin{pmatrix} 1 & -1 & 0 & 0 & 0 & 0 \\ 0 & 0 & 1 & -1 & 0 & 0 \\ 0 & 0 & 0 & 0 & 1 & -1 \end{pmatrix} \frac{1}{30} \begin{pmatrix} 1 \\ -5 \\ 1 \\ 1 \\ 1 \\ 1 \end{pmatrix} = \frac{3}{5\sqrt{2}} \begin{pmatrix} 1 \\ 0 \\ 0 \end{pmatrix}. \qquad \text{(SS 45)}$$

Hence, we obtain

$$\widehat{\rho}_{\text{LIN}} = \frac{1}{2} \left(1 + \frac{3}{5} \sigma_x \right), \qquad \text{(SS 46)}$$

which is a positive operator that is inconsistent with the data. For instance, $\mathrm{tr}\{\widehat{\rho}_{\mathrm{LIN}}\,\Pi_1\} = 4/15$ and this contradicts with its frequency of 0.2.

After measuring $N = 18$ copies, one has

$$
f = \frac{1}{18}\begin{pmatrix} 6 \\ 0 \\ 3 \\ 3 \\ 2 \\ 4 \end{pmatrix}, \quad \text{and } f' = \frac{1}{18}\begin{pmatrix} 3 \\ -3 \\ 0 \\ 0 \\ -1 \\ 1 \end{pmatrix},
\tag{SS 47}
$$

giving us

$$
t = \frac{3}{\sqrt{2}}\begin{pmatrix} 1 & -1 & 0 & 0 & 0 & 0 \\ 0 & 0 & 1 & -1 & 0 & 0 \\ 0 & 0 & 0 & 0 & 1 & -1 \end{pmatrix}\frac{1}{18}\begin{pmatrix} 3 \\ -3 \\ 0 \\ 0 \\ -1 \\ 1 \end{pmatrix} = \frac{1}{\sqrt{2}}\begin{pmatrix} 1 \\ 0 \\ -\frac{1}{3} \end{pmatrix}.
\tag{SS 48}
$$

This leads to the linear-inversion estimator

$$
\widehat{\rho}_{\mathrm{LIN}} = \frac{1}{2}\left(1 + \sigma_x - \frac{1}{3}\sigma_z\right),
\tag{SS 49}
$$

which is a nonpositive operator that is consistent with the data.

Sample solution to Problem 1.15

For the first part of this problem, it is enough to see that the scalar

$$
\begin{aligned}
\boldsymbol{y} \cdot (\boldsymbol{\Pi} - \boldsymbol{G}) \cdot \boldsymbol{y} &= \sum_{j=1}^{M}\mathrm{tr}\{\Pi_j\}y_j^2 - \sum_{j=1}^{M}\sum_{k=1}^{M}\mathrm{tr}\{\Pi_j\Pi_k\}y_jy_k \\
&= \sum_{j=1}^{M}\sum_{k=1}^{M}\mathrm{tr}\{\Pi_j\Pi_k\}\left(y_j^2 - y_jy_k\right) \\
&= \frac{1}{2}\sum_{j=1}^{M}\sum_{k=1}^{M}\mathrm{tr}\{\Pi_j\Pi_k\}\underbrace{\left(y_j^2 - 2y_jy_k + y_k^2\right)}_{= (y_j - y_k)^2} \geq 0.
\end{aligned}
\tag{SS 50}
$$

For the second part, using the operator basis $\{1/\sqrt{D}, \Omega_l\}$, we have

$$\mathbf{G}_{jk} = \mathrm{tr}\{\Pi_j \Pi_k\} = \langle\!\langle \Pi_j | \Pi_k \rangle\!\rangle$$

$$= \langle\!\langle \Pi_j | 1 \rangle\!\rangle \frac{1}{D} \langle\!\langle 1 | \Pi_k \rangle\!\rangle + \sum_{l=1}^{D^2-1} \langle\!\langle \Pi_j | \Omega_l \rangle\!\rangle \langle\!\langle \Omega_l | \Pi_k \rangle\!\rangle$$

$$= \frac{1}{D} \Pi_j \Pi_k + \sum_{l=1}^{D^2-1} \mathbf{Q}_{jl} \mathbf{Q}_{lk}^{\mathsf{T}}, \qquad \text{(SS 51)}$$

or

$$\mathbf{G} = \frac{1}{D} \Pi 1 1^{\mathsf{T}} \Pi + \mathbf{Q}\mathbf{Q}^{\mathsf{T}}. \qquad \text{(SS 52)}$$

Since $1^{\mathsf{T}}\Pi 1 = D$ and $\mathbf{Q}^{\mathsf{T}} 1 = 0$, we have $\mathbf{G}1 = \Pi 1$, which means that 1 is the null eigenvector of $\Pi - \mathbf{G}$.

Sample solution to Problem 1.16

One example would be the positive ML estimator $\widehat{\rho}_{\mathrm{ML}}$ we have just discussed, where the average operator $\overline{\widehat{\rho}_{\mathrm{ML}}} \neq \rho_{\mathrm{true}}$ in general, yet $\widehat{\rho}_{\mathrm{ML}}$ approaches ρ_{true} as the number of copies N approaches infinity. Another example would be the Bayesian estimator $\widehat{\rho}_{\mathrm{B}}$, which is clearly biased but approaches $\widehat{\rho}_{\mathrm{ML}}$ and, ultimately, ρ_{true} as N increases.

Sample solution to Problem 1.17

This follows from the fact that

$$\mathbf{F}(\rho_{\mathrm{true}}) = \overline{\left[\frac{\partial}{\partial r} \log \mathcal{L}(\mathbb{D}; \rho_{\mathrm{true}})\right]\left[\frac{\partial}{\partial r} \log \mathcal{L}(\mathbb{D}; \rho_{\mathrm{true}})\right]}$$

$$= \int (\mathrm{d}\mathbb{D}) \left\{ \mathcal{L}(\mathbb{D}; \rho_{\mathrm{true}}) \left[\frac{\partial}{\partial r} \log \mathcal{L}(\mathbb{D}; \rho_{\mathrm{true}})\right]\left[\frac{\partial}{\partial r} \log \mathcal{L}(\mathbb{D}; \rho_{\mathrm{true}})\right] \right\}$$

$$= \int (\mathrm{d}\mathbb{D}) \left\{ \left[\frac{\partial}{\partial r} \mathcal{L}(\mathbb{D}; \rho_{\mathrm{true}})\right]\left[\frac{\partial}{\partial r} \log \mathcal{L}(\mathbb{D}; \rho_{\mathrm{true}})\right] \right\}$$

$$= \frac{\partial}{\partial r} \left\{ \int (d\mathbb{D}) \left[\mathcal{L}(\mathbb{D}; \rho_{\text{true}}) \frac{\partial}{\partial r} \log \mathcal{L}(\mathbb{D}; \rho_{\text{true}}) \right] \right\}$$

$$- \frac{\partial}{\partial r} \frac{\partial}{\partial r} \int (d\mathbb{D}) \mathcal{L}(\mathbb{D}; \rho_{\text{true}}) \log \mathcal{L}(\mathbb{D}; \rho_{\text{true}}). \tag{SS 53}$$

It is clear that the first term in the final equality is zero since

$$\int (d\mathbb{D}) \left[\mathcal{L}(\mathbb{D}; \rho_{\text{true}}) \frac{\partial}{\partial r} \log \mathcal{L}(\mathbb{D}; \rho_{\text{true}}) \right] = \frac{\partial}{\partial r} \int (d\mathbb{D}) \mathcal{L}(\mathbb{D}; \rho_{\text{true}}) = \boldsymbol{0}.$$
$$\tag{SS 54}$$

Sample solution to Problem 1.18

For multinomial statistics, the likelihood is given by

$$\mathcal{L}(\mathbb{D}; \rho) = \prod_k p_k^{n_k} \tag{SS 55}$$

and

$$\int (d\mathbb{D}) = \sum_{\{n_j\}} \frac{N!}{\prod_j n_j!}, \tag{SS 56}$$

such that

$$\int (d\mathbb{D}) \mathcal{L}(\mathbb{D}; \rho) = \sum_{\{n_j\}} \left(\frac{N!}{\prod_j n_j!} \prod_k p_k^{n_k} \right)$$

$$= \left(\sum_k p_k \right)^N = 1. \tag{SS 57}$$

In terms of the column $\boldsymbol{r}$ of coefficients r_j, one can evaluate the term

$$\frac{\partial}{\partial r} \log \mathcal{L}(\mathbb{D}; \rho) = \sum_k n_k \frac{\partial}{\partial r} \log p_k$$

$$= \sum_k \frac{n_k}{p_k} \frac{\partial p_k}{\partial r}. \tag{SS 58}$$

The Fisher information dyadic for ρ can then be found to be

$$
\begin{aligned}
\mathbf{F}(\rho) &= \overline{\left(\sum_k \frac{n_k}{p_k} \frac{\partial p_k}{\partial \boldsymbol{r}} \right) \left(\sum_l \frac{n_l}{p_l} \frac{\partial p_l}{\partial \boldsymbol{r}} \right)} \\
&= \sum_{k,l} \frac{\overline{n_k n_l}}{p_k p_l} \frac{\partial p_k}{\partial \boldsymbol{r}} \frac{\partial p_l}{\partial \boldsymbol{r}} \\
&= (N^2 - N) \underbrace{\sum_{k,l} \frac{\partial p_k}{\partial \boldsymbol{r}} \frac{\partial p_l}{\partial \boldsymbol{r}}}_{= \, \mathbf{0}} + N \sum_l \frac{1}{p_l} \frac{\partial p_l}{\partial \boldsymbol{r}} \frac{\partial p_l}{\partial \boldsymbol{r}} \\
&= N \sum_l \frac{1}{p_l} \frac{\partial p_l}{\partial \boldsymbol{r}} \frac{\partial p_l}{\partial \boldsymbol{r}}.
\end{aligned}
\tag{SS 59}
$$

In terms of the operator basis $\left\{ 1/\sqrt{D}, \Omega_k \right\}$ that defines $\boldsymbol{r}$,

$$
\frac{\partial p_l}{\partial \boldsymbol{r}} = \frac{\partial}{\partial \boldsymbol{r}} \operatorname{tr}\{\rho \Pi_l\} = \operatorname{tr}\{\Pi_l \boldsymbol{\Omega}\},
\tag{SS 60}
$$

where $\boldsymbol{\Omega}$ is the column of Ω_ks, $\mathbf{F}(\rho)$ can be written as

$$
\mathbf{F}(\rho) = \mathbf{Q}^{\mathsf{T}} \mathbf{P}^{-1} \mathbf{Q}.
\tag{SS 61}
$$

Sample solution to Problem 1.19

It is more convenient to start with the result $\mathbf{F}(\rho) = \mathbf{Q}^{\mathsf{T}} \mathbf{P}^{-1} \mathbf{Q}$ from **Problem 1.18**. Then, by rewriting the function

$$
\begin{aligned}
\mathbf{F}(\lambda \rho_1 + (1 - \lambda)\rho_2) &= \mathbf{Q}^{\mathsf{T}} \mathbf{P}^{-1} \mathbf{Q} \\
&\equiv \mathbf{Q}^{\mathsf{T}} \left[\lambda \mathbf{P}_1 + (1 - \lambda)\mathbf{P}_2 \right]^{-1} \mathbf{Q},
\end{aligned}
\tag{SS 62}
$$

involving $0 \le \lambda \le 1$, $\mathbf{P}_1$ for ρ_1 and $\mathbf{P}_2$ for ρ_2, one arrives at the derivatives

$$
\frac{\partial}{\partial \lambda} \mathbf{F}(\lambda \rho_1 + (1 - \lambda)\rho_2) = -\mathbf{Q}^{\mathsf{T}} \mathbf{P}^{-2} (\mathbf{P}_1 - \mathbf{P}_2) \mathbf{Q},
$$

$$
\left(\frac{\partial}{\partial \lambda} \right)^2 \mathbf{F}(\lambda \rho_1 + (1 - \lambda)\rho_2) = 2 \mathbf{Q}^{\mathsf{T}} \mathbf{P}^{-3} (\mathbf{P}_1 - \mathbf{P}_2)^2 \mathbf{Q}.
\tag{SS 63}
$$

Upon losing the arguments of $\mathbf{F}$ on the basis of a common ground of notational understanding, the first-order derivative of $\mathrm{Sp}\left\{\mathbf{F}^{-1}\right\}$ reads

$$\frac{\partial}{\partial\lambda}\mathrm{Sp}\left\{\mathbf{F}^{-1}\right\} = -\mathrm{Sp}\left\{\mathbf{F}^{-1}\left(\frac{\partial}{\partial\lambda}\mathbf{F}\right)\mathbf{F}^{-1}\right\} \tag{SS 64}$$

and the second-order derivative is given by

$$\left(\frac{\partial}{\partial\lambda}\right)^2 \mathrm{Sp}\left\{\mathbf{F}^{-1}\right\}$$

$$= 2\,\mathrm{Sp}\left\{\mathbf{F}^{-1}\left(\frac{\partial}{\partial\lambda}\mathbf{F}\right)\mathbf{F}^{-1}\left(\frac{\partial}{\partial\lambda}\mathbf{F}\right)\mathbf{F}^{-1}\right\} - \mathrm{Sp}\left\{\mathbf{F}^{-1}\left[\left(\frac{\partial}{\partial\lambda}\right)^2\mathbf{F}\right]\mathbf{F}^{-1}\right\}$$

$$= 2\Big[\mathrm{Sp}\left\{\mathbf{F}^{-1}\mathbf{Q}^{\mathsf{T}}\mathbf{P}^{-2}\left(\mathbf{P}_1 - \mathbf{P}_2\right)\mathbf{Q}\mathbf{F}^{-1}\mathbf{Q}^{\mathsf{T}}\mathbf{P}^{-2}\left(\mathbf{P}_1 - \mathbf{P}_2\right)\mathbf{Q}\mathbf{F}^{-1}\right\}$$

$$- \mathrm{Sp}\left\{\mathbf{F}^{-1}\mathbf{Q}^{\mathsf{T}}\mathbf{P}^{-3}\left(\mathbf{P}_1 - \mathbf{P}_2\right)^2\mathbf{Q}\mathbf{F}^{-1}\right\}\Big]. \tag{SS 65}$$

The derivatives are carried out by noting the simple identity $\delta\mathbf{F}^{-1} = -\mathbf{F}^{-1}\delta\mathbf{F}\mathbf{F}^{-1}$.

The fact that

$$\mathrm{Sp}\left\{\mathbf{A}\mathbf{B}\right\} \leq \|\mathbf{A}\|_2 \,\mathrm{Sp}\left\{\mathbf{B}\right\} \text{ for any } \mathbf{A} \geq \mathbf{0} \text{ and } \mathbf{B} \geq \mathbf{0} \tag{SS 66}$$

means that one can find the upper bound of the term

$$\mathrm{Sp}\left\{\mathbf{F}^{-1}\mathbf{Q}^{\mathsf{T}}\mathbf{P}^{-2}\left(\mathbf{P}_1 - \mathbf{P}_2\right)\mathbf{Q}\mathbf{F}^{-1}\mathbf{Q}^{\mathsf{T}}\mathbf{P}^{-2}\left(\mathbf{P}_1 - \mathbf{P}_2\right)\mathbf{Q}\mathbf{F}^{-1}\right\}$$

$$= \mathrm{Sp}\left\{\mathbf{P}^{-\frac{1}{2}}\mathbf{Q}\mathbf{F}^{-1}\mathbf{Q}^{\mathsf{T}}\mathbf{P}^{-\frac{1}{2}}\mathbf{P}^{-\frac{3}{2}}\left(\mathbf{P}_1 - \mathbf{P}_2\right)\mathbf{Q}\mathbf{F}^{-2}\mathbf{Q}^{\mathsf{T}}\mathbf{P}^{-\frac{3}{2}}\left(\mathbf{P}_1 - \mathbf{P}_2\right)\right\}$$

$$\leq \mathrm{Sp}\left\{\mathbf{P}^{-\frac{3}{2}}\left(\mathbf{P}_1 - \mathbf{P}_2\right)\mathbf{Q}\mathbf{F}^{-2}\mathbf{Q}^{\mathsf{T}}\mathbf{P}^{-\frac{3}{2}}\left(\mathbf{P}_1 - \mathbf{P}_2\right)\right\}$$

$$= \mathrm{Sp}\left\{\mathbf{F}^{-1}\mathbf{Q}^{\mathsf{T}}\mathbf{P}^{-3}\left(\mathbf{P}_1 - \mathbf{P}_2\right)^2\mathbf{Q}\mathbf{F}^{-1}\right\}, \tag{SS 67}$$

where one would have noticed that $\mathbf{P}^{-\frac{1}{2}}\mathbf{Q}\mathbf{F}^{-1}\mathbf{Q}^{\mathsf{T}}\mathbf{P}^{-\frac{1}{2}}$ is a projector. Thus, $\left(\frac{\partial}{\partial\lambda}\right)^2 \mathrm{Sp}\left\{\mathbf{F}^{-1}\right\} \leq 0$ for any λ, $\mathbf{P}_1$ and $\mathbf{P}_2$, and concavity of $\mathrm{Sp}\left\{\mathbf{F}^{-1}\right\}$ follows.

Deriving the inequality in Eq. (SS 66) simply requires the spectral decomposition of one of the positive dyadics:

$$\mathrm{Sp}\left\{\mathbf{A}\mathbf{B}\right\} = \sum_j b_j \underbrace{\boldsymbol{b}_j \cdot \mathbf{A} \cdot \boldsymbol{b}_j}_{\leq \|\mathbf{A}\|_2} \leq \|\mathbf{A}\|_2\,\mathrm{Sp}\left\{\mathbf{B}\right\}. \tag{SS 68}$$

Sample solution to Problem 1.20

One can either make use of the usual set of basis operators $\left\{1/\sqrt{D}, \Omega_k\right\}$ or simply take the results of **Problem 1.15** to write $\mathrm{Sp}\left\{\mathbf{F_0}\right\}$ in terms of the Π_js. Here, we choose the latter to quickly obtain Eq. (1.3.18),

$$\mathrm{Sp}\left\{\mathbf{F_0}\right\} = D\,\mathrm{Sp}\left\{\mathbf{QQ^T\Pi^{-1}}\right\}$$

$$= D\,\mathrm{Sp}\left\{\left(\mathbf{G} - \frac{1}{D}\mathbf{\Pi 1 1^T\Pi}\right)\mathbf{\Pi}^{-1}\right\}$$

$$= D\left(\sum_{j=1}^{M}\frac{\mathrm{tr}\{\Pi_j^2\}}{\mathrm{tr}\{\Pi_j\}} - 1\right) \leq D\left(\sum_{j=1}^{M}\mathrm{tr}\{\Pi_j\} - 1\right) = D(D-1),$$

$$\text{(SS 69)}$$

where the inequality

$$\mathrm{tr}\{\mathcal{AB}\} \leq \mathrm{tr}\{\mathcal{A}\}\mathrm{tr}\{\mathcal{B}\} \text{ for any } \mathcal{A} \geq 0 \text{ and } \mathcal{B} \geq 0 \qquad \text{(SS 70)}$$

is used. This inequality may be obtained simply with the spectral decomposition of one of the positive operators inasmuch as

$$\mathrm{tr}\{\mathcal{AB}\} = \sum_{j}b_j\underbrace{\langle b_j|\,\mathcal{A}\,|b_j\rangle}_{\leq\,\mathrm{tr}\{\mathcal{A}\}} \leq \mathrm{tr}\{\mathcal{A}\}\mathrm{tr}\{\mathcal{B}\}. \qquad \text{(SS 71)}$$

The next inequality in Eq. (1.3.19) may be derived by first looking at the product $\mathrm{Sp}\left\{\mathcal{A}\right\}\mathrm{Sp}\left\{\mathcal{A}^{-1}\right\}$ for a positive operator $\mathcal{A}$. In terms of its eigenvalues, we have

$$\mathrm{Sp}\left\{\mathcal{A}\right\}\mathrm{Sp}\left\{\mathcal{A}^{-1}\right\} = \sum_{j}\sum_{k}\frac{a_j}{a_k}$$

$$= \frac{1}{2}\sum_{j}\sum_{k}\underbrace{\left(\frac{a_j}{a_k} + \frac{a_k}{a_j}\right)}_{= \frac{a_j^2 + a_k^2}{a_j a_k} \geq 2}$$

$$\geq \dim\{\mathcal{A}\}^2. \qquad \text{(SS 72)}$$

Hence,

$$\mathrm{Sp}\left\{\mathbf{F_0^{-1}}\right\} \geq \frac{(D^2-1)^2}{D(D-1)} = \frac{(D+1)(D^2-1)}{D}. \qquad \text{(SS 73)}$$

Sample solution to Problem 2.1

Given that

$$\Pi_j = \frac{1 + \boldsymbol{v}_j \cdot \boldsymbol{\sigma}}{a} \qquad \text{(SS 74)}$$

for positive a, the entries of the square matrix that represents the Gram dyadic **G** are

$$\text{tr}\{\Pi_j \Pi_k\} = \frac{2}{a^2}(1 + \boldsymbol{v}_j \cdot \boldsymbol{v}_k). \qquad \text{(SS 75)}$$

For completeness's sake, we shall demonstrate informational completeness for both the six-outcome POM and the tetrahedron POM by inspecting their respective Gram matrices.

For the six-outcome POM, by taking $a = 6$ and ordering the vectors

$$\boldsymbol{v}_1 \mathrel{\hat{=}} \begin{pmatrix} 1 \\ 0 \\ 0 \end{pmatrix}, \qquad \boldsymbol{v}_2 \mathrel{\hat{=}} -\begin{pmatrix} 1 \\ 0 \\ 0 \end{pmatrix}, \qquad \boldsymbol{v}_3 \mathrel{\hat{=}} \begin{pmatrix} 0 \\ 1 \\ 0 \end{pmatrix},$$

$$\boldsymbol{v}_4 \mathrel{\hat{=}} -\begin{pmatrix} 0 \\ 1 \\ 0 \end{pmatrix}, \qquad \boldsymbol{v}_5 \mathrel{\hat{=}} \begin{pmatrix} 0 \\ 0 \\ 1 \end{pmatrix}, \qquad \boldsymbol{v}_6 \mathrel{\hat{=}} -\begin{pmatrix} 0 \\ 0 \\ 1 \end{pmatrix}, \qquad \text{(SS 76)}$$

the Gram dyadic is given by

$$\mathbf{G}_{\text{six}} \mathrel{\hat{=}} \frac{1}{18} \begin{pmatrix} 2 & 0 & 1 & 1 & 1 & 1 \\ 0 & 2 & 1 & 1 & 1 & 1 \\ 1 & 1 & 2 & 0 & 1 & 1 \\ 1 & 1 & 0 & 2 & 1 & 1 \\ 1 & 1 & 1 & 1 & 2 & 0 \\ 1 & 1 & 1 & 1 & 0 & 2 \end{pmatrix}, \qquad \text{(SS 77)}$$

which can be written as a sum of **21** and a multiple of the matrix

$$\begin{pmatrix} 0 & 0 & 1 & 1 & 1 & 1 \\ 0 & 0 & 1 & 1 & 1 & 1 \\ 1 & 1 & 0 & 0 & 1 & 1 \\ 1 & 1 & 0 & 0 & 1 & 1 \\ 1 & 1 & 1 & 1 & 0 & 0 \\ 1 & 1 & 1 & 1 & 0 & 0 \end{pmatrix}. \qquad \text{(SS 78)}$$

It is clear that this matrix has three nonzero eigenvalues and one can look for them easily by noting that

$$
\begin{pmatrix}
0 & 0 & 1 & 1 & 1 & 1 \\
0 & 0 & 1 & 1 & 1 & 1 \\
1 & 1 & 0 & 0 & 1 & 1 \\
1 & 1 & 0 & 0 & 1 & 1 \\
1 & 1 & 1 & 1 & 0 & 0 \\
1 & 1 & 1 & 1 & 0 & 0
\end{pmatrix}
\begin{pmatrix} 1 \\ 1 \\ 1 \\ 1 \\ 1 \\ 1 \end{pmatrix}
= 4
\begin{pmatrix} 1 \\ 1 \\ 1 \\ 1 \\ 1 \\ 1 \end{pmatrix}
\tag{SS 79}
$$

and

$$
\begin{pmatrix}
0 & 0 & 1 & 1 & 1 & 1 \\
0 & 0 & 1 & 1 & 1 & 1 \\
1 & 1 & 0 & 0 & 1 & 1 \\
1 & 1 & 0 & 0 & 1 & 1 \\
1 & 1 & 1 & 1 & 0 & 0 \\
1 & 1 & 1 & 1 & 0 & 0
\end{pmatrix}
\begin{pmatrix} 1 \\ 1 \\ 0 \\ 0 \\ -1 \\ -1 \end{pmatrix}
= -2
\begin{pmatrix} 1 \\ 1 \\ 0 \\ 0 \\ -1 \\ -1 \end{pmatrix}.
\tag{SS 80}
$$

So, two of the eigenvalues are respectively 4 and -2. The third eigenvalue is -2 since the trace of the dyadic in Eq. (SS 78) is zero. Thus, there are four positive eigenvalues for $\mathbf{G}_{\text{six}}$ given in the set $\{1/3, 1/9, 1/9, 1/9\}$, which implies that the six-outcome POM is informationally complete. For the tetrahedron POM, we have $a = 4$ and $v_j = a_j$ and the entries of the Gram dyadic $\mathbf{G}_{\text{tetra}}$ is simply

$$
\text{tr}\{\Pi_j \Pi_k\} = \frac{1}{12}\left(2\delta_{j,k} + 1\right)
\tag{SS 81}
$$

using Eq. (2.1.3). The resulting matrix representation for $\mathbf{G}_{\text{tetra}}$ is then

$$
\mathbf{G}_{\text{tetra}} \;\hat{=}\; \frac{1}{12}
\begin{pmatrix}
3 & 1 & 1 & 1 \\
1 & 3 & 1 & 1 \\
1 & 1 & 3 & 1 \\
1 & 1 & 1 & 3
\end{pmatrix}
$$

$$
= \frac{1}{6}\mathbf{1} + \frac{1}{12}
\begin{pmatrix} 1 \\ 1 \\ 1 \\ 1 \end{pmatrix}
\begin{pmatrix} 1 & 1 & 1 & 1 \end{pmatrix}.
\tag{SS 82}
$$

The eigenvalues of $\mathbf{G}_{\text{tetra}}$ are therefore $\{1/3, 1/6, 1/6, 1/6\}$, which also implies that the tetrahedron POM is informationally complete.

Sample solution to Problem 2.2

By writing

$$\mathcal{A} = U\Sigma V^\dagger = \sum_j |u_j\rangle \Sigma_{jj} \langle v_j| \qquad \text{(SS 83)}$$

in terms of its singular-value decomposition, where $U = \sum_j |u_j\rangle e^{i\theta_j} \langle u_j|$ and $V = \sum_j |v_j\rangle e^{i\phi_j} \langle v_j|$ are unitary operators and the rectangular Σ has diagonal elements equal to the eigenvalues of $\sqrt{\mathcal{A}^\dagger \mathcal{A}}$, we note that the $|v_j\rangle$s corresponding to zero singular values span the kernel of $\mathcal{A}$, the $|v_j\rangle$s corresponding to nonzero singular values span its row space, the $|u_j\rangle$s corresponding to zero singular values span its cokernel, and the $|u_j\rangle$s corresponding to nonzero singular values span its column space. It is clear that the row and column spaces have the same dimensions. This means that the rank of $\mathcal{A}$ is simply the number of nonzero singular values. With that observation, the final result is immediate.

Sample solution to Problem 2.3

The fact that $R_{\text{ML}} = 1$ means that $\hat{\rho}_{\text{ML}}$ is invertible and so the actual peak of the likelihood function lies inside the state space. Therefore $\hat{\rho}_{\text{ML}} = \hat{\rho}_{\text{LIN}}$. On the other hand, $\hat{\rho}_{\text{ML}} = \hat{\rho}_{\text{LIN}}$ implies that

$$R_{\text{ML}} = \sum_j \frac{f_j}{\text{tr}\{\hat{\rho}_{\text{LIN}}\Pi_j\}} \Pi_j$$

$$= \sum_j \Pi_j = 1. \qquad \text{(SS 84)}$$

Sample solution to Problem 2.4

The vector identities given in the hints are useful for this problem. To derive the first identity, in the notational spirit of this problem, the well-known commutation relation $\sigma_x \sigma_y = i\sigma_z$ for cyclically permutable subscripts immediately gives

$$\boldsymbol{\sigma}\boldsymbol{\sigma}^{\text{T}} = \mathbf{1} - i\boldsymbol{\sigma} \times \mathbf{1}, \qquad \text{(SS 85)}$$

where **1** in this equation is understood as the operator unit dyadic. Therefore, dot products with a and b gives

$$a \cdot \sigma\sigma \cdot b = a \cdot b - ia \cdot \sigma \times b$$

$$= a \cdot b + ia \cdot b \times \sigma$$

$$= a \cdot b + ia \times b \cdot \sigma. \qquad \text{(SS 86)}$$

To arrive at the double-cross-product identity, one may either calculate every component of its left-hand side, or observe that $a \times (b \times c)$ is coplanar with b and c, so that we have

$$a \times (b \times c) = \beta b + \chi c, \qquad \text{(SS 87)}$$

and linearity in a, b and c of the left-hand side imposes that $\beta = \kappa_1 a \cdot c$ and $\chi = \kappa_2 a \cdot b$, where κ_1 and κ_2 are proportionality constants that are independent of the three vectors. Asymmetry in $b \times c$ further implies that $\kappa_1 = -\kappa_2$,

$$a \times (b \times c) = \kappa_1 (ba \cdot c - ca \cdot b). \qquad \text{(SS 88)}$$

Since this equation is true for all sorts of vectors, a special case where $a = e_1$, $b = e_2$ and $c = e_3$ which are the three right-handed position unit vectors gives $\kappa_1 = 1$. Consequently, one can learn that

$$\mathbf{1} \times (b \times c) = bc^{\mathrm{T}} - cb^{\mathrm{T}}. \qquad \text{(SS 89)}$$

Using those two vector identities,

$$Uv \cdot \sigma U^{\dagger} = (\cos \vartheta + in \cdot \sigma \sin \vartheta) \, v \cdot \sigma \, (\cos \vartheta - in \cdot \sigma \sin \vartheta)$$

$$= (\cos \vartheta)^2 \, n \cdot \sigma + (\sin \vartheta)^2 \, n \cdot \sigma v \cdot \sigma n \cdot \sigma$$

$$+ i \sin \vartheta \cos \vartheta \, (n \cdot \sigma v \cdot \sigma - v \cdot \sigma n \cdot \sigma), \qquad \text{(SS 90)}$$

where

$$n \cdot \sigma v \cdot \sigma n \cdot \sigma = n \cdot \sigma \, (v \cdot n + iv \times n \cdot \sigma)$$

$$= v \cdot nn \cdot \sigma - n \times (v \times n) \cdot \sigma$$

$$= 2v \cdot nn \cdot \sigma - v \cdot \sigma \qquad \text{(SS 91)}$$

and

$$n \cdot \sigma v \cdot \sigma - v \cdot \sigma n \cdot \sigma = 2 in \times v \cdot \sigma. \qquad \text{(SS 92)}$$

Hence,

$$Uv \cdot \sigma U^\dagger$$

$$= (\cos \vartheta)^2\, v \cdot \sigma + (\sin \vartheta)^2 \left(2v \cdot nn \cdot \sigma - v \cdot \sigma \right) - 2 \sin \vartheta \cos \vartheta n \times v \cdot \sigma$$

$$= \cos(2\vartheta) v \cdot \sigma + \left[1 - \cos(2\vartheta) \right] v \cdot nn \cdot \sigma + 2 \sin(2\vartheta) v \times n \cdot \sigma$$

$$= v \cdot \left[\cos(2\vartheta)\left(1 - nn \right) + \sin(2\vartheta) 1 \times n + nn \right] \cdot \sigma$$

$$= \left(v^\mathsf{T} \mathbf{O}^\mathsf{T} \right) \sigma. \tag{SS 93}$$

So,

$$\mathbf{O} = \cos(2\vartheta)\left(1 - nn^\mathsf{T} \right) + \sin(2\vartheta) n \times 1 + nn^\mathsf{T}. \tag{SS 94}$$

To show that $\mathbf{O}$ is orthogonal, we note that

$$\mathbf{O}^\mathsf{T}\mathbf{O} = \left[\cos(2\vartheta)\left(1 - nn^\mathsf{T} \right) + \sin(2\vartheta) 1 \times n^\mathsf{T} + nn^\mathsf{T} \right]$$

$$\times \left[\cos(2\vartheta)\left(1 - nn^\mathsf{T} \right) + \sin(2\vartheta) n \times 1 + nn^\mathsf{T} \right]$$

$$= \left[\cos(2\vartheta) \right]^2 \left(1 - nn^\mathsf{T} \right)^2 + \left[\sin(2\vartheta) \right]^2 \left(1 \times n^\mathsf{T} \right)\left(n \times 1 \right) + nn^\mathsf{T}, \tag{SS 95}$$

recognizing that the cross terms do not contribute to the product expansion. The dyadic $1 - nn^\mathsf{T}$ is a projector as

$$\left(1 - nn^\mathsf{T} \right)^2 = 1 - nn^\mathsf{T}. \tag{SS 96}$$

The product $\left(1 \times n^\mathsf{T} \right)\left(n \times 1 \right)$, on the other hand, gives a positive dyadic and to find out what this dyadic is, we start by evaluating the scalar product $a \times n \cdot b \times n$ for arbitrary vectors a and b. The expression for this is

$$a \times n \cdot b \times n = a \cdot n \times \left(b \times n \right)$$

$$= a \cdot b - a \cdot nn \cdot b$$

$$= a^\mathsf{T}\left(1 - nn^\mathsf{T} \right) b. \tag{SS 97}$$

The left-hand side of the above equation can be rewritten as

$$a \times n \cdot b \times n = \left(n \times a \right)^\mathsf{T}\left(n \times b \right)$$

$$= a^\mathsf{T}\left(n \times 1 \right)^\mathsf{T}\left(n \times 1 \right) b$$

$$= a^\mathsf{T}\left(1 \times n^\mathsf{T} \right)\left(n \times 1 \right) b, \tag{SS 98}$$

which implies that

$$(1 \times n^{\mathrm{T}})(n \times 1) = 1 - nn^{\mathrm{T}} = (n \times 1)(1 \times n^{\mathrm{T}}). \qquad \text{(SS 99)}$$

Combining the terms together, we finally have

$$\begin{aligned}
\mathbf{O}^{\mathrm{T}}\mathbf{O} &= [\cos(2\vartheta)]^2 \left(1 - nn^{\mathrm{T}}\right)^2 + [\sin(2\vartheta)]^2 (n \times 1)(1 \times n^{\mathrm{T}}) + nn^{\mathrm{T}} \\
&= \left\{ [\cos(2\vartheta)]^2 + [\sin(2\vartheta)]^2 \right\} (1 - nn^{\mathrm{T}}) + nn^{\mathrm{T}} \\
&= 1. \qquad\qquad\qquad\qquad\qquad\qquad\qquad\qquad\qquad \text{(SS 100)}
\end{aligned}$$

Sample solution to Problem 2.5

Since $\widehat{\rho}_{\mathrm{ML}} = \widehat{\sigma}$ for von Neumann measurements, the frequencies $f_\pm$ are related to the outcomes $\Pi_\pm$ by

$$f_\pm = \frac{1 \pm \frac{1}{\sqrt{2}}(s_1 + s_3)}{2}, \qquad \text{(SS 101)}$$

where s_1, s_2 and s_3 are coordinates of the Bloch vector of $\widehat{\rho}_{\mathrm{ML}}$. We then have

$$s_3 = \sqrt{2}\,(2f_+ - 1) - s_1 \qquad \text{(SS 102)}$$

and the positivity criterion for all $\widehat{\rho}_{\mathrm{ML}}$ dictates that

$$\begin{aligned}
s_1^2 + s_2^2 + s_3^2 &= s_1^2 + s_2^2 + \left[\sqrt{2}\,(2f_+ - 1) - s_1\right]^2 \\
&= 2\left[s_1 - \frac{1}{\sqrt{2}}(2f_+ - 1)\right]^2 + s_2^2 + (2f_+ - 1)^2 \\
&\leq 1. \qquad\qquad\qquad\qquad\qquad\qquad\qquad \text{(SS 103)}
\end{aligned}$$

Thus, the apparent geometry of the convex set is an ellipse. This means that the circular disc of estimators is tilted with respect to the (s_1, s_2, s_3) coordinate axes.

If one is astute, then one would immediately compare Eq. (SS 101) with Eq. (2.1.22) and deduce that the third coordinate has to be transformed to

$$s_3' = \frac{1}{\sqrt{2}}(s_1 + s_3) \qquad \text{(SS 104)}$$

so that the two equations have the same form. To make the overall transformation on all three coordinates orthogonal, a possible transformation for the first coordinate would be

$$s_1' = \frac{1}{\sqrt{2}}(-s_1 + s_3). \tag{SS 105}$$

Once these two transformations are fixed, in order to maintain the right-handedness of the primed coordinate system, we must have

$$s_2' = -s_2. \tag{SS 106}$$

Thus, in this new coordinate system, we would obtain the inequality Eq. (2.1.24).

To recover the actual circular geometry in a systematic way, we note that the outcomes $\Pi_\pm$ are related to the outcomes $\Pi_{0,1}$ in Eq. (2.1.19) by a unitary transformation $\Pi_\pm = U\Pi_{0,1}U^\dagger$, so that

$$\begin{aligned} f_\pm &= \mathrm{tr}\{\widehat{\rho}_{\mathrm{ML}}\,\Pi_\pm\} \\ &= \mathrm{tr}\{\widehat{\rho}_{\mathrm{ML}}\,U\Pi_{0,1}U^\dagger\} \\ &= \mathrm{tr}\{U^\dagger\widehat{\rho}_{\mathrm{ML}}U\,\Pi_{0,1}\}. \end{aligned} \tag{SS 107}$$

So, we can recover Eq. (2.1.24) by switching to the coordinates of the new Bloch vector of $U^\dagger\widehat{\rho}_{\mathrm{ML}}U$. The operator U is obtained by noting that

$$\begin{aligned} \Pi_\pm &\widehat{=} \frac{1}{2}\begin{pmatrix} 1 \pm \frac{1}{\sqrt{2}} & \pm\frac{1}{\sqrt{2}} \\ \pm\frac{1}{\sqrt{2}} & 1 \mp \frac{1}{\sqrt{2}} \end{pmatrix} \\ &= \frac{1}{\sqrt{2}}\begin{pmatrix} \sqrt{1 \pm \frac{1}{\sqrt{2}}} \\ \pm\sqrt{1 \mp \frac{1}{\sqrt{2}}} \end{pmatrix}\begin{pmatrix} \sqrt{1 \pm \frac{1}{\sqrt{2}}} & \pm\sqrt{1 \mp \frac{1}{\sqrt{2}}} \end{pmatrix}\frac{1}{\sqrt{2}}. \end{aligned} \tag{SS 108}$$

The unitary operator U is then given by

$$\begin{aligned} U &\widehat{=} \frac{1}{\sqrt{2}}\begin{pmatrix} \sqrt{1 + \frac{1}{\sqrt{2}}} \\ \sqrt{1 - \frac{1}{\sqrt{2}}} \end{pmatrix}\begin{pmatrix} 1 & 0 \end{pmatrix} + \frac{1}{\sqrt{2}}\begin{pmatrix} \sqrt{1 - \frac{1}{\sqrt{2}}} \\ -\sqrt{1 + \frac{1}{\sqrt{2}}} \end{pmatrix}\begin{pmatrix} 0 & 1 \end{pmatrix} \\ &= \frac{\sqrt{2 + \sqrt{2}}}{2}\begin{pmatrix} 1 & \sqrt{2} - 1 \\ \sqrt{2} - 1 & -1 \end{pmatrix} \\ &= \frac{\sqrt{2 - \sqrt{2}}}{2}\sigma_x + \frac{\sqrt{2 + \sqrt{2}}}{2}\sigma_z, \end{aligned} \tag{SS 109}$$

which corresponds to $\vartheta = \pi/2$, $\phi = -\pi/2$ and

$$n \triangleq \frac{1}{2} \begin{pmatrix} \sqrt{2-\sqrt{2}} \\ 0 \\ \sqrt{2+\sqrt{2}} \end{pmatrix} \tag{SS 110}$$

in **Problem 2.4**. This gives

$$nn^T \triangleq \frac{1}{4} \begin{pmatrix} 2-\sqrt{2} & 0 & \sqrt{2} \\ 0 & 0 & 0 \\ \sqrt{2} & 0 & 2+\sqrt{2} \end{pmatrix}, \tag{SS 111}$$

which leads to the orthogonal dyadic

$$\mathbf{O} \triangleq \begin{pmatrix} -\frac{1}{\sqrt{2}} & 0 & \frac{1}{\sqrt{2}} \\ 0 & -1 & 0 \\ \frac{1}{\sqrt{2}} & 0 & \frac{1}{\sqrt{2}} \end{pmatrix} \tag{SS 112}$$

required to reveal the circular-planar geometry of the convex set.

Sample solution to Problem 2.6

Since the superoperator $\mathcal{T}$ is Hermitian and $\mathcal{T}^2 = \mathcal{I}$, upon denoting the matrix

$$\mathbf{U} = \frac{1}{\sqrt{2}} \begin{pmatrix} \mathcal{I} & i\mathcal{T} \\ \mathcal{T} & -i\mathcal{T} \end{pmatrix}, \tag{SS 113}$$

we have

$$\mathbf{U}^\dagger \mathbf{U} = \frac{1}{2} \begin{pmatrix} \mathcal{I} & \mathcal{T} \\ -i\mathcal{I} & i\mathcal{T} \end{pmatrix} \begin{pmatrix} \mathcal{I} & i\mathcal{T} \\ \mathcal{T} & -i\mathcal{T} \end{pmatrix} = \begin{pmatrix} \mathcal{I} & 0 \\ 0 & \mathcal{I} \end{pmatrix}. \tag{SS 114}$$

This matrix $\mathbf{U}$ has dimension $2D^2 \times 2D^2$ that is divided into a 2×2 block matrix. Based on the structure of the blocks, it is suggestive to consider two relevant 2×2 numerical matrices

$$\mathbf{U}_1 = \frac{1}{\sqrt{2}} \begin{pmatrix} 1 & i \\ 1 & -i \end{pmatrix}, \quad \mathbf{U}_2 = \frac{1}{\sqrt{2}} \begin{pmatrix} 1 & i \\ -1 & i \end{pmatrix}. \tag{SS 115}$$

The matrix $\mathbf{U}_1$ has eigenvectors

$$\boldsymbol{u}_{11} = \frac{1}{\sqrt{3+\sqrt{3}}}\begin{pmatrix} \frac{1}{\sqrt{2}}e^{i\frac{\pi}{4}}(1+\sqrt{3}) \\ 1 \end{pmatrix}, \; \boldsymbol{u}_{12} = \frac{1}{\sqrt{3-\sqrt{3}}}\begin{pmatrix} \frac{1}{\sqrt{2}}e^{i\frac{\pi}{4}}(1-\sqrt{3}) \\ 1 \end{pmatrix}$$

$$\text{(SS 116)}$$

and eigenvalues

$$u_{11} = \frac{1-i+e^{i\frac{\pi}{4}}\sqrt{6}}{2\sqrt{2}} = e^{i\frac{\pi}{12}}, \quad u_{12} = \frac{1-i-e^{i\frac{\pi}{4}}\sqrt{6}}{2\sqrt{2}} = e^{-i\frac{7\pi}{12}}, \quad \text{(SS 117)}$$

and the matrix $\mathbf{U}_2$ has eigenvectors

$$\boldsymbol{u}_{21} = \frac{1}{\sqrt{3-\sqrt{3}}}\begin{pmatrix} \frac{1}{\sqrt{2}}e^{i\frac{3\pi}{4}}(1-\sqrt{3}) \\ 1 \end{pmatrix}, \; \boldsymbol{u}_{22} = \frac{1}{\sqrt{3+\sqrt{3}}}\begin{pmatrix} \frac{1}{\sqrt{2}}e^{-i\frac{3\pi}{4}}(1+\sqrt{3}) \\ 1 \end{pmatrix}$$

$$\text{(SS 118)}$$

and eigenvalues

$$u_{21} = \frac{1+i+e^{i\frac{3\pi}{4}}\sqrt{6}}{2\sqrt{2}} = e^{i\frac{7\pi}{12}}, \quad u_{22} = \frac{1+i-e^{i\frac{3\pi}{4}}\sqrt{6}}{2\sqrt{2}} = e^{-i\frac{\pi}{12}}. \quad \text{(SS 119)}$$

The purpose of considering these matrices becomes clear after realizing that the eigenvectors of $\mathbf{U}$ are essentially the columns analogous to those in Eq. (SS 116) and Eq. (SS 118), namely

$$\boldsymbol{u}_{11} = \frac{1}{\sqrt{3+\sqrt{3}}}\begin{pmatrix} |S\rangle\!\rangle \frac{1}{\sqrt{2}}e^{i\frac{\pi}{4}}(1+\sqrt{3}) \\ |S\rangle\!\rangle \end{pmatrix},$$

$$\boldsymbol{u}_{12} = \frac{1}{\sqrt{3-\sqrt{3}}}\begin{pmatrix} |S\rangle\!\rangle \frac{1}{\sqrt{2}}e^{i\frac{\pi}{4}}(1-\sqrt{3}) \\ |S\rangle\!\rangle \end{pmatrix},$$

$$\boldsymbol{u}_{21} = \frac{1}{\sqrt{3-\sqrt{3}}}\begin{pmatrix} |AS\rangle\!\rangle \frac{1}{\sqrt{2}}e^{i\frac{3\pi}{4}}(1-\sqrt{3}) \\ |AS\rangle\!\rangle \end{pmatrix},$$

$$\boldsymbol{u}_{22} = \frac{1}{\sqrt{3+\sqrt{3}}}\begin{pmatrix} |AS\rangle\!\rangle \frac{1}{\sqrt{2}}e^{i\frac{3\pi}{4}}(1+\sqrt{3}) \\ |AS\rangle\!\rangle \end{pmatrix}, \quad \text{(SS 120)}$$

where $|S\rangle\!\rangle$ and $|AS\rangle\!\rangle$ are respectively superkets of a symmetric and antisymmetric operator such that $\mathcal{T}|S\rangle\!\rangle = |S\rangle\!\rangle$ and $\mathcal{T}|AS\rangle\!\rangle = -|AS\rangle\!\rangle$.

From linear algebra, we know that there are $D(D + 1)/2$ trace-orthonormal symmetric basis operators and $D(D-1)/2$ trace-orthonormal antisymmetric basis operators for D-dimensional matrices. These two sets of basis operators respectively span the mutually orthogonal symmetric and antisymmetric subspaces for the matrix space. The corresponding eigenvalues that go with the eigenvectors therefore exist with multiplicities; $D(D+1)/2$ of u_{11} and u_{12}, and $D(D-1)/2$ of u_{21} and u_{22}. By recalling that $\text{Tr}\{\mathcal{I}\} = D^2$ and $\text{Tr}\{\mathcal{T}\} = D$, one can check for consistency by summing all these eigenvalues:

$$\frac{D(D+1)}{2}(u_{11} + u_{12}) + \frac{D(D-1)}{2}(u_{21} + u_{22})$$

$$= \frac{D(D+1)}{2}\frac{1-\mathrm{i}}{\sqrt{2}} + \frac{D(D-1)}{2}\frac{1+\mathrm{i}}{\sqrt{2}}$$

$$= \frac{1}{\sqrt{2}}(D^2 - \mathrm{i}D) = \text{Sp}\{\mathbf{U}\}. \tag{SS 121}$$

The superkets $|S\rangle\rangle$ and $|AS\rangle\rangle$ may be generated from the symmetric and antisymmetric part of a set of D^2 trace-orthonormal basis operators, that is $|S\rangle\rangle = (|\Gamma_j\rangle\rangle + |\Gamma_j^T\rangle\rangle)/2$ and $|AS\rangle\rangle = (|\Gamma_j\rangle\rangle - |\Gamma_j^T\rangle\rangle)/2$. These form the entire set of $2D^2$ orthonormal eigenvectors for $\mathbf{U}$.

Sample solution to Problem 2.7

Taking

$$\frac{1}{\lambda} = \int_0^\infty \mathrm{d}y\, e^{-\lambda y}, \tag{SS 122}$$

the operator

$$\left(-\frac{\mathrm{d}}{\mathrm{d}\lambda}\right)^{v\frac{\partial}{\partial v}} \frac{1}{\lambda}\bigg|_{\lambda=1} = \int_0^\infty \mathrm{d}y\left(-\frac{\mathrm{d}}{\mathrm{d}\lambda}\right)^{v\frac{\partial}{\partial v}} e^{-\lambda y}\bigg|_{\lambda=1}$$

$$= \int_0^\infty \mathrm{d}y\, y^{v\frac{\partial}{\partial v}} e^{-y}$$

$$= \left(v\frac{\partial}{\partial v}\right)!. \tag{SS 123}$$

Since

$$\left(v\frac{\partial}{\partial v}\right)! \, v = D! \, v, \qquad \text{(SS 124)}$$

the operator in Eq. (2.1.51) simply gives $D!$ when applied to v.

Sample solution to Problem 2.8

Since $\log \mathcal{L}'_{\widehat{N}_t}(\{n_j\}; \rho) \geq \log \mathcal{L}'_{\widehat{N}_t+1}(\{n_j\}; \rho)$, one has the inequality

$$\widehat{N}_t - N + 1 \geq (1-\eta)\left(\widehat{N}_t + 1\right). \qquad \text{(SS 125)}$$

Also, $\log \mathcal{L}'_{\widehat{N}_t}(\{n_j\}; \rho) \geq \log \mathcal{L}'_{\widehat{N}_t-1}(\{n_j\}; \rho)$ implies the inequality

$$\widehat{N}_t(1-\eta) \geq \widehat{N}_t - N. \qquad \text{(SS 126)}$$

The two inequalities together give the condition

$$\widehat{N}_t(1-\eta) - \eta \leq \widehat{N}_t - N \leq \widehat{N}_t(1-\eta), \qquad \text{(SS 127)}$$

or

$$\widehat{N}_t = \left\lfloor \widehat{N}_t(1-\eta) \right\rfloor + N. \qquad \text{(SS 128)}$$

Sample solution to Problem 2.9

Taking $x = n$ to be a positive integer, from the recurrence relation in Eq. (2.2.17), we get

$$F(n) = F(0) + H_n, \qquad \text{(SS 129)}$$

where

$$F(0) = \int_0^\infty ds \, e^{-s} \, \log s. \qquad \text{(SS 130)}$$

Using the hint, $F(0)$ may be expressed as a double integral inasmuch as

$$F(0) = \int_0^\infty \frac{dt}{t}\, e^{-t} \underbrace{\int_0^\infty ds\, e^{-s}}_{=\,1} - \int_0^\infty \frac{dt}{t} \underbrace{\int_0^\infty ds\, e^{-s(t+1)}}_{=\,\frac{1}{t+1}}$$

$$= \log m \Big|_{m \to \infty} - \int_0^\infty \frac{dt}{t(t+1)}. \qquad \text{(SS 131)}$$

The second integral term can be evaluated by introducing a new integration variable $e^y = t + 1$,

$$\int_0^\infty \frac{dt}{t(t+1)} = \int_0^\infty dy\, \frac{1}{e^y - 1}$$

$$= \sum_{k=0}^\infty \int_0^\infty dy\, e^{-(k+1)y}$$

$$= \sum_{k=0}^\infty \frac{1}{k+1} = H_m \Big|_{m \to \infty}. \qquad \text{(SS 132)}$$

Putting everything together, we find that

$$F(0) = (\log m - H_m) \Big|_{m \to \infty} = -\gamma. \qquad \text{(SS 133)}$$

Sample solution to Problem 2.10

Using Eq. (2.1.8), the right-hand side of Eq. (2.2.32) simply yields

$$\frac{N}{\mathrm{tr}\{A^\dagger A\}} \mathrm{tr}\left\{ \delta A^\dagger A \left[\left(R - \frac{1}{\eta}G \right) + \mathrm{tr}\left\{ \left(R - \frac{1}{\eta}G \right) \rho \right\} \right] \right.$$

$$\left. + \left[\left(R - \frac{1}{\eta}G \right) + \mathrm{tr}\left\{ \left(R - \frac{1}{\eta}G \right) \rho \right\} \right] A^\dagger \delta A \right\}. \qquad \text{(SS 134)}$$

Since $\text{tr}\left\{\left(R - \dfrac{1}{\eta}G\right)\rho\right\}$ is clearly zero, maximal likelihood function implies that

$$\frac{\mathcal{A}}{\text{tr}\{\mathcal{A}^\dagger\mathcal{A}\}}\left(R_{\text{ML}} - \frac{1}{\eta_{\text{ML}}}G\right) = 0 = \left(R_{\text{ML}} - \frac{1}{\eta_{\text{ML}}}G\right)\frac{\mathcal{A}^\dagger}{\text{tr}\{\mathcal{A}^\dagger\mathcal{A}\}}, \qquad \text{(SS 135)}$$

or

$$\widehat{\rho}_{\text{ML}}\left(R_{\text{ML}} - \frac{1}{\eta_{\text{ML}}}G\right) = 0 = \left(R_{\text{ML}} - \frac{1}{\eta_{\text{ML}}}G\right)\widehat{\rho}_{\text{ML}}. \qquad \text{(SS 136)}$$

Problem 2.11

As $\eta_j = \eta_0$, the fact that

$$G = \sum_j \Pi'_j = \eta_0 \sum_j \Pi_j = \eta_0 \qquad \text{(SS 137)}$$

and

$$\eta = \sum_j p_j = \eta_0 \qquad \text{(SS 138)}$$

means that $G/\eta = 1$. Therefore, the algorithm defined by Eq. (2.2.34) reduces to that defined by Eq. (2.1.64). The lesson here is that if all detectors have the same efficiencies, there is no relative difference between any two outcomes. Therefore such an imperfection translates to having perfect detectors with a reduced number of detected copies. This is the reason why experimentalists prefer to introduce additional attenuators in front of detectors in order to ensure that they have equal efficiencies.

Sample solution to Problem 3.1

In terms of the computational basis $\{e_j\}$ and the eigencolumns $\{m_j\}$ of a D-dimensional positive operator M, we have

$$M = \sum_{l=1}^{D} m_l m_l m_l$$

$$= \sum_{j',k'} e_{j'} e_{j'} \cdot \sum_{l=1}^{D} m_l m_l m_l \cdot e_{k'} e_{k'}. \qquad \text{(SS 139)}$$

Suppose that the matrix representation $\mathcal{M}$ for M in the computational basis has the diagonal elements $\mathcal{M}_{jj}$ equal to zero. Then

$$0 = \mathcal{M}_{jj} = e_j \cdot \sum_{l=1}^{D} m_l m_l m_l \cdot e_j$$

$$= \begin{pmatrix} e_j \cdot m_1 \sqrt{m_1} \\ e_j \cdot m_2 \sqrt{m_2} \\ \vdots \\ e_j \cdot m_D \sqrt{m_D} \end{pmatrix} \cdot \begin{pmatrix} e_j \cdot m_1 \sqrt{m_1} \\ e_j \cdot m_2 \sqrt{m_2} \\ \vdots \\ e_j \cdot m_D \sqrt{m_D} \end{pmatrix}$$

$$\Rightarrow e_j \cdot m_l = 0 \text{ for all } l, \tag{SS 140}$$

which implies that e_j is an eigenvector of M for the eigenvalue zero. It then follows that the row and column that contain $\mathcal{M}_{jj}$ are both zero.

Sample solution to Problem 3.2

If ρ is pure, the purity

$$\text{tr}\{(\rho + H)^2\} = \underbrace{\text{tr}\{\rho^2\}}_{=1} + \underbrace{2\text{tr}\{\rho H\}}_{=0} + \text{tr}\{H^2\} > 1, \tag{SS 141}$$

telling us that $\rho + H$ is always nonpositive. If ρ is now mixed, the first necessary condition can be obtained again from the purity of $\rho + H$,

$$\text{tr}\{(\rho + H)^2\} = \text{tr}\{\rho^2\} + \text{tr}\{H^2\}$$

$$= \text{tr}\{\rho^2\} + \sum_{j=1}^{D} h_j^2$$

$$\leq 1, \tag{SS 142}$$

or

$$\sum_{j=1}^{D} h_j^2 \leq 1 - \text{tr}\{\rho^2\}. \tag{SS 143}$$

The second necessary condition can come directly from the fact that

$$\rho + H \geq 0$$

$$\Rightarrow \langle h_j | \rho | h_j \rangle + h_j \geq 0, \tag{SS 144}$$

where $\{|h_j\rangle\}$ is the set of eigenkets of H. Since

$$\langle h_j| \rho |h_j\rangle \leq \|\rho\|_2 , \qquad\qquad \text{(SS 145)}$$

we then have

$$h_j \geq -\|\rho\|_2 . \qquad\qquad \text{(SS 146)}$$

These necessary inequalities are, however, weak inequalities in the sense that the bounds are too conservative and may never be reached by many H. Obtaining tighter inequalities is a challenge and is only feasible when much more restrictions are imposed on the set of ρs. This further illustrates the complexity of the state-space boundary.

Sample solution to Problem 3.3

It is obvious to start with a set of D^2 positive operators with $\Pi_1 = 1$ and the rest of $\Pi_{j\neq 1} = V_{j\neq 1}$ being random operators that are linearly independent, so that

$$\tilde{\Gamma}_1 = 1,$$

$$\tilde{\Gamma}_2 = V_2 - \frac{\text{tr}\{V_2\}}{D},$$

$$\tilde{\Gamma}_3 = V_3 - \frac{\text{tr}\{V_3\}}{D} - \frac{\text{tr}\{V_3\tilde{\Gamma}_2\}}{\text{tr}\{\tilde{\Gamma}_2^2\}}\tilde{\Gamma}_2,$$

$$\text{(SS 147)}$$

and so forth. It is clear that all operators $\tilde{\Gamma}_j$ generated this way are traceless and trace-orthogonal. All that is left to do is to normalize these operators to obtain the proper basis operators Γ_j.

Sample solution to Problem 3.4

We can start by noting that any state ρ can be written as a convex sum of two other reference states ρ_1 and ρ_2, that is

$$\rho = t\rho_1 + (1 - t)\rho_2 \qquad\qquad \text{(SS 148)}$$

with $0 \leq t \leq 1$. So,

$$\delta S(\rho) = -\text{tr}\{\delta \rho \log \rho\}$$
$$= -\delta t \, \text{tr}\{(\rho_1 - \rho_2) \log \rho\}, \tag{SS 149}$$

or

$$\frac{\partial}{\partial t} S(\rho) = -\text{tr}\{(\rho_1 - \rho_2) \log \rho\}. \tag{SS 150}$$

We then vary the first-order derivative

$$\delta \frac{\partial}{\partial t} S(\rho) = -\text{tr}\{(\rho_1 - \rho_2) \, \delta \log \rho\}$$
$$= -\delta t \int_0^\infty dy \, \text{tr}\left\{ \left[(\rho_1 - \rho_2) \frac{1}{y + \rho} \right]^2 \right\} \tag{SS 151}$$

to obtain the second-order derivative

$$\left(\frac{\partial}{\partial t} \right)^2 S(\rho) = -\int_0^\infty dy \, \text{tr}\left\{ \left[(\rho_1 - \rho_2) \frac{1}{y + \rho} \right]^2 \right\}. \tag{SS 152}$$

To show that this second-order derivative is a strictly negative function for any ρ_1 and ρ_2, we note that for the operators $\mathcal{A} = \mathcal{A}^\dagger$ and $\mathcal{B} \geq 0$, where $\mathcal{A} = \rho_1 - \rho_2$ and $\mathcal{B} = 1/(y + \rho)$ in our context,

$$\text{tr}\{(\mathcal{AB})^2\} = \text{tr}\{\mathcal{ABAB}\}$$
$$= \text{tr}\left\{ \left(\sqrt{\mathcal{B}} \mathcal{A} \sqrt{\mathcal{B}} \right) \left(\sqrt{\mathcal{B}} \mathcal{A} \sqrt{\mathcal{B}} \right) \right\}$$
$$\geq 0. \tag{SS 153}$$

Thus, $S(\rho)$ is a concave function for all ρs.

Sample solution to Problem 3.5

Reminding ourselves that an MLME estimator is, of course, also an ML estimator, the operator

$$R_{\text{MLME}} = 1_{\widehat{\rho}_{\text{MLME}}} + R_0, \tag{SS 154}$$

where $R_0\widehat{\rho}_{\text{MLME}} = 0$. Hence,

$$\int_0^1 dx\, e^{x\sum_j \lambda_j \Pi_j}\, R\, e^{-x\sum_j \lambda_j \Pi_j} \bigg|_{\rho = \widehat{\rho}_{\text{MLME}}}$$

$$= \int_0^1 dx\, \widehat{\rho}_{\text{MLME}}^{\,x}\, R_{\text{MLME}}\, \widehat{\rho}_{\text{MLME}}^{\,-x}$$

$$= \int_0^1 dx\, \widehat{\rho}_{\text{MLME}}^{\,x}\, 1_{\widehat{\rho}_{\text{MLME}}}\, \widehat{\rho}_{\text{MLME}}^{\,-x} = 1_{\widehat{\rho}_{\text{MLME}}} \int_0^1 dx = 1_{\widehat{\rho}_{\text{MLME}}}. \tag{SS 155}$$

Sample solution to Problem 3.6

To obtain Eq. (3.1.33), we first verify that indeed

$$\int_0^\infty dt\, \frac{1}{t+1} - \int_0^\infty dt\, \frac{1}{t+a} = \log\left(\frac{t+1}{t+a}\right)\bigg|_{t=0}^{t=\infty} = \log a. \tag{SS 156}$$

Upon invoking $\delta\mathcal{A} = -\mathcal{A}^{-1}\,\delta\mathcal{A}\,\mathcal{A}^{-1}$, Eq. (3.1.33) is straightforwardly established.

To obtain Eq. (3.1.34), we shall first evaluate the following relevant integral provided by the hints using a simple substitution:

$$t = a(\tan\vartheta)^2, \quad dt = 2a\tan\vartheta(\sec\vartheta)^2. \tag{SS 157}$$

Then, this integral simplifies to

$$\int_0^\infty \frac{dt}{\pi\sqrt{t}}\, \frac{a}{a+t} = 2\sqrt{a} \int_0^{\frac{\pi}{2}} \frac{d\vartheta}{\pi} = \sqrt{a}, \tag{SS 158}$$

so that an application of the operator variation identity gives

$$
\begin{aligned}
\delta\sqrt{\mathcal{A}} &= \int_0^\infty \frac{dt}{\pi\sqrt{t}}\, \delta\frac{1}{1+t\mathcal{A}^{-1}} \\[2mm]
&= \int_0^\infty \frac{dt}{\pi\sqrt{t}}\, \frac{1}{1+t\mathcal{A}^{-1}} t\mathcal{A}^{-1}\,\delta\mathcal{A}\, t\mathcal{A}^{-1}\frac{1}{1+t\mathcal{A}^{-1}} \\[2mm]
&= \int_0^\infty \frac{dt}{\pi}\, \sqrt{t}\frac{\mathcal{A}}{t+\mathcal{A}}\,\delta\mathcal{A}\,\frac{\mathcal{A}}{t+\mathcal{A}}.
\end{aligned}
\tag{SS 159}
$$

Sample solution to Problem 3.7

This can be shown by rewriting

$$
\mathrm{tr}\left\{ \rho\,\Pi_k \int_0^1 dx\, \left(e^{x\sum_j \lambda_j\Pi_j}\, R\, e^{-x\sum_j \lambda_j\Pi_j} \right) \right\} = \mathrm{tr}\left\{ \Pi_k \int_0^1 dx\, \rho^x\, R\, \rho^{1-x} \right\},
\tag{SS 160}
$$

where

$$
\int_0^1 dx\, \rho^x\, R\, \rho^{1-x} = \int_0^1 dx\, \rho^{1-x}\, R\, \rho^x.
\tag{SS 161}
$$

We thus have

$$
\begin{aligned}
&\mathrm{tr}\left\{ \Pi_k \int_0^1 \frac{dx}{2}\, \left(\rho^x\, R\, \rho^{1-x} + \rho^{1-x}\, R\, \rho^x \right) \right\} \\[2mm]
&= \int_0^1 \frac{dx}{2}\, \mathrm{tr}\left\{ \sqrt{\Pi_k}\, \left(\rho^x\, R\, \rho^{1-x} + \rho^{1-x}\, R\, \rho^x \right) \sqrt{\Pi_k} \right\},
\end{aligned}
\tag{SS 162}
$$

which is an integral of the trace of a Hermitian operator.

Sample solution to Problem 3.8

By denoting the eigenvalues of a single-qubit state ρ by λ_+ and λ_-, such that $\lambda_+ + \lambda_- = 1$,

$$S(\rho) = -[\lambda_+ \log \lambda_+ + (1 - \lambda_+) \log (1 - \lambda_+)]. \qquad \text{(SS 163)}$$

The purity of ρ is related to the eigenvalues by

$$\text{tr}\{\rho^2\} = \lambda_+^2 + \lambda_-^2$$
$$= 2\lambda_+^2 - 2\lambda_+ + 1. \qquad \text{(SS 164)}$$

Thus,

$$\lambda_+ = \frac{1}{2} + \frac{1}{2}\sqrt{2\text{tr}\{\rho^2\} - 1} = \frac{1 + \mathcal{K}}{2}. \qquad \text{(SS 165)}$$

In terms of $\mathcal{K}$,

$$S(\rho) = -\left[\left(\frac{1 + \mathcal{K}}{2}\right) \log \left(\frac{1 + \mathcal{K}}{2}\right) + \left(\frac{1 - \mathcal{K}}{2}\right) \log \left(\frac{1 - \mathcal{K}}{2}\right)\right]$$
$$= -\frac{1}{2}\left[\log \left(\frac{1 - \mathcal{K}^2}{4}\right) + \mathcal{K} \log \left(\frac{1 + \mathcal{K}}{1 - \mathcal{K}}\right)\right]. \qquad \text{(SS 166)}$$

For $\rho = |\ \rangle\langle\ |$, $\mathcal{K} = 1$ and

$$S(|\ \rangle\langle\ |) = -\frac{1}{2}\left[\log \left(\frac{1 - \mathcal{K}}{2}\right) + \log \left(\frac{2}{1 - \mathcal{K}}\right)\right]\bigg|_{\mathcal{K}=1} = 0. \qquad \text{(SS 167)}$$

For $\rho = 1/D$, $\mathcal{K} = 0$, so

$$S\left(\frac{1}{D}\right) = -\frac{1}{2}\log \left(\frac{1}{4}\right). \qquad \text{(SS 168)}$$

Sample solution to Problem 3.9

The ML probabilities for these outcomes are given in terms of the Bloch vector components of the ML estimators $\widehat{\rho}_{\mathrm{ML}}$ by

$$\widehat{p}_1 = \frac{1}{2}\left(1 - s_2 \sin \vartheta + s_3 \cos \vartheta\right),$$

$$\widehat{p}_2 = \frac{1}{2}\left[1 + \frac{\sqrt{3}}{2}s_1 + \frac{1}{2}\left(s_2 \sin \vartheta - s_3 \cos \vartheta\right)\right],$$

$$\widehat{p}_3 = \frac{1}{2}\left[1 - \frac{\sqrt{3}}{2}s_1 + \frac{1}{2}\left(s_2 \sin \vartheta - s_3 \cos \vartheta\right)\right], \qquad \text{(SS 169)}$$

where we can express s_1 and s_3 in terms of the probabilities as

$$s_1 = \frac{2}{\sqrt{3}}\left(\widehat{p}_2 - \widehat{p}_3\right),$$

$$s_3 = \frac{2\widehat{p}_1 - 1 + s_2 \sin \vartheta}{\cos \vartheta} \qquad \text{(SS 170)}$$

and leave s_2 as a free variable. From here, there are two ways to look for the value of s_2 that maximizes the entropy of the ML estimator.

One way to go about this is to see that the vectors b_j are actually rotated Bloch vectors of the standard trine outcomes we have discussed, with the corresponding orthogonal dyadic represented as

$$\mathbf{O} \triangleq \begin{pmatrix} 1 & 0 & 0 \\ 0 & \cos \vartheta & -\sin \vartheta \\ 0 & \sin \vartheta & \cos \vartheta \end{pmatrix} \qquad \text{(SS 171)}$$

that comes from a unitary transformation $U\Pi_j U^\dagger$ on the standard trine outcomes, a result of **Problem 2.4**. Therefore, the estimator $\widehat{\rho}'_{\mathrm{ML}} = U^\dagger \widehat{\rho}_{\mathrm{ML}} U^\dagger$ with the Bloch vector

$$\begin{pmatrix} s'_1 \\ s'_2 \\ s'_3 \end{pmatrix} = \begin{pmatrix} 1 & 0 & 0 \\ 0 & \cos \vartheta & \sin \vartheta \\ 0 & -\sin \vartheta & \cos \vartheta \end{pmatrix}\begin{pmatrix} s_1 \\ s_2 \\ s_3 \end{pmatrix} = \begin{pmatrix} s_1 \\ s_2 \cos \vartheta + s_3 \sin \vartheta \\ -s_2 \sin \vartheta + s_3 \cos \vartheta \end{pmatrix} \qquad \text{(SS 172)}$$

possesses the largest possible entropy if $s_2' = 0$, or

$$s_2 = -\frac{s_3 \sin \vartheta}{\cos \vartheta}$$

$$= -\frac{(2\widehat{p}_1 - 1) \sin \vartheta + s_2 (\sin \vartheta)^2}{(\cos \vartheta)^2}$$

$$\Rightarrow s_2 = -(2\widehat{p}_1 - 1) \sin \vartheta, \qquad \text{(SS 173)}$$

so that

$$\boldsymbol{s}_{\text{MLME}} \widehat{=} \begin{pmatrix} \frac{2}{\sqrt{3}} (\widehat{p}_2 - \widehat{p}_3) \\ -(2\widehat{p}_1 - 1) \sin \vartheta \\ \frac{2\widehat{p}_1 - 1 + s_2 \sin \vartheta}{\cos \vartheta} \end{pmatrix}. \qquad \text{(SS 174)}$$

One can also obtain this result by the usual method of purity minimization on the ML estimator

$$\widehat{\rho}_{\text{ML}} \widehat{=} \frac{1}{2} \begin{pmatrix} 1 + \frac{2\widehat{p}_1 - 1 + s_2 \sin \vartheta}{\cos \vartheta} & \frac{2}{\sqrt{3}} (\widehat{p}_2 - \widehat{p}_3) - is_2 \\ \frac{2}{\sqrt{3}} (\widehat{p}_2 - \widehat{p}_3) + is_2 & 1 - \frac{2\widehat{p}_1 - 1 + s_2 \sin \vartheta}{\cos \vartheta} \end{pmatrix}, \qquad \text{(SS 175)}$$

whose purity is given by

$$\text{tr}\{\widehat{\rho}_{\text{ML}}^2\} = \frac{1}{4}\left[\left(1 + \frac{2\widehat{p}_1 - 1 + s_2 \sin \vartheta}{\cos \vartheta}\right)^2 + \left(1 - \frac{2\widehat{p}_1 - 1 + s_2 \sin \vartheta}{\cos \vartheta}\right)^2 \right.$$

$$\left. + \frac{8}{3}(\widehat{p}_2 - \widehat{p}_3)^2 + 2s_2^2 \right]$$

$$= \frac{1}{2}\left[1 + \left(\frac{2\widehat{p}_1 - 1 + s_2 \sin \vartheta}{\cos \vartheta}\right)^2 + \frac{4}{3}(\widehat{p}_2 - \widehat{p}_3)^2 + s_2^2 \right]. \qquad \text{(SS 176)}$$

The minimum of $\text{tr}\{\widehat{\rho}_{\text{ML}}^2\}$ is found by setting the derivative

$$\frac{\partial}{\partial s_2} \text{tr}\{\widehat{\rho}_{\text{ML}}^2\} = \frac{(2\widehat{p}_1 - 1) \sin \vartheta + s_2}{(\cos \vartheta)^2} \qquad \text{(SS 177)}$$

to zero, out of which one also gets Eq. (SS 173).

Sample solution to Problem 3.10

The maximum-entropy form of $\widehat{\rho}_{\mathrm{MLME}}$ can be simplified into

$$\widehat{\rho}_{\mathrm{MLME}} = \frac{e^{\sum_j |j\rangle \lambda_j \langle j|}}{\mathrm{tr}\left\{e^{\sum_j |j\rangle \lambda_j \langle j|}\right\}} \,\widehat{=}\, \frac{1}{\displaystyle\sum_{j=1}^{D} e^{\lambda_j}} \sum_{j=1}^{D} |j\rangle \, e^{\lambda_j} \, \langle j| \qquad \text{(SS 178)}$$

for the von Neumann measurements $\{|j\rangle \langle j|\}$, and this form has to be equal to the operator $\sum_j |j\rangle f_j \langle j|$. Therefore, the most general form for λ_j is

$$\lambda_j = \log f_j + \alpha, \qquad \text{(SS 179)}$$

where α is an arbitrary additive constant.

Sample solution to Problem 3.11

A sample routine for the ML algorithm that is written for MATLAB is given below:

```
1    function [rho,prob_ML,maxloglike,time_ML] = ...
        ML_SA(pom,freqs,epsilon,prec,ini_state)

3        % Inputs of the ML_SA (ML steepest-ascent) routine:
4        % pom:       a cell (array of matrices) of POM outcomes
5        % freqs:     an array of frequencies of POM outcomes
6        % epsilon:   step size
7        % prec:      accuracy/precision of the ML state estimator
8        % ini_state: starting state for the routine

10       % Outputs of the ML_SA routine:
11       % rho:        ML state estimator
12       % prob_ML:    an array of ML probabilities
13       % maxloglike: maximum log-likelihood value
14       % time_ML:    duration of routine execution

16       numpom=length(freqs); % number of POM outcomes
17       dim=length(pom{1});   % dimension of the Hilbert space
18       gop=zeros(dim,dim);   % sum of all POM outcomes
19       for j=1:numpom
20           gop=gop+pom{j};
21       end

23       rho=ini_state;        % routine initialization
24       dist=1.0;
25       probs=zeros(numpom,1);
```

```
27      tic;
28      while dist>prec                 % loop till precision is reached
29          for j=1:numpom
30              probs(j)=real(trace(rho*pom{j})); % probabilities
31          end
32          rmatrix=zeros(dim,dim);  % R computation
33          for j=1:numpom
34              rmatrix=rmatrix+freqs(j).*pom{j}./probs(j);
35          end
36          eta=sum(probs);           % effective efficiency
37          grad=rmatrix-gop./eta;    % operator gradient
38          dist=norm(grad*rho);      % accuracy of ML state estimator

40          tmp_mat=eye(dim)+epsilon.*grad; % iterative equations
41          rho=tmp_mat*rho*tmp_mat';
42          rho=rho./trace(rho);
43      end
44      time_ML=toc;                    % duration of routine execution

46      maxloglike=0.0;
47      for j=1:numpom
48          maxloglike=maxloglike+freqs(j).*log(probs(j)./eta);
49      end
50      maxloglike=real(maxloglike);% maximum log-likelihood value

52      prob_ML=probs;                  % ML probabilities

54  end
```

For the parameters specified in this problem, the resulting ML estimators that are obtained by repeating the routine at sufficiently high precision with different starting states should always be equal to

$$\widehat{\rho}_{ML} = \begin{pmatrix} 0.985483 & 0.0725539 + 0.0950889\,i \\ 0.0725539 - 0.0950889\,i & 0.0145167 \end{pmatrix}, \quad \text{(SS 180)}$$

which is a rank-one operator that gives the ML probabilities $\{0.0860887, 0.492674, 0.421237\}$.

Sample solution to Problem 3.12

To perform this constrained function optimization, one may design a barrier function $b(x,y)$ for the region $x^2 + (y-1)^2 < 1$ as

$$b(x,y) = \frac{\tau}{2}\left\{1 - \tanh\left(\kappa\left[1 - x^2 - (y-1)^2\right]\right)\right\}, \quad \text{(SS 181)}$$

where the parameters τ and κ are set to reasonably large values during the numerical iteration. This barrier function $b(x, y)$ approaches a tall, inverted step function in the limit of large parameters, and any part of $f(x, y)$ that lies outside this barrier will play no role in the function minimization.

The variation of this function gives

$$\delta\, b(x, y) = \tau \left\{ \operatorname{sech} \left(\kappa \left[1 - x^2 - (y-1)^2 \right] \right) \right\}^2 \begin{pmatrix} x \\ y-1 \end{pmatrix} \cdot \begin{pmatrix} \delta x \\ \delta y \end{pmatrix}. \quad \text{(SS 182)}$$

Together with the variation of $f(x, y)$,

$$\delta f(x, y) = 2(x - 1)\, \delta x + 3y^2\, \delta y - 2\delta x\, y - 2x\, \delta y$$

$$= \begin{pmatrix} 2(x - y - 1) \\ 3y^2 - x \end{pmatrix} \cdot \begin{pmatrix} \delta x \\ \delta y \end{pmatrix}, \quad \text{(SS 183)}$$

the combined variation of the relevant Lagrange function is given by

$$\delta \mathcal{D}_{\mathrm{L}}(x, y; \tau, \kappa, \lambda) = \begin{pmatrix} w(x, y; \tau, \kappa)\, x + 2(x - y - 1) \\ w(x, y; \tau, \kappa)\, (y - 1) + 3y^2 - 2x \end{pmatrix} \cdot \begin{pmatrix} \delta x \\ \delta y \end{pmatrix}, \quad \text{(SS 184)}$$

where

$$w(x, y; \tau, \kappa) = \tau \left\{ \operatorname{sech} \left(\kappa \left[1 - x^2 - (y-1)^2 \right] \right) \right\}^2. \quad \text{(SS 185)}$$

Since we are minimizing $f(x, y)$, we may set

$$\delta x = -\epsilon \left[w(x, y; \tau, \kappa)\, x + 2(x - y - 1) \right],$$

$$\delta y = -\epsilon \left[w(x, y; \tau, \kappa)\, (y - 1) + 3y^2 - 2x \right] \quad \text{(SS 186)}$$

for a small step size ϵ. These iterative equations constitutes the steepest-ascent algorithm for the barrier function in Eq. (SS 181). Its output should yield a minimum function value of about -1.066 at $x = 0.99$ and $y = 0.86$.

Sample solution to Problem 4.1

For SIC POMs,

$$\operatorname{tr}\{\Theta_j^2\} = (D + 1)^2 + D - 2(D + 1) = D^2 + D - 1. \quad \text{(SS 187)}$$

Hence,

$$\overline{\mathcal{D}_{\mathrm{H\text{-}S}}(\hat{\rho}_{\mathrm{LIN}}, \rho_{\mathrm{true}})} = \frac{1}{2N} \left(D^2 + D - 1 - \operatorname{tr}\{\rho_{\mathrm{true}}^2\} \right). \quad \text{(SS 188)}$$

For $D = 2$, the coefficient of the measure is equal to $5 - \operatorname{tr}\{\rho_{\mathrm{true}}^2\}$.

For the dual operators of the six-outcome POM,

$$\text{tr}\{\Theta_j^2\} = \text{tr}\left\{\left(\frac{1}{2} \pm \frac{3}{2}\sigma_{x,y,z}\right)^2\right\} = 5. \qquad \text{(SS 189)}$$

Hence, for the LS LIN estimators, the tomographic accuracies for both SIC POMs and the six-outcome POM are equal.

Sample solution to Problem 4.2

For minimally complete POMs, the orthonormality relation $\langle\!\langle\Pi_j|\Theta_k\rangle\!\rangle = \delta_{j,k}$ apply. So for this class of POMs, apart from the canonical dual superkets, both the exact and approximate optimal dual superkets also give

$$\langle\!\langle\Pi_j|\,\mathcal{F}(\rho_{\text{true}})^{-1}\,|\Pi_k\rangle\!\rangle\,\frac{1}{p_k} = \langle\!\langle\Pi_j|\,\mathcal{F}(\{f_k\})^{-1}\,|\Pi_k\rangle\!\rangle\,\frac{1}{f_k} = \delta_{j,k}. \qquad \text{(SS 190)}$$

Now, consider a basis consisting of superkets spanned by these D^2 linearly independent POM outcomes,

$$|\Gamma_j\rangle\!\rangle = \sum_{l=1}^{D^2} |\Pi_l\rangle\!\rangle\, c_{jl}. \qquad \text{(SS 191)}$$

Then we find that

$$\langle\!\langle\Gamma_j|\mathcal{F}(\rho_{\text{true}})^{-1}|\Pi_k\rangle\!\rangle\,\frac{1}{p_k} = \sum_{l=1}^{D^2} c_{jl}\,\langle\!\langle\Pi_l|\mathcal{F}(\rho_{\text{true}})^{-1}|\Pi_k\rangle\!\rangle\,\frac{1}{p_k} = c_{jk},$$

$$\langle\!\langle\Gamma_j|\mathcal{F}(\{f_k\})^{-1}|\Pi_k\rangle\!\rangle\,\frac{1}{f_k} = \sum_{l=1}^{D^2} c_{jl}\,\langle\!\langle\Pi_l|\mathcal{F}(\{f_k\})^{-1}|\Pi_k\rangle\!\rangle\,\frac{1}{f_k} = c_{jk},$$

$$\langle\!\langle\Gamma_j|\mathcal{F}^{-1}|\Pi_k\rangle\!\rangle\,\frac{1}{\text{tr}\{\Pi_k\}} = \sum_{l=1}^{D^2} c_{jl}\,\langle\!\langle\Pi_l|\mathcal{F}^{-1}|\Pi_k\rangle\!\rangle\,\frac{1}{\text{tr}\{\Pi_k\}} = c_{jk}. \qquad \text{(SS 192)}$$

It then follows that all three superkets are the same.

Sample solution to Problem 4.3

Upon comparing the expression for $\overline{\mathcal{D}_{\text{H-S}}\left(\widehat{\rho}_{\text{LIN}}, \rho_{\text{true}}\right)}$ in Eq. (4.1.9) with that for $\overline{\mathcal{D}_{\text{H-S}}\left(\{\Pi_j\}; \{\Theta_j\}\right)}$ in Eq. (4.1.22), the optimal dual operators can be found immediately by simply replacing p_j with $\text{tr}\{\Pi_j\}/D$ in Eq. (4.1.15).

Therefore, the optimal operators are nothing else but the canonical dual operators.

Sample solution to Problem 4.4

We suppose, for the moment, that $b = 0$. Using the decomposition

$$\mathbf{M} = \sum_{j=1}^{D} m_j\, \boldsymbol{m}_j \boldsymbol{m}_j^{\mathsf{T}} \tag{SS 193}$$

for a positive real dyadic $\mathbf{M}$, with $\boldsymbol{m}_j^{\mathsf{T}} \boldsymbol{m}_k = \delta_{j,k}$, we have

$$\int (d\boldsymbol{x})\, e^{-\frac{1}{2}\boldsymbol{x}^{\mathsf{T}}\mathbf{M}\boldsymbol{x}} = \int \left(\prod_{j=1}^{\kappa} dx_j \right) e^{-\frac{1}{2}\sum_k m_j (\boldsymbol{m}_j^{\mathsf{T}}\boldsymbol{x})^2}. \tag{SS 194}$$

Upon introducing a new set of integration variables $x_j' = \boldsymbol{m}_j^{\mathsf{T}}\boldsymbol{x}$, or

$$\boldsymbol{x} = \sum_{j=1}^{D} \boldsymbol{m}_j x_j' = \begin{pmatrix} \boldsymbol{m}_1 & \boldsymbol{m}_2 & \cdots & \boldsymbol{m}_\kappa \end{pmatrix} \boldsymbol{x}', \tag{SS 195}$$

where

$$\mathbf{O} = \begin{pmatrix} \boldsymbol{m}_1 & \boldsymbol{m}_2 & \cdots & \boldsymbol{m}_\kappa \end{pmatrix} \tag{SS 196}$$

is an orthogonal dyadic which is of unit determinant in magnitude. Therefore, $dx_j' = dx_j$ and the integral turns into a product of single-variable Gaussian integrals inasmuch as

$$\prod_{j=1}^{\kappa} \left(\int dx_j'\, e^{-\frac{1}{2}m_j x_j'^2} \right) = \prod_{j=1}^{\kappa} \left(\sqrt{\frac{2\pi}{m_j}} \right) = \sqrt{\frac{(2\pi)^\kappa}{\det\{\mathbf{M}\}}}. \tag{SS 197}$$

Now consider $b \neq 0$. Equation (4.1.31) can be transformed into the first integral of Eq. (SS 194) with the variable transformation $\boldsymbol{x} \longrightarrow \boldsymbol{x} - \mathbf{M}^{-1}b/2$, thereby incurring an additional exponential factor $e^{\frac{1}{4}b^{\mathsf{T}}\mathbf{M}^{-1}b}$.

Sample solution to Problem 4.5

Since μ and $\mathbf{C}$ are functions of the column of κ parameters a, we have, after suppressing the well-understood arguments using Eq. (4.1.28) and

Eq. (3.1.28),

$$\log p = -\frac{1}{2}\log\left(\det\{\mathbf{C}\}\right) - \frac{1}{2}(t-\mu)^{\mathrm{T}}\mathbf{C}^{-1}(t-\mu),$$

$$\delta\log p = -\frac{1}{2}\mathrm{Sp}\left\{\mathbf{C}^{-1}\delta\mathbf{C}\right\} + \frac{1}{2}\mathrm{Sp}\left\{\mathbf{C}^{-1}\delta\mathbf{C}\mathbf{C}^{-1}(t-\mu)(t-\mu)^{\mathrm{T}}\right\}$$

$$+ (t-\mu)^{\mathrm{T}}\mathbf{C}^{-1}\delta\mu. \tag{SS 198}$$

In the subsequent calculation of the Fisher information dyadic, we need the definition $\overline{(t-\mu)(t-\mu)^{\mathrm{T}}} = \mathbf{C}$ and the fourth-order moment, which can be extracted by the general identity

$$\overline{(t-\mu)^{\mathrm{T}}\mathbf{A}(t-\mu)(t-\mu)^{\mathrm{T}}\mathbf{B}(t-\mu)} = \mathrm{Sp}\left\{\mathbf{A}\mathbf{C}\right\}\mathrm{Sp}\left\{\mathbf{B}\mathbf{C}\right\} + 2\,\mathrm{Sp}\left\{\mathbf{A}\mathbf{C}\mathbf{B}\mathbf{C}\right\}$$

$$\tag{SS 199}$$

for any dyadics $\mathbf{A}$ and $\mathbf{B}$. The latter can be shown straightforwardly by following very similar calculations performed in the same section:

$$\overline{(t-\mu)^{\mathrm{T}}\mathbf{A}(t-\mu)(t-\mu)^{\mathrm{T}}\mathbf{B}(t-\mu)}$$

$$= \mathcal{N}\int (\mathrm{d}t)\left[(t-\mu)^{\mathrm{T}}\mathbf{A}(t-\mu)\,\mathrm{Sp}\left\{\mathbf{B}(t-\mu)(t-\mu)^{\mathrm{T}}\right\}\right.$$

$$\left. \times \exp\left(-\frac{1}{2}(t-\mu)^{\mathrm{T}}\mathbf{C}^{-1}(t-\mu)\right)\right]$$

$$= -2\mathcal{N}\int (\mathrm{d}t)\,(t-\mu)^{\mathrm{T}}\mathbf{A}(t-\mu)\,\mathrm{Sp}\left\{\mathbf{B}\frac{\delta}{\delta\mathbf{C}^{-1}}\exp\left(\cdots\right)\right\}$$

$$= -2\mathcal{N}\mathrm{Sp}\left\{\mathbf{B}\frac{\delta}{\delta\mathbf{C}^{-1}}\int (\mathrm{d}t)\,\mathrm{Sp}\left\{\mathbf{A}(t-\mu)(t-\mu)^{\mathrm{T}}\right\}\exp\left(\cdots\right)\right\}$$

$$= 4\mathcal{N}\mathrm{Sp}\left\{\mathbf{B}\frac{\delta}{\delta\mathbf{C}^{-1}}\mathrm{Sp}\left\{\mathbf{A}\frac{\delta}{\delta\mathbf{C}^{-1}}\sqrt{\frac{(2\pi)^{\kappa}}{\det\{\mathbf{C}^{-1}\}}}\right\}\right\}$$

$$= -2\sqrt{(2\pi)^{\kappa}}\mathcal{N}\mathrm{Sp}\left\{\mathbf{B}\frac{\delta}{\delta\mathbf{C}^{-1}}\left(\mathrm{Sp}\left\{\mathbf{A}\det\{\mathbf{C}^{-1}\}^{-\frac{1}{2}}\right\}\mathbf{C}\right)\right\}$$

$$= 2\sqrt{(2\pi)^{\kappa}}\mathcal{N}\mathrm{Sp}\left\{\mathbf{B}\mathbf{C}\left(\frac{1}{2}\det\{\mathbf{C}^{-1}\}^{-\frac{1}{2}}\mathrm{Sp}\left\{\mathbf{A}\mathbf{C}\right\}\right) + \mathbf{B}\det\{\mathbf{C}^{-1}\}^{-\frac{1}{2}}\mathbf{C}\mathbf{A}\mathbf{C}\right\}$$

$$= \mathrm{Sp}\left\{\mathbf{A}\mathbf{C}\right\}\mathrm{Sp}\left\{\mathbf{B}\mathbf{C}\right\} + 2\,\mathrm{Sp}\left\{\mathbf{A}\mathbf{C}\mathbf{B}\mathbf{C}\right\}. \tag{SS 200}$$

With these two identities, it is now straightforward to compute the Fisher information dyadic by taking advantage of the fact that *all* odd moments of a Gaussian distribution are zero, so that

$$\overline{\delta \log p \, \delta' \log p} = \delta\boldsymbol{\mu}^{\mathsf{T}}\mathbf{C}\delta'\boldsymbol{\mu} - \frac{1}{4}\operatorname{Sp}\left\{\mathbf{C}^{-1}\delta\mathbf{C}\right\}\operatorname{Sp}\left\{\mathbf{C}^{-1}\delta'\mathbf{C}\right\}$$

$$+ \frac{1}{4}\left(\operatorname{Sp}\left\{\mathbf{C}^{-1}\delta\mathbf{C}\right\}\operatorname{Sp}\left\{\mathbf{C}^{-1}\delta'\mathbf{C}\right\} + 2\operatorname{Sp}\left\{\mathbf{C}^{-1}\delta\mathbf{C}\mathbf{C}^{-1}\delta'\mathbf{C}\right\}\right)$$

$$= \delta\boldsymbol{\mu}^{\mathsf{T}}\mathbf{C}\delta'\boldsymbol{\mu} + \frac{1}{2}\operatorname{Sp}\left\{\mathbf{C}^{-1}\delta\mathbf{C}\mathbf{C}^{-1}\delta'\mathbf{C}\right\}, \qquad \text{(SS 201)}$$

where the distinction between the two variations $\delta\cdot$ and $\delta'\cdot$ simply serves to avoid confusion in the objects that are varied. The answer to this problem immediately follows hereafter.

Sample solution to Problem 4.6

Since

$$\delta \left|\mathbf{A}^{-1}\right\rangle\!\rangle = -\left|\mathbf{A}^{-1}\delta\mathbf{A}\mathbf{A}^{-1}\right\rangle\!\rangle = -\mathbf{A}^{-1}\otimes\mathbf{A}^{\mathsf{T}^{-1}}\left|\delta\mathbf{A}\right\rangle\!\rangle, \qquad \text{(SS 202)}$$

we have

$$\frac{\delta}{\delta\mathbf{A}}\mathbf{A}^{-1} \equiv \frac{\delta}{\delta\left|\mathbf{A}\right\rangle\!\rangle}\left|\mathbf{A}^{-1}\right\rangle\!\rangle = -\mathbf{A}^{-1}\otimes\mathbf{A}^{\mathsf{T}^{-1}}. \qquad \text{(SS 203)}$$

Equation (4.1.44) comes from the product rule,

$$\delta\left|\boldsymbol{x}\boldsymbol{x}^{\mathsf{T}}\right\rangle\!\rangle = \left|\delta\boldsymbol{x}\boldsymbol{x}^{\mathsf{T}}\right\rangle\!\rangle + \underbrace{\left|\boldsymbol{x}\delta\boldsymbol{x}^{\mathsf{T}}\right\rangle\!\rangle}_{= \mathcal{T}\left|\delta\boldsymbol{x}\boldsymbol{x}^{\mathsf{T}}\right\rangle\!\rangle}$$

$$= (\mathbf{1}\otimes\boldsymbol{x} + \mathcal{T}\mathbf{1}\otimes\boldsymbol{x})\left|\delta\boldsymbol{x}\right\rangle\!\rangle, \qquad \text{(SS 204)}$$

where the action of the transposition superoperator on $\mathbf{1}\otimes\boldsymbol{x}$ simply gives $\boldsymbol{x}\otimes\mathbf{1}$ in some computational basis $\{\boldsymbol{e}_j\}$:

$$\mathcal{T}\mathbf{1}\otimes\boldsymbol{x} = \sum_{j,k}\boldsymbol{e}_k\otimes\boldsymbol{e}_j\boldsymbol{e}_j^{\mathsf{T}}\otimes\boldsymbol{e}_k^{\mathsf{T}}\sum_{j'}\boldsymbol{e}_{j'}\boldsymbol{e}_{j'}^{\mathsf{T}}\otimes\sum_{k'}\boldsymbol{e}_{k'}x_{k'}$$

$$= \sum_{j,k}\boldsymbol{e}_k\boldsymbol{e}_j\boldsymbol{e}_j^{\mathsf{T}}\boldsymbol{e}_k^{\mathsf{T}}\sum_{j',k'}\boldsymbol{e}_{j'}\boldsymbol{e}_{k'}x_{k'}\boldsymbol{e}_{j'}^{\mathsf{T}}$$

$$= \sum_{j,k}\boldsymbol{e}_k\boldsymbol{e}_j x_k\boldsymbol{e}_j^{\mathsf{T}} = \sum_k\boldsymbol{e}_k x_k\otimes\sum_j\boldsymbol{e}_j\boldsymbol{e}_j^{\mathsf{T}} = \boldsymbol{x}\otimes\mathbf{1}. \qquad \text{(SS 205)}$$

Sample solution to Problem 4.7

The eight POM outcomes for a three-port device with efficiencies given by T_1, T_2 and T_3 are obtained as follows:

$$\Pi_{000} = \sum_{n=0}^{D_{\text{rec}}-1} |n\rangle \, (1 - T_1 - T_2 - T_3)^n \, \langle n| \, ,$$

$$\Pi_{001} = \Pi_{00\forall} - \Pi_{000}$$
$$= \sum_{n=0}^{D_{\text{rec}}-1} |n\rangle \left[(1 - T_1 - T_2)^n - (1 - T_1 - T_2 - T_3)^n \right] \langle n| \, ,$$

$$\Pi_{010} = \Pi_{0\forall 0} - \Pi_{000}$$
$$= \sum_{n=0}^{D_{\text{rec}}-1} |n\rangle \left[(1 - T_1 - T_3)^n - (1 - T_1 - T_2 - T_3)^n \right] \langle n| \, ,$$

$$\Pi_{011} = \Pi_{0\forall\forall} - \Pi_{001} - \Pi_{010} - \Pi_{000}$$
$$= \sum_{n=0}^{D_{\text{rec}}-1} |n\rangle \left[(1 - T_1)^n - (1 - T_1 - T_2)^n - (1 - T_1 - T_3)^n \right.$$
$$\left. + (1 - T_1 - T_2 - T_3)^n \right] \langle n| \, ,$$

$$\Pi_{100} = \Pi_{\forall 00} - \Pi_{000}$$
$$= \sum_{n=0}^{D_{\text{rec}}-1} |n\rangle \left[(1 - T_2 - T_3)^n - (1 - T_1 - T_2 - T_3)^n \right] \langle n| \, ,$$

$$\Pi_{101} = \Pi_{\forall 0\forall} - \Pi_{001} - \Pi_{100} - \Pi_{000}$$
$$= \sum_{n=0}^{D_{\text{rec}}-1} |n\rangle \left[(1 - T_2)^n - (1 - T_1 - T_2)^n - (1 - T_2 - T_3)^n \right.$$
$$\left. + (1 - T_1 - T_2 - T_3)^n \right] \langle n| \, ,$$

$$\Pi_{110} = \Pi_{\forall\forall 0} - \Pi_{010} - \Pi_{100} - \Pi_{000}$$
$$= \sum_{n=0}^{D_{\text{rec}}-1} |n\rangle \left[(1 - T_3)^n - (1 - T_1 - T_3)^n - (1 - T_2 - T_3)^n \right.$$
$$\left. + (1 - T_1 - T_2 - T_3)^n \right] \langle n| \, ,$$

$$\Pi_{111} = 1 - \Pi_{000} - \Pi_{001} - \Pi_{010} - \Pi_{011} - \Pi_{100} - \Pi_{101} - \Pi_{110}. \quad \text{(SS 206)}$$

One can verify that these expressions make sense by realizing that when $D_{\rm rec} \leq \mathcal{K}$, all POM outcomes referring to the detection of more than $D_{\rm rec}-1$ pulses vanish, as they should. For instance, for $D_{\rm rec} = 3 = \mathcal{K}$, we have $\Pi_{111} = 0$, and for $D_{\rm rec} = 2 < 3 = \mathcal{K}$, the outcomes Π_{011}, Π_{101}, Π_{110} and Π_{111} all vanish.

For a general $\mathcal{K}$-port device, a sample routine written for MATLAB is given below.

```matlab
1  function pom = Multiport_Dev(dim,trans_coeff)

3      % Inputs of the Multiport_Dev (Multiport Device) routine:
4      % dim:          dimension of the reconstruction subspace
5      % trans_coeff: array of transmission coefficients

7      % Output of the Multiport_Dev routine:
8      % pom:          a cell (array of matrices) of POM outcomes

10     if isrow(trans_coeff)    % converting a row array to a ...
            column array
11         trans=trans_coeff';
12     else
13         trans=trans_coeff;
14     end
15     numports=length(trans); % number of ports
16     numconfig=2.^numports;  % number of configurations

18     pom=cell(numconfig,1);  % POM initialization

20     for j=1:numconfig
21         binstring=dec2bin(j-1,numports); % binary string ...
              generation
22         binstring=str2num(binstring(:));

24         numfa=0; % counting and locating all "1"s in the string
25         for k=1:numports
26             if binstring(k)==1
27                 numfa=numfa+1;
28                 if numfa==1
29                     falist=[k];
30                 else
31                     falist=[falist;k];
32                 end
33             end
34         end
```

```
35          numfa

37          p_forall=1+sum(-mod(binstring+1,2).*trans); % for-all...
                probabilities

39          pom_outcome=zeros(dim,dim); % generating the POM outcome
40          for k=1:dim
41              pom_outcome(k,k)=p_forall.^(k-1);
42          end
43          for k=0:2.^numfa-2
44              tstrg1=dec2bin(k,numfa);
45              tstrg1=str2num(tstrg1(:));
46              tstrg2=zeros(1,numports);
47              for l=1:numfa
48                  tstrg2(falist(l))=tstrg1(l);
49              end
50              tstrg2=num2str(tstrg2);
51              pom_outcome=pom_outcome-pom{bin2dec(tstrg2)+1};
52          end
53          pom{j}=pom_outcome;
54      end

56  end
```

Sample solution to Problem 4.8

(1) By multiplying $|n\rangle$ to the right of the commutation relation $[A, A^\dagger] = 1$, we get

$$AA^\dagger |n\rangle = |n\rangle (n+1), \quad (\text{SS } 207)$$

from which we recognize that $A^\dagger |n\rangle \propto |n+1\rangle$ after another multiplication of $A^\dagger$ from the left of Eq. (SS 207). The real normalization constant $\sqrt{n+1}$, on the other hand, is obtained by multiplying $\langle n|$ on the left of Eq. (SS 207). To derive the relation in Eq. (4.2.39), we simply note the eigenvalue equation

$$A^\dagger A |n\rangle = |n\rangle n \quad (\text{SS } 208)$$

and repeat the same procedures as before.

(2) By multiplying $\langle n|$ on the left of the eigenvalue equation for $|\alpha\rangle$, one has

$$\langle n| A |\alpha\rangle = \sqrt{n+1} \langle n+1|\alpha\rangle = \alpha \langle n|\alpha\rangle. \quad (\text{SS } 209)$$

This simple iterative equation has the simple solution

$$\langle n|\alpha\rangle = C\,\frac{\alpha^n}{\sqrt{n!}}, \tag{SS 210}$$

where C is some constant that may be taken to be real, and the expression for $|\alpha\rangle$ in terms of $|n\rangle$ is thus obtained by multiplying $|n\rangle$ on both sides of the above equation and taking the sum over all nonnegative n. The value $C = e^{-\frac{1}{2}|\alpha|^2}$ is found by normalizing $|\alpha\rangle$.

(3) Using Eq. (4.2.38),

$$\langle \alpha^*|\alpha'\rangle = e^{-\frac{1}{2}\left(|\alpha|^2+|\alpha'|^2\right)} \sum_{n=0}^{\infty}\sum_{n'=0}^{\infty} \frac{\alpha^{*n}\alpha'^{n'}}{\sqrt{n!\,n'!}} \underbrace{\langle n|n'\rangle}_{=\,\delta_{n,n'}}$$

$$= e^{-\frac{1}{2}\left(|\alpha|^2+|\alpha'|^2\right)} \sum_{n=0}^{\infty} \frac{(\alpha^*\alpha')^n}{n!}$$

$$= e^{-\frac{1}{2}\left(|\alpha|^2+|\alpha'|^2\right)+\alpha^*\alpha'}. \tag{SS 211}$$

Sample solution to Problem 4.9

Upon multiplying the position eigenbra $\langle x|$ from the left to the eigenvalue equation for $|\alpha\rangle$, one obtains

$$\langle x|\,A\,|\alpha\rangle = \frac{1}{\sqrt{2}}\,\langle x|\,X+iP\,|\alpha\rangle$$

$$= \frac{1}{\sqrt{2}}\left(x+\frac{\partial}{\partial x}\right)\langle x|\alpha\rangle$$

$$= \alpha\,\langle x|\alpha\rangle\,. \tag{SS 212}$$

This yields the differential equation

$$\frac{\partial}{\partial x}\,\langle x|\alpha\rangle = \left(\sqrt{2}\alpha - x\right)\langle x|\alpha\rangle\,, \tag{SS 213}$$

whose solution is given by

$$\langle x|\alpha\rangle = C_\alpha\,e^{-\frac{1}{2}\left(x-\sqrt{2}\alpha\right)^2}, \tag{SS 214}$$

where C_α is a complex normalization constant that is a function of α. To look for C, we can make use of the result in Part (3) of **Problem 4.8** in as much as

$$\langle \alpha^*|\alpha'\rangle = \mathcal{C}_{\alpha^*}\mathcal{C}_{\alpha'} \int dx\, e^{-\frac{1}{2}\left(x-\sqrt{2}\alpha^*\right)^2} e^{-\frac{1}{2}\left(x-\sqrt{2}\alpha'\right)^2}$$

$$= \mathcal{C}_{\alpha^*}\mathcal{C}_{\alpha'}\, e^{-\left(\alpha^{*2}+\alpha'^2\right)} \underbrace{\int dx\, e^{-x^2+\sqrt{2}\left(\alpha^*+\alpha'\right)x}}_{=\sqrt{\pi}\, e^{\frac{1}{2}\left(\alpha^*+\alpha'\right)^2}}$$

$$= \mathcal{C}_{\alpha^*}\mathcal{C}_{\alpha'}\sqrt{\pi}\, e^{-\frac{1}{2}\alpha^{*2}-\frac{1}{2}\alpha'^2+\alpha^*\alpha'}$$

$$\equiv e^{-\frac{1}{2}|\alpha|^2-\frac{1}{2}|\alpha'|^2+\alpha^*\alpha'}, \tag{SS 215}$$

so that

$$\mathcal{C}_\alpha = \frac{e^{-\frac{1}{2}|\alpha|^2+\frac{1}{2}\alpha^2}}{\pi^{\frac{1}{4}}}. \tag{SS 216}$$

Finally, the full wave function is given by

$$\langle x|\alpha\rangle = \frac{e^{-\frac{1}{2}|\alpha|^2+\frac{1}{2}\alpha^2}}{\pi^{\frac{1}{4}}} e^{-\frac{1}{2}\left(x-\sqrt{2}\alpha\right)^2}$$

$$= \frac{1}{\pi^{\frac{1}{4}}} e^{-\frac{1}{2}|\alpha|^2-\frac{1}{2}\alpha^2-\frac{1}{2}x^2+\sqrt{2}\alpha x}. \tag{SS 217}$$

Sample solution to Problem 4.10

Assuming the eigenvalue equation

$$\langle \alpha^*|A = \alpha^*\langle \alpha^*| \tag{SS 218}$$

for the eigenbras $\langle \alpha^*|$, we can multiply the position eigenket $|x\rangle$ from the right, so that

$$\langle \alpha^*|A|x\rangle = \frac{1}{\sqrt{2}}\langle \alpha^*|X+iP|x\rangle$$

$$= \frac{1}{\sqrt{2}}\left(x-\frac{\partial}{\partial x}\right)\langle \alpha^*|x\rangle$$

$$= \alpha^*\langle \alpha^*|x\rangle, \tag{SS 219}$$

or

$$\frac{\partial}{\partial x} \langle \alpha^* | x \rangle = \left(x - \sqrt{2} \alpha^* \right) \langle \alpha^* | x \rangle .$$

(SS 220)

Solving this differential equation yields the function

$$\langle \alpha^* | x \rangle \propto \mathrm{e}^{\frac{1}{2}\left(x - \sqrt{2} \alpha^*\right)^2} ,$$

(SS 221)

which is strictly not normalizable as the integral

$$\int \mathrm{d}x \, |\langle \alpha^* | x \rangle|^2$$

(SS 222)

diverges. Consequently, the complex conjugate of Eq. (SS 220) is also not a physical wave function. Therefore, A has no eigenbras and $A^\dagger$ has no eigenkets.

Sample solution to Problem 4.11

We can investigate the analyticity of both kets by inspecting their components in the Fock basis.

The quick way is to adopt the polar representation of the Cauchy–Riemann equations that is compatible with the α-description of the kets. The simplest route to go about this is to say that since x and p are labels of one specific pair of orthogonal coordinate axes along e_x and e_p, the equations would also thus hold for any pair of coordinate unit vectors. In particular, for polar coordinates s and φ corresponding to the unit vectors e_s and e_φ, the equations become

$$s\frac{\partial u}{\partial s} = \frac{\partial v}{\partial \varphi} ,$$

$$\frac{\partial u}{\partial \varphi} = -s\frac{\partial v}{\partial s} .$$

(SS 223)

In this coordinate system, we have $\alpha = s\mathrm{e}^{\mathrm{i}\varphi}$ and entirety of $|\alpha\rangle \, \mathrm{e}^{\frac{1}{2}|\alpha|^2} = s^n \mathrm{e}^{\mathrm{i}n\varphi}/\sqrt{n!}$ follows because

$$s\frac{\partial u}{\partial s} = n\frac{s^n \cos(n\varphi)}{\sqrt{n!}} = \frac{\partial v}{\partial \varphi} ,$$

$$\frac{\partial u}{\partial \varphi} = -n\frac{s^n \sin(n\varphi)}{\sqrt{n!}} = -s\frac{\partial v}{\partial s} .$$

(SS 224)

On the other hand $|\alpha\rangle = s^n e^{\frac{1}{2}s^2} e^{in\varphi}/\sqrt{n!}$ is not entire since

$$s\frac{\partial u}{\partial s} = \frac{(n+s^2)s^n e^{\frac{1}{2}s^2}\cos(n\varphi)}{\sqrt{n!}} \neq \frac{\partial v}{\partial \varphi} = n\frac{s^n e^{\frac{1}{2}s^2}\cos(n\varphi)}{\sqrt{n!}}, \qquad \text{(SS 225)}$$

for instance.

One can also carry out the calculations in the x-p coordinate system. For the ket $|\alpha\rangle\, e^{\frac{1}{2}|\alpha|^2}$, the components are

$$\langle n|\alpha\rangle\, e^{\frac{1}{2}|\alpha|^2} = \frac{\alpha^n}{\sqrt{n!}}$$

$$= \frac{(x+ip)^n}{2^n\sqrt{n!}}$$

$$= \frac{1}{2^n\sqrt{n!}}\sum_{k=0}^{n}\binom{n}{k}x^{n-k}\,(ip)^k, \qquad \text{(SS 226)}$$

where the sum can be written as a linear combination of two parts,

$$\sum_{k=0}^{n}\binom{n}{k}x^{n-k}\,(ip)^k$$

$$= \underbrace{\sum_{\text{even }k}\binom{n}{k}(-1)^{\frac{k}{2}}x^{n-k}p^k}_{\equiv\, u(x,p)} + i\underbrace{\sum_{\text{odd }k}\binom{n}{k}(-1)^{\frac{k-1}{2}}x^{n-k}p^k}_{\equiv\, v(x,p)}. \qquad \text{(SS 227)}$$

The relations

$$\frac{\partial u}{\partial x} = \sum_{\text{even }k}\binom{n}{k}(-1)^{\frac{k}{2}}(n-k)\,x^{n-k-1}p^k \qquad (k'=k+1)$$

$$= \sum_{\text{odd }k'}\binom{n}{k'-1}(-1)^{\frac{k'-1}{2}}(n-k'+1)\,x^{n-k'}p^{k'-1}$$

$$= \sum_{\text{odd }k'}\binom{n}{k'}(-1)^{\frac{k'-1}{2}}k'\,x^{n-k'}p^{k'-1}$$

$$= \frac{\partial v}{\partial p} \qquad\qquad\qquad\qquad\qquad\qquad \text{(SS 228)}$$

and

$$\frac{\partial u}{\partial p} = \sum_{\text{even } k} \binom{n}{k} (-1)^{\frac{k}{2}} k\, x^{n-k} p^{k-1} \qquad (k' = k - 1)$$

$$= \sum_{\text{odd } k'} \binom{n}{k'+1} (-1)^{\frac{k'+1}{2}} (k'+1)\, x^{n-k'-1} p^{k'}$$

$$= -\sum_{\text{odd } k'} \binom{n}{k'} (-1)^{\frac{k'-1}{2}} (n - k')\, x^{n-k'-1} p^{k'}$$

$$= -\frac{\partial v}{\partial x} \qquad\qquad\qquad\qquad \text{(SS 229)}$$

among the partial derivatives imply that all components $\langle n | \alpha \rangle\, e^{\frac{1}{2}|\alpha|^2}$ are entire. It then follows that the ket $|\alpha\rangle\, e^{\frac{1}{2}|\alpha|^2}$ is entire. Things are different for the components of $|\alpha\rangle$, which are given by

$$\langle n | \alpha \rangle = e^{-\frac{1}{2}|\alpha|^2} \frac{\alpha^n}{\sqrt{n!}}$$

$$= e^{-\frac{1}{2}(x^2+p^2)} \frac{(x+ip)^n}{2^n \sqrt{n!}}$$

$$= \frac{1}{2^n \sqrt{n!}} [u(x,p) + iv(x,p)], \qquad \text{(SS 230)}$$

where the functions $u(x,p)$ and $v(x,p)$ are now defined as

$$u(x,p) = e^{-\frac{1}{2}(x^2+p^2)} \sum_{\text{even } k} \binom{n}{k} (-1)^{\frac{k}{2}} x^{n-k} p^k,$$

$$v(x,p) = e^{-\frac{1}{2}(x^2+p^2)} \sum_{\text{odd } k} \binom{n}{k} (-1)^{\frac{k-1}{2}} x^{n-k} p^k. \qquad \text{(SS 231)}$$

Owing to the exponential factor, the partial derivatives $\partial u / \partial x$ and $\partial v / \partial p$ are now different as

$$\frac{\partial u}{\partial x} = e^{-\frac{1}{2}(x^2+p^2)} \sum_{\text{even } k} \binom{n}{k} (-1)^{\frac{k}{2}} (n - k - x^2)\, x^{n-k-1} p^k \qquad (k' = k + 1)$$

$$= e^{-\frac{1}{2}(x^2+p^2)} \sum_{\text{odd } k'} \binom{n}{k'-1} (-1)^{\frac{k'-1}{2}} (n - k' + 1 - x^2)\, x^{n-k'} p^{k'-1},$$

$$\text{(SS 232)}$$

whereas

$$\frac{\partial v}{\partial p} = \mathrm{e}^{-\frac{1}{2}(x^2+p^2)} \sum_{\text{odd } k'} \binom{n}{k'}(-1)^{\frac{k'-1}{2}}\left(k'-p^2\right)x^{n-k'}p^{k'-1} \neq \frac{\partial u}{\partial x}.$$

(SS 233)

Sample solution to Problem 4.12

Upon the insertion of Eq. (4.2.59) into the right-hand side of Eq. (4.2.38), we have

$$\mathrm{e}^{-\frac{1}{2}|\alpha|^2} \sum_{n=0}^{\infty} \frac{\alpha^n}{\sqrt{n!}} \left(\frac{\sqrt{n!}}{2\pi\mathrm{i}} \oint_{\substack{\text{unit}\\\text{circle}}} \mathrm{d}\alpha' \, \frac{|\alpha'\rangle\, \mathrm{e}^{\frac{1}{2}|\alpha'|^2}}{\alpha'^{n+1}} \right)$$

$$= \frac{\mathrm{e}^{-\frac{1}{2}|\alpha|^2}}{2\pi\mathrm{i}} \sum_{n=0}^{\infty} \alpha^n \left[2\pi\mathrm{i}\,\mathrm{Res}\left(|\alpha'\rangle\, \mathrm{e}^{\frac{1}{2}|\alpha'|^2}, \{\alpha'=0, n+1\} \right) \right]$$

$$= \mathrm{e}^{-\frac{1}{2}|\alpha|^2} \sum_{n=0}^{\infty} \frac{\alpha^n}{n!} \left(\frac{\partial}{\partial\alpha'} \right)^n \left(|\alpha'\rangle\, \mathrm{e}^{\frac{1}{2}|\alpha'|^2} \right) \Bigg|_{\alpha'=0}$$

$$= \mathrm{e}^{-\frac{1}{2}|\alpha|^2}\, \mathrm{e}^{\alpha\frac{\partial}{\partial\alpha'}} \left(|\alpha'\rangle\, \mathrm{e}^{\frac{1}{2}|\alpha'|^2} \right) \Bigg|_{\alpha'=0}$$

$$= |\alpha'+\alpha\rangle\, \mathrm{e}^{-\frac{1}{2}|\alpha|^2}\, \mathrm{e}^{\frac{1}{2}|\alpha'+\alpha|^2} \Bigg|_{\alpha'=0}$$

$$= |\alpha\rangle .$$

(SS 234)

Sample solution to Problem 4.13

Using Eq. (4.2.38),

$$\int \frac{(\mathrm{d}\alpha')}{\pi} |\alpha'\rangle\langle\alpha'^*| = \sum_{m=0}^{\infty}\sum_{n=0}^{\infty} \frac{|m\rangle\langle n|}{\sqrt{m!\,n!}} \int \frac{(\mathrm{d}\alpha')}{\pi} \alpha'^m \alpha'^{*n}\, \mathrm{e}^{-|\alpha'|^2} \quad (\alpha'=s\,\mathrm{e}^{\mathrm{i}\varphi})$$

$$= \sum_{m=0}^{\infty}\sum_{n=0}^{\infty} \frac{|m\rangle\langle n|}{\sqrt{m!\,n!}} \underbrace{\int_0^{2\pi} \frac{\mathrm{d}\varphi}{\pi}\, \mathrm{e}^{\mathrm{i}(m-n)\varphi} \int_0^{\infty} \mathrm{d}s\, s^{m+n+1}\, \mathrm{e}^{-s^2}}_{= 2\delta_{m,n}}$$

$$= \sum_{n=0}^{\infty} \frac{|n\rangle \langle n|}{n!} \, 2 \int_0^{\infty} \mathrm{d}s \, s^{2n+1} \, \mathrm{e}^{-s^2} \quad (y = s^2)$$

$$= \sum_{n=0}^{\infty} \frac{|n\rangle \langle n|}{n!} \underbrace{\int_0^{\infty} \mathrm{d}y \, y^n \, \mathrm{e}^{-y}}_{= \, n!}$$

$$= 1. \tag{SS 235}$$

To show the overcompleteness relation, we may first express a coherent ket $|\alpha\rangle$ in the basis of position eigenkets $|x\rangle$, so that

$$|\alpha\rangle = \int \mathrm{d}x' \, |x'\rangle \, \langle x'|\alpha\rangle$$

$$= \frac{1}{\pi^{\frac{1}{4}}} \, \mathrm{e}^{-\frac{1}{2}|\alpha|^2 - \frac{1}{2}\alpha^2} \int \mathrm{d}x' \, |x'\rangle \, \mathrm{e}^{-\frac{1}{2}x'^2 + \sqrt{2}\alpha x'}. \tag{SS 236}$$

Thus,

$$\int \frac{\mathrm{d}x \, \mathrm{d}p}{2\pi} \, \frac{|\alpha'\rangle \langle \alpha^*|}{\langle \alpha^*|\alpha'\rangle} \bigg|_{\substack{\alpha'=x \\ \alpha=\mathrm{i}p}}$$

$$= \int \frac{\mathrm{d}x \, \mathrm{d}p}{2\pi^{\frac{3}{2}}} \, \mathrm{e}^{-\alpha^*\alpha' - \frac{1}{2}\alpha^2 - \frac{1}{2}\alpha^{*2}}$$

$$\times \int \mathrm{d}x' \, \mathrm{d}x'' \, |x'\rangle \, \mathrm{e}^{-\frac{1}{2}(x'^2 + x''^2) + \sqrt{2}(\alpha' x' + \alpha^* x'')} \, \langle x''| \bigg|_{\substack{\alpha'=x \\ \alpha=\mathrm{i}p}}$$

$$= \int \mathrm{d}x' \, \mathrm{d}x'' \, |x'\rangle \, \mathrm{e}^{-\frac{1}{2}(x'^2 + x''^2)} \, \langle x''|$$

$$\times \int \frac{\mathrm{d}p}{2\pi} \, \mathrm{e}^{\frac{1}{2}p^2 - \mathrm{i}\sqrt{2}px''} \int \frac{\mathrm{d}x}{\sqrt{\pi}} \, \mathrm{e}^{-\frac{1}{2}x^2 + (\mathrm{i}p + \sqrt{2}x')x}. \tag{SS 237}$$

The double integral in x and p is evaluated as follows:

$$\int \frac{\mathrm{d}p}{2\pi} \, \mathrm{e}^{\frac{1}{2}p^2 - \mathrm{i}\sqrt{2}px''} \int \frac{\mathrm{d}x}{\sqrt{\pi}} \, \mathrm{e}^{-\frac{1}{2}x^2 + (\mathrm{i}p + \sqrt{2}x')x}$$

$$= \int \frac{\mathrm{d}p}{\sqrt{2\pi}} \, \mathrm{e}^{\frac{1}{2}p^2 - \mathrm{i}\sqrt{2}px''} \, \mathrm{e}^{\frac{1}{2}(\mathrm{i}p + \sqrt{2}x')^2}$$

$$= e^{x'^2} \int \frac{dp}{\sqrt{2\pi}} \, e^{i\sqrt{2}p(x'-x'')}$$

$$= e^{x'^2} \, \delta(x' - x''). \tag{SS 238}$$

In the end, we have

$$\int \frac{dx\,dp}{2\pi} \frac{|\alpha'\rangle\langle\alpha^*|}{\langle\alpha^*|\alpha'\rangle}\bigg|_{\substack{\alpha'=x\\\alpha=ip}} = \int dx'\,dx''\,|x'\rangle\,e^{\frac{1}{2}\left(x'^2-x''^2\right)}\,\delta(x'-x'')\,\langle x''|$$

$$= \int dx'\,|x'\rangle\,\langle x'|$$

$$= 1. \tag{SS 239}$$

Sample solution to Problem 4.14

One can write $\left[F\left(A^\dagger A\right)\right]_{\mathrm{N}}$ as the linear combination

$$\left[F\left(A^\dagger A\right)\right]_{\mathrm{N}} = \sum_{m=0}^{\infty} f_m A^{\dagger\,m} A^m \tag{SS 240}$$

in terms of the operators $A^{\dagger\,m}A^m$. It is straightforward to see that the Fock ket $|n\rangle$ is an eigenket of $A^{\dagger\,m}A^m$ since there are as many creation operators as there are annihilation operators in this product, and so

$$A^{\dagger\,m}A^m\,|n\rangle = A^{\dagger\,m}\,|n-m\rangle\,\sqrt{\frac{n!}{(n-m)!}}$$

$$= |n\rangle\,\frac{n!}{(n-m)!}. \tag{SS 241}$$

If that is the case, by invoking the identity in Eq. (4.2.57) the relations

$$\left[F\left(A^\dagger A\right)\right]_{\mathrm{N}}|n\rangle = |n\rangle\sum_{m=0}^{\infty} f_m \frac{n!}{(n-m)!}$$

$$= |n\rangle\sum_{m=0}^{\infty} f_m \left(\frac{d}{dx}\right)^m x^n\bigg|_{x=1}$$

$$= \sum_{m=0}^{\infty} f_m \left(\frac{d}{dx}\right)^m x^{A^\dagger A} \Big|_{x=1} |n\rangle$$

$$= F\left(\frac{d}{dx}\right) x^{A^\dagger A} \Big|_{x=1} |n\rangle \qquad \text{(SS 242)}$$

must be true for all $|n\rangle$, thus implying Eq. (4.2.92).

It is now easy to prove the two identities in Eq. (4.2.94). For the first identity,

$$\left[e^{-\lambda A^\dagger A}\right]_N = e^{-\lambda \frac{d}{dx}} x^{A^\dagger A} \Big|_{x=1}$$

$$= (x-\lambda)^{A^\dagger A} \Big|_{x=1}$$

$$= (1-\lambda)^{A^\dagger A}, \qquad \text{(SS 243)}$$

and for the second identity,

$$A^{\dagger n} A^n = \left(\frac{d}{dx}\right)^n x^{A^\dagger A} \Big|_{x=1}$$

$$= \frac{(A^\dagger A)!}{(A^\dagger A - n)!} x^{A^\dagger A} \Big|_{x=1}$$

$$= \frac{(A^\dagger A)!}{(A^\dagger A - n)!}. \qquad \text{(SS 244)}$$

Sample solution to Problem 4.15

It is enough to show the first of the two identities. Multiplying the left-hand side by $|n\rangle$ from the right,

$$f\left(A^\dagger A\right) A |n\rangle = f\left(A^\dagger A\right) |n-1\rangle \sqrt{n}$$

$$= |n-1\rangle \sqrt{n} f(n-1)$$

$$= A |n\rangle f(n-1)$$

$$= A f\left(A^\dagger A - 1\right) |n\rangle. \qquad \text{(SS 245)}$$

The second identity is simply obtained by taking the adjoint of the first identity and replacing $A^\dagger A$ with $A^\dagger A + 1$.

Sample solution to Problem 4.16

One may expand the characteristic function for I'_{diff} into the power series

$$e^{ikI'_{\text{diff}}} = \sum_{l=0}^{\infty} \frac{1}{l!} \left(\frac{ik}{\sqrt{2}|\beta|\eta_{\text{L}}} \right)^l$$

$$\times \Big[\eta_{\text{L}} (C^\dagger C - D^\dagger D) + \sqrt{\eta_{\text{L}}(1-\eta_{\text{L}})} (C^\dagger C_{\text{L}} + C_{\text{L}}^\dagger C - D^\dagger D_{\text{L}} - D_{\text{L}}^\dagger D)$$

$$+ (1-\eta_{\text{L}}) (C_{\text{L}}^\dagger C_{\text{L}} - D_{\text{L}}^\dagger D_{\text{L}}) \Big]^l \tag{SS 246}$$

and realize that all terms with B absent contribute to expansion terms of order lower than l in $|\beta|$ after taking the average with the reference coherent state. Thus, these terms do not contribute in the high-energy limit since they give corrections in orders of $1/|\beta|$.

Sample solution to Problem 4.17

Using Eq. (4.2.122),

$$\sum_{n=0}^{\infty} \frac{t^n}{n!} H_n(x) = e^{x^2} \int \frac{dy}{\sqrt{4\pi}} e^{-\frac{1}{4}y^2 + ixy} \sum_{n=0}^{\infty} \frac{t^n}{n!} (-iy)^n$$

$$= e^{x^2} \int \frac{dy}{\sqrt{4\pi}} e^{-\frac{1}{4}y^2 + i(x-t)y}$$

$$= e^{x^2} e^{-(x-t)^2}$$

$$= e^{2xt - t^2}. \tag{SS 247}$$

The orthogonality property can then be shown using this generating function. We first compute the integral of the product of e^{-x^2} and the square of the generating function, that is

$$\int dx\, e^{-x^2} e^{4xt - 2t^2} = \sqrt{\pi}\, e^{2t^2}. \tag{SS 248}$$

Using the identity for the generating function, the above integral is equal to

$$\int \mathrm{d}x \, e^{-x^2} \left[\sum_{n=0}^{\infty} \frac{t^n}{n!} H_n(x) \right]^2$$

$$= \sum_{m=0}^{\infty} \sum_{n=0}^{\infty} \frac{t^{m+n}}{m! \, n!} \int \mathrm{d}x \, e^{-x^2} H_m(x) H_n(x) . \qquad \text{(SS 249)}$$

Since the double sum must converge to

$$\sqrt{\pi} \, e^{2t^2} = \sqrt{\pi} \sum_{n=0}^{\infty} \frac{2^n t^{2n}}{n!}$$

$$= \sum_{m=0}^{\infty} \sum_{n=0}^{\infty} \frac{t^{m+n}}{m! \, n!} \left(2^n \, n! \, \sqrt{\pi} \, \delta_{m,n} \right), \qquad \text{(SS 250)}$$

we must have

$$\int \mathrm{d}x \, e^{-x^2} H_m(x) H_n(x) = 2^n \, n! \, \sqrt{\pi} \, \delta_{m,n} \qquad \text{(SS 251)}$$

from comparing the coefficients in both sums.

Sample solution to Problem 4.18

By inserting the resolution of the identity

$$\int \mathrm{d}x' \, |x'\rangle \langle x'| = 1, \qquad \text{(SS 252)}$$

we have

$$\langle p' | x_\vartheta \rangle = \int \mathrm{d}x' \, \langle p' | x' \rangle \langle x' | x_\vartheta \rangle$$

$$= \int \mathrm{d}x' \, \frac{e^{-ix'p'}}{\sqrt{2\pi}} \, \frac{1}{\sqrt{-2\pi i \, e^{i\vartheta} \sin \vartheta}} \, e^{-\frac{i}{2} \cot \vartheta \left(x_\vartheta^2 + x'^2 \right) + \frac{ix_\vartheta x'}{\sin \vartheta}}$$

$$= \frac{1}{\sqrt{-2\pi i\, e^{i\vartheta}\sin\vartheta}}\, e^{-\frac{i}{2}\cot\vartheta\, x_\vartheta^2} \underbrace{\int \frac{dx'}{\sqrt{2\pi}}\, e^{-\frac{i}{2}\cot\vartheta\, x'^2 + i\left(\frac{x_\vartheta}{\sin\vartheta} - p'\right)x'}}_{= \sqrt{\frac{1}{i\cot\vartheta}}\, e^{\frac{i}{2}\tan\vartheta\left(\frac{x_\vartheta}{\sin\vartheta} - p'\right)^2}}$$

$$= \frac{1}{\sqrt{2\pi\, e^{i\vartheta}\cos\vartheta}}\, e^{\frac{i}{2}\tan\vartheta\left(x_\vartheta^2 + p'^2\right) - \frac{i x_\vartheta p'}{\cos\vartheta}}. \tag{SS 253}$$

Sample solution to Problem 4.19

To prove the first relation,

$$\langle x' | p' \rangle$$

$$= \int dx_\vartheta\, \langle x' | x_\vartheta \rangle \langle x_\vartheta | p' \rangle$$

$$= \int dx_\vartheta \left[\frac{1}{\sqrt{-2\pi i\, e^{i\vartheta}\sin\vartheta}}\, e^{-\frac{i}{2}\cot\vartheta\left(x_\vartheta^2 + x'^2\right) + \frac{i x_\vartheta x'}{\sin\vartheta}} \right]$$

$$\times \left[\frac{1}{\sqrt{2\pi\, e^{-i\vartheta}\cos\vartheta}}\, e^{-\frac{i}{2}\tan\vartheta\left(x_\vartheta^2 + p'^2\right) + \frac{i x_\vartheta p'}{\cos\vartheta}} \right]$$

$$= \frac{1}{\pi\sqrt{-2\,i\sin(2\vartheta)}}\, e^{-\frac{i}{2}\left(\cot\vartheta\, x'^2 + \tan\vartheta\, p'^2\right)} \underbrace{\int dx_\vartheta\, e^{-\frac{i}{2\sin\vartheta\cos\vartheta}x_\vartheta^2 + i\left(\frac{x'}{\sin\vartheta} + \frac{p'}{\cos\vartheta}\right)x_\vartheta}}_{= \sqrt{\frac{\pi\sin(2\vartheta)}{i}}\, e^{\frac{i}{2}\sin\vartheta\cos\vartheta\left(\frac{x'}{\sin\vartheta} + \frac{p'}{\cos\vartheta}\right)^2}}$$

$$= \frac{1}{\sqrt{2\pi}}\, e^{i x' p'}. \tag{SS 254}$$

To prove the second relation,

$$\langle x' | x'' \rangle = \int dx_\vartheta\, \langle x' | x_\vartheta \rangle \langle x_\vartheta | x'' \rangle$$

$$= \int dx_\vartheta \left[\frac{1}{\sqrt{-2\pi i\, e^{i\vartheta}\sin\vartheta}}\, e^{-\frac{i}{2}\cot\vartheta\left(x_\vartheta^2 + x'^2\right) + \frac{i x_\vartheta x'}{\sin\vartheta}} \right]$$

$$\times \left[\frac{1}{\sqrt{2\pi i\, e^{-i\vartheta}\sin\vartheta}}\, e^{\frac{i}{2}\cot\vartheta\left(x_\vartheta^2 + x''^2\right) - \frac{i x_\vartheta x''}{\sin\vartheta}} \right]$$

$$= e^{\frac{i}{2}\cot\vartheta\left(x''^2 - x'^2\right)} \underbrace{\int \frac{dx_\vartheta}{2\pi|\sin\vartheta|} e^{\frac{i}{\sin\vartheta}\left(x' - x''\right)x_\vartheta}}_{=\,\delta(x' - x'')}$$

$$= \delta(x' - x''). \tag{SS 255}$$

Finally, the third relation may also be proven likewise.

Sample solution to Problem 4.20

The relation in Eq. (4.2.133) can be shown by doing the same type of calculation in **Problem 4.18**,

$$\langle x_\vartheta | x'_{\vartheta'} \rangle = \int dx \, \langle x_\vartheta | x \rangle \, \langle x | x'_{\vartheta'} \rangle$$

$$= \int dx \left[\frac{1}{\sqrt{2\pi i\, e^{-i\vartheta}\sin\vartheta}} e^{\frac{i}{2}\cot\vartheta\left(x_\vartheta^2 + x^2\right) - \frac{ix_\vartheta x}{\sin\vartheta}} \right]$$

$$\times \left[\frac{1}{\sqrt{-2\pi i\, e^{i\vartheta'}\sin\vartheta'}} e^{-\frac{i}{2}\cot\vartheta'\left(x_{\vartheta'}'^2 + x^2\right) + \frac{ix'_{\vartheta'} x}{\sin\vartheta'}} \right]$$

$$= \frac{e^{\frac{i}{2}\left(\vartheta - \vartheta'\right)}}{2\pi\sqrt{\sin\vartheta \sin\vartheta'}} e^{\frac{i}{2}\left(\cot\vartheta\, x_\vartheta^2 - \cot\vartheta'\, x_{\vartheta'}'^2\right)}$$

$$\times \underbrace{\int dx\, e^{-\frac{i}{2}\left(\cot\vartheta' - \cot\vartheta\right)x^2 + i\left(\frac{x'_{\vartheta'}}{\sin\vartheta'} - \frac{x_\vartheta}{\sin\vartheta}\right)x}}_{=\sqrt{\dfrac{2\pi}{i\left(\cot\vartheta' - \cot\vartheta\right)}}\, e^{\frac{i}{2\left(\cot\vartheta' - \cot\vartheta\right)}\left(\frac{x'_{\vartheta'}}{\sin\vartheta'} - \frac{x_\vartheta}{\sin\vartheta}\right)^2}} .$$

$$\tag{SS 256}$$

Manipulating the trigonometric functions, we find that

$$\sin\vartheta \sin\vartheta' \left(\cot\vartheta' - \cot\vartheta\right) = \sin\vartheta \cos\vartheta' - \cos\vartheta \sin\vartheta'$$

$$= \sin\left(\vartheta - \vartheta'\right), \tag{SS 257}$$

$$\frac{1}{(\cot \vartheta' - \cot \vartheta)(\sin \vartheta')^2} - \cot \vartheta'$$

$$= \frac{1}{\sin \vartheta' \cos \vartheta' - \dfrac{\cos \vartheta (\sin \vartheta')^2}{\sin \vartheta}} - \cot \vartheta'$$

$$= \frac{\sin \vartheta}{\sin \vartheta \sin \vartheta' \cos \vartheta' - \cos \vartheta (\sin \vartheta')^2} - \frac{\cos \vartheta'}{\sin \vartheta'}$$

$$= \frac{\sin \vartheta - \sin \vartheta (\cos \vartheta')^2 + \cos \vartheta \sin \vartheta' \cos \vartheta'}{\sin \vartheta' \sin (\vartheta - \vartheta')}$$

$$= \cot (\vartheta - \vartheta') \qquad\qquad \text{(SS 258)}$$

and

$$\cot \vartheta + \frac{1}{(\cot \vartheta' - \cot \vartheta)(\sin \vartheta)^2} = -\frac{1}{(\cot \vartheta - \cot \vartheta')(\sin \vartheta)^2} + \cot \vartheta$$

$$= -\left[\frac{1}{(\cot \vartheta - \cot \vartheta')(\sin \vartheta)^2} - \cot \vartheta \right]$$

$$= -\cot (\vartheta' - \vartheta)$$

$$= \cot (\vartheta - \vartheta'). \qquad\qquad \text{(SS 259)}$$

Therefore, inserting all the simplifications yields

$$\langle x_\vartheta | x'_{\vartheta'} \rangle = \frac{e^{\frac{i}{2}(\vartheta - \vartheta')}}{\sqrt{2\pi i \sin(\vartheta - \vartheta')}} \, e^{\frac{i}{2\sin(\vartheta - \vartheta')}\left[\left(x_\vartheta^2 + x_{\vartheta'}'^2\right)\cos(\vartheta - \vartheta') - 2x_\vartheta x'_{\vartheta'}\right]}.$$

$$\text{(SS 260)}$$

Now, as $\vartheta - \vartheta' \longrightarrow 0$,

$$\langle x_\vartheta | x'_{\vartheta'} \rangle \Big|_{\vartheta - \vartheta' \longrightarrow 0} = \frac{1}{\sqrt{2\pi i \sin(\vartheta - \vartheta')}} \, e^{\frac{i}{2\sin(\vartheta - \vartheta')}\left(x_\vartheta - x'_{\vartheta'}\right)^2} \Bigg|_{\vartheta - \vartheta' \longrightarrow 0}$$

$$= \frac{1}{\sqrt{2\pi\gamma}} \, e^{-\frac{\left(x_\vartheta - x'_{\vartheta'}\right)^2}{2\gamma}} \Bigg|_{\vartheta - \vartheta' \longrightarrow 0} \qquad (\gamma = i \sin(\vartheta - \vartheta'))$$

$$= \delta \left(x_\vartheta - x'_{\vartheta'}\right). \qquad\qquad \text{(SS 261)}$$

When $\vartheta - \vartheta' = \pi/2$,

$$\left\langle x_\vartheta \,\middle|\, x'_{\vartheta - \frac{\pi}{2}} \right\rangle = \frac{e^{i\frac{\pi}{4}}}{\sqrt{2\pi i}} \, e^{\frac{i}{2}\left(-2x_\vartheta x'_\vartheta - \frac{\pi}{2}\right)}$$

$$= \frac{1}{\sqrt{2\pi}} \, e^{-i\, x_\vartheta \, x'_{\vartheta - \frac{\pi}{2}}}. \qquad \text{(SS 262)}$$

Sample solution to Problem 4.21

We can start from the relation between the Hermite polynomials and their generating function in Eq. (4.2.125). After computing the Fourier transform of the product of the generating function and $e^{-\frac{1}{2}x^2}$, namely

$$F_k \left\{ e^{-\frac{1}{2}x^2 + 2xt - t^2} \right\} = e^{-t^2} \int \frac{dx}{\sqrt{2\pi}} \, e^{-\frac{1}{2}x^2 + (2t - ik)x}$$

$$= e^{-t^2} \, e^{\frac{1}{2}(2t - ik)^2}$$

$$= e^{-\frac{1}{2}k^2} \, e^{2k(-it) - (it)^2}$$

$$= e^{-\frac{1}{2}k^2} \sum_{n=0}^{\infty} \frac{t^n}{n!} (-i)^n H_n(k)$$

$$= \sum_{n=0}^{\infty} \frac{t^n}{n!} (-i)^n \, g(n; k), \qquad \text{(SS 263)}$$

and considering this is equal to

$$F_k \left\{ e^{-\frac{1}{2}x^2} \sum_{n=0}^{\infty} \frac{t^n}{n!} H_n(x) \right\} = \sum_{n=0}^{\infty} \frac{t^n}{n!} F_k \{g(n; x)\} \qquad \text{(SS 264)}$$

we realize that

$$F_k \{g(n; x)\} = (-i)^n \, g(n; k). \qquad \text{(SS 265)}$$

Thus, the eigenfunctions and eigenvalues of F_k are respectively $\left\{ e^{-\frac{1}{2}x^2} H_n(x) \right\}$ and $\{(-i)^n\}$, which confirms that F_k is unitary.

Sample solution to Problem 5.1

From the definition in Eq. (5.1.2),

$$\text{tr}\{W_{x,p}\,W_{x'p'}\}$$

$$= \text{tr}\left\{\left(2\int dy\,|x+y\rangle\,e^{2\,\mathrm{i}py}\,\langle x-y|\right)\left(2\int dy'\,|x'+y'\rangle\,e^{2\,\mathrm{i}p'y'}\,\langle x'-y'|\right)\right\}$$

$$= 4\int dy\,dy'\,e^{2\,\mathrm{i}(py+p'y')}\,\underbrace{\langle x-y|x'+y'\rangle\,\langle x'-y'|x+y\rangle}_{=\,\delta(x-x'-y-y')\delta(x'-x-y-y')}$$

$$= 2\,\underbrace{e^{2\,\mathrm{i}p'\left(x-x'\right)}}_{=\,1}\,\delta(x-x')\int dy\,e^{2\,\mathrm{i}(p-p')y}$$

$$= 2\pi\,\delta(x-x')\delta(p-p'). \tag{SS 266}$$

In the language of superkets/superbras, this means that

$$\langle\!\langle W_{x,p}|W_{x',p'}\rangle\!\rangle = 2\pi\,\delta(x-x')\delta(p-p') \tag{SS 267}$$

or

$$\int \frac{dx\,dp}{2\pi}\,|W_{x,p}\rangle\!\rangle\,\langle\!\langle W_{x,p}| = \mathcal{I}, \tag{SS 268}$$

which says that the Wigner operators form a complete set of trace-orthonormal basis operators that span the infinite-dimensional space of Hermitian operators. This permits another simple relationship between the state ρ and its Wigner function $\rho_{\text{w}}(x,p)$, that is

$$\rho = \int \frac{dx\,dp}{2\pi}\,\rho_{\text{w}}(x,p)\,W_{x,p}. \tag{SS 269}$$

Along with Eq. (5.1.2), this implies that every ρ has a unique Wigner function.

Sample solution to Problem 5.2

It is now a trivial affair to prove Eq. (5.1.25), since

$$
\int \frac{dx\,dp}{2\pi}\, \rho_{\mathrm{W}}(x,p)\, f_{\mathrm{W}}(x,p) = \int \frac{dx\,dp}{2\pi}\, \langle\!\langle \rho | W_{x,p} \rangle\!\rangle \, \langle\!\langle W_{x,p} | F \rangle\!\rangle
$$

$$
= \langle\!\langle \rho | F \rangle\!\rangle = \mathrm{tr}\{\rho F\}. \tag{SS 270}
$$

Sample solution to Problem 5.3

Upon noting that

$$
\langle \beta^* | (-1)^{\left(A^\dagger - \alpha^*\right)(A - \alpha)} | \beta' \rangle = \langle \beta^* | \left[e^{-2\left(A^\dagger - \alpha^*\right)(A - \alpha)} \right]_{\mathrm{N}} | \beta' \rangle
$$

$$
= e^{-2|\alpha|^2 - \frac{1}{2}|\beta|^2 - \frac{1}{2}|\beta'|^2 - \beta^* \beta' + 2\alpha^* \beta' + 2\alpha\beta^*}, \tag{SS 271}
$$

the Wigner function may be turned into the double integral

$$
\rho_{\mathrm{W}}^{\mathrm{sq}}(\alpha, \alpha'; z) = 2 \operatorname{sech} r \, e^{-2|\alpha|^2 - |\alpha'|^2 - \mathrm{Re}\left\{\alpha'^2 e^{-i\theta}\right\} \tanh r}
$$

$$
\times \int \frac{(d\beta)}{\pi} \, e^{-|\beta|^2 + \alpha'\beta \operatorname{sech} r + 2\alpha\beta^* + \frac{1}{2}e^{-i\theta} \tanh r \beta^2}
$$

$$
\times \int \frac{(d\beta')}{\pi} \, e^{-|\beta'|^2 + (2\alpha^* - \beta^*)\beta' + \alpha'\beta'^* \operatorname{sech} r + \frac{1}{2}e^{i\theta} \tanh r \beta'^{*2}}. \tag{SS 272}
$$

The β' integral involves the parameters $a = 1$, $b_1 = 2\alpha^* - \beta^*$, $b_2 = \alpha' \operatorname{sech} r$, $c_1 = 0$ and $c_2 = \frac{1}{2}e^{i\theta} \tanh r$, where $a^2 - 4c_1 c_2 = 1$, so that

$$
\rho_{\mathrm{W}}^{\mathrm{sq}}(\alpha, \alpha'; z) = 2 \operatorname{sech} r \, e^{-2|\alpha|^2 - |\alpha'|^2 - \mathrm{Re}\left\{\alpha'^2 e^{-i\theta}\right\} \tanh r} \, e^{2\alpha^* \alpha' \operatorname{sech} r + 2\alpha^{*2} e^{i\theta} \tanh r}
$$

$$
\times \int \frac{(d\beta)}{\pi} \left[e^{-|\beta|^2 + \alpha'\beta \operatorname{sech} r + \left(2\alpha - 2\alpha^* e^{i\theta} \tanh r - \alpha' \operatorname{sech} r\right)\beta^*} \right.
$$

$$
\left. \times e^{\frac{1}{2}e^{-i\theta} \tanh r \beta^2 + \frac{1}{2}e^{i\theta} \tanh r \beta^{*2}} \right], \tag{SS 273}
$$

where now $a = 1$, $b_1 = \alpha' \operatorname{sech} r$, $b_2 = 2\alpha - 2\alpha^* e^{i\theta} \tanh r - \alpha' \operatorname{sech} r$, $c_1 = \frac{1}{2} e^{-i\theta} \tanh r$ and $c_2 = \frac{1}{2} e^{i\theta} \tanh r$, with $a^2 - 4c_1 c_2 = (\operatorname{sech} r)^2$.

After some straightforward simplification, one arrives at

$$\rho_{\mathrm{W}}^{\mathrm{sq}}(\alpha, \alpha'; z) = 2\, e^{-2\left\{\cosh(2r)|\alpha|^2 + |\alpha'|^2\right\}}$$

$$\times\, e^{4\,\mathrm{Re}\left\{\alpha\left(\alpha'^* \cosh r - \alpha' e^{-i\theta} \sinh r\right)\right\} + 2\sinh(2r)\mathrm{Re}\left\{\alpha^2 e^{-i\theta}\right\}}.$$

$$\text{(SS 274)}$$

It is clear that when $z = 0$, the Wigner function reduces to that for the coherent state.

As a final check, we see if this Wigner function normalizes to unity over the entire phase space. For that, we make use of Eq. (5.1.130) one more time to evaluate the phase-space integral

$$\int \frac{(\mathrm{d}\alpha)}{\pi}\, \rho_{\mathrm{W}}^{\mathrm{sq}}(\alpha, \alpha'; z)$$

$$= 2\, e^{-2|\alpha'|^2}\, I_{\mathrm{gauss}}\left(2\cosh(2r), 2\zeta^*, 2\zeta, e^{-i\theta}\sinh(2r), e^{i\theta}\sinh(2r)\right), \quad \text{(SS 275)}$$

where $\zeta = \alpha' \cosh r - \alpha'^* e^{i\theta} \sinh r$ and $a^2 - 4c_1 c_2 = 4$ in this case. It is then a simple matter to show that the right-hand side is indeed equal to one.

Sample solution to Problem 5.4

Substituting the formula in Eq. (5.1.62) into the two Hermite polynomials that are present in Eq. (5.1.48),

$$I_{m,n} = m!\, n! \underbrace{\oint_{\substack{\text{unit}\\\text{circle}}} \frac{\mathrm{d}z}{2\pi i} \frac{e^{2xz - z^2}}{z^{m+1}} \oint_{\substack{\text{unit}\\\text{circle}}} \frac{\mathrm{d}z'}{2\pi i} \frac{e^{2xz' - z'^2}}{z'^{n+1}} \int \mathrm{d}y\, e^{-y^2 + 2\mathrm{i}py + 2(z' - z)y}}_{= \sqrt{\pi}\, e^{(\mathrm{i}p + z' - z)^2}}$$

$$= m!\, n!\, \sqrt{\pi}\, e^{-p^2} \oint_{\substack{\text{unit}\\\text{circle}}} \frac{\mathrm{d}z}{2\pi i} \frac{1}{z^{m+1}} \oint_{\substack{\text{unit}\\\text{circle}}} \frac{\mathrm{d}z'}{2\pi i} \frac{1}{z'^{n+1}} e^{-2zz' + 2(x - \mathrm{i}p)z + 2(x + \mathrm{i}p)z'}.$$

$$\text{(SS 276)}$$

For the case $m \geq n$, we can take $x = \sqrt{2}\,z$, $y = \sqrt{2}\,z'$ and $\alpha = \sqrt{2}\,(x-ip) = \beta^*$, following which

$$I_{m \geq n} = m!\,n!\,\sqrt{\pi}\,e^{-p^2} \sum_{j=0}^{\infty}\sum_{k=0}^{\infty} \frac{(-1)^k}{j!} \left(\sqrt{2}\alpha^*\right)^{j-k} L_k^{(j-k)}\left(4|\alpha|^2\right)$$

$$\times \underbrace{\oint_{\substack{\text{unit} \\ \text{circle}}} \frac{dz}{2\pi i} \frac{(\sqrt{2}\,z)^j}{z^{m+1}} \oint_{\substack{\text{unit} \\ \text{circle}}} \frac{dz'}{2\pi i} \frac{(\sqrt{2}\,z')^k}{z'^{n+1}}}_{= (\sqrt{2})^{j+k}\delta_{j,m}\delta_{k,n}}$$

$$= (-1)^n\,2^m\,n!\,\sqrt{\pi}\,e^{-p^2}\,(x-ip)^{m-n}\,L_n^{(m-n)}\left(2x^2+2p^2\right). \quad \text{(SS 277)}$$

For the case $m < n$, we may repeat the above calculation by taking $x = \sqrt{2}\,z'$, $y = \sqrt{2}\,z$ and $\alpha = \sqrt{2}\,(x+ip) = \beta^*$, and obtain Eq. (5.1.59) as a result.

Sample solution to Problem 5.5

By expanding the exponential function e^{-y} in Eq. (5.1.57) into its power series, we get

$$L_n^{(\nu)}(y) = \frac{1}{n!}\,y^{-\nu}\,e^y \left(\frac{d}{dy}\right)^n \left(y^{n+\nu}\,e^{-y}\right)$$

$$= \frac{1}{n!}\,y^{-\nu}\,e^y \sum_{k=0}^{\infty} \frac{(-1)^k}{k!} \left(\frac{d}{dy}\right)^n y^{k+n+\nu}$$

$$= e^y \sum_{k=0}^{\infty} \frac{(-1)^k}{k!}\,y^k \binom{k+\nu+n}{n}, \quad \text{(SS 278)}$$

with which the series

$$\sum_{n=0}^{\infty} t^n L_n^{(\nu)}(y) = e^y \sum_{k=0}^{\infty} \frac{(-1)^k}{k!}\,y^k \underbrace{\sum_{n=0}^{\infty} \binom{k+\nu+n}{n} t^n}_{= (1-t)^{-k-\nu-1}}$$

$$= \frac{e^y\,e^{-\frac{y}{1-t}}}{(1-t)^{\nu+1}}$$

$$= \frac{e^{-\frac{yt}{1-t}}}{(1-t)^{\nu+1}}. \quad \text{(SS 279)}$$

To continue, we follow the calculations for **Problem 4.17** and compute the integral involving the square of the generating function for the associated Laguerre polynomials,

$$\int_0^\infty dy\, y^\nu\, e^{-y} \left[\frac{e^{-\frac{yt}{1-t}}}{(1-t)^{\nu+1}}\right]^2 = \frac{1}{(1-t)^{2\nu+2}} \int_0^\infty dy\, y^\nu\, e^{-\left(\frac{1+t}{1-t}\right)y}$$

$$\underbrace{\qquad\qquad\qquad\qquad}_{} = \left(\frac{1-t}{1+t}\right)^{\nu+1} \nu!$$

$$= \frac{\nu!}{(1-t^2)^{\nu+1}}$$

$$= \sum_{n=0}^\infty t^{2n}\frac{(n+\nu)!}{n!}$$

$$= \sum_{m=0}^\infty \sum_{n=0}^\infty t^{m+n}\frac{(n+\nu)!}{n!}\delta_{m,n}, \qquad \text{(SS 280)}$$

and compare this result with the integral

$$\int_0^\infty dy\, y^\nu\, e^{-y}\left[\sum_{n=0}^\infty t^n \mathrm{L}_n^{(\nu)}(y)\right]^2$$

$$= \sum_{m=0}^\infty \sum_{n=0}^\infty t^{m+n}\int_0^\infty dy\, y^\nu\, e^{-y}\mathrm{L}_m^{(\nu)}(y)\,\mathrm{L}_n^{(\nu)}(y) \qquad \text{(SS 281)}$$

to find that

$$\int_0^\infty dy\, y^\nu\, e^{-y}\mathrm{L}_m^{(\nu)}(y)\,\mathrm{L}_n^{(\nu)}(y) = \frac{(n+\nu)!}{n!}\delta_{m,n}. \qquad \text{(SS 282)}$$

Sample solution to Problem 5.6

Using the second formula in Eq. (5.1.23), we easily see that

$$\rho_{\mathrm{W}}^{\mathrm{disp\ Fock}}(\alpha;\alpha';n) = 2\,\mathrm{tr}\Big\{[D(\alpha')\,|n\rangle\langle n|\,D(-\alpha')]\,D(\alpha)\,(-1)^{A^\dagger A}\,D(-\alpha)\Big\}$$

$$= 2\,\langle n|\,D(\alpha-\alpha')\,(-1)^{A^\dagger A}\,D(-\alpha+\alpha')\,|n\rangle$$

$$= 2\,(-1)^n\,e^{-2|\alpha-\alpha'|^2}\,\mathrm{L}_n\big(4\,|\alpha-\alpha'|^2\big). \qquad \text{(SS 283)}$$

Sample solution to Problem 5.7

We can rewrite $p(x_\vartheta)$ as

$$p(x_\vartheta) = \frac{1}{1+\epsilon} \left(1 - 2\epsilon \frac{\partial}{\partial t}\right) \frac{e^{-t x_\vartheta^2}}{\sqrt{\pi}} \Bigg|_{t=1}, \qquad \text{(SS 284)}$$

which essentially means that we just need to invert a Gaussian Radon transform. Evaluating the right-hand side of Eq. (5.1.37),

$$\int_0^\pi \frac{d\vartheta}{2\pi} \int dk\, |k|\, e^{ik(x\cos\vartheta + p\sin\vartheta)} \int dx_\vartheta\, e^{-ikx_\vartheta} \frac{e^{-t x_\vartheta^2}}{\sqrt{\pi}}$$

$$= \sqrt{\frac{1}{t}} \int_0^\pi \frac{d\vartheta}{2\pi} \int dk\, |k|\, e^{ik(x\cos\vartheta + p\sin\vartheta) - \frac{k^2}{4t}}$$

$$= \sqrt{\frac{1}{t}} \int \frac{dx'dp'}{2\pi}\, e^{i(xx' + pp') - \frac{1}{4t}(x'^2 + p'^2)}$$

$$= 2\sqrt{t}\, e^{-t(x^2 + p^2)}. \qquad \text{(SS 285)}$$

The Wigner function is therefore equal to

$$\rho_{\mathrm{W}}(\alpha) = \frac{2}{1+\epsilon} \left(1 - 2\epsilon \frac{\partial}{\partial t}\right) \sqrt{t}\, e^{-2t|\alpha|^2} \Bigg|_{t=1}$$

$$= \frac{2}{1+\epsilon} \left[e^{-2|\alpha|^2} + \epsilon \left(4|\alpha|^2 - 1\right) e^{-2|\alpha|^2}\right]. \qquad \text{(SS 286)}$$

According to Eq. (5.1.71), the first term corresponds to the vacuum state $|0\rangle\langle 0|$ and the second term corresponds to the Fock state $|1\rangle\langle 1|$ since $L_1(x) = -x + 1$. Thus, the unknown state is

$$\rho_{\mathrm{true}} = |0\rangle \frac{1}{1+\epsilon} \langle 0| + |1\rangle \frac{\epsilon}{1+\epsilon} \langle 1|. \qquad \text{(SS 287)}$$

Sample solution to Problem 5.8

With Eq. (4.2.59) and Eq. (4.2.49), the operator elements

$$\langle m| D(\alpha) |n\rangle$$

$$= \sqrt{m!\,n!} \oint_{\substack{\text{unit}\\\text{circle}}} \frac{d\alpha'^{*}}{-2\pi i} \frac{e^{\frac{1}{2}|\alpha'|^2}}{\alpha'^{*\,m+1}} \oint_{\substack{\text{unit}\\\text{circle}}} \frac{d\alpha''}{2\pi i} \frac{e^{\frac{1}{2}|\alpha''|^2}}{\alpha''^{\,n+1}} \underbrace{\langle \alpha'^{*}| D(\alpha) |\alpha''\rangle}_{= e^{-\frac{1}{2}|\alpha'|^2 - \frac{1}{2}|\alpha''+\alpha|^2 + \alpha'^{*}(\alpha''+\alpha)} e^{i\,\text{Im}\{\alpha\alpha''^{*}\}}}$$

$$= -\sqrt{m!\,n!} \oint_{\substack{\text{unit}\\\text{circle}}} \frac{d\alpha'^{*} d\alpha''}{(2\pi i)^2} \frac{e^{\alpha'^{*}\alpha'' + \alpha'^{*}\alpha - \alpha''\alpha^{*}}}{\alpha'^{*\,m+1}\alpha''^{\,n+1}}. \tag{SS 288}$$

Considering the case where $m \geq n$, we first evaluate the integral in α'^{*} and then the integral in α''. This gives

$$\langle m| D(\alpha) |n\rangle \Big|_{m\geq n}$$

$$= \sqrt{\frac{n!}{m!}} e^{-\frac{1}{2}|\alpha|^2} \oint_{\substack{\text{unit}\\\text{circle}}} \frac{d\alpha''}{2\pi i} \frac{e^{-\alpha''\alpha^{*}}}{\alpha''^{\,n+1}} \underbrace{\left(\frac{\partial}{\partial\alpha'^{*}}\right)^{m} e^{\alpha'^{*}(\alpha''+\alpha)} \Big|_{\alpha'^{*}=0}}_{= (\alpha''+\alpha)^{m}}$$

$$= \frac{e^{-\frac{1}{2}|\alpha|^2}}{\sqrt{m!\,n!}} \left(\frac{\partial}{\partial\alpha''}\right)^{n} \left[(\alpha''+\alpha)^{m} e^{-\alpha''\alpha^{*}}\right]\Big|_{\alpha''=0} \qquad (t = \alpha^{*}(\alpha''+\alpha))$$

$$= \frac{e^{-\frac{1}{2}|\alpha|^2}}{\sqrt{m!\,n!}} \alpha^{*\,n-m} \left(\frac{d}{dt}\right)^{n} \left(t^{m} e^{-t}\right)\Big|_{t=|\alpha|^2}$$

$$= \alpha^{m-n} e^{-\frac{1}{2}|\alpha|^2} \sqrt{\frac{n!}{m!}} L_n^{(m-n)}\left(|\alpha|^2\right). \tag{SS 289}$$

For the case where $m < n$, we simply replace α with $-\alpha^{*}$ and interchange m and n to get

$$\langle m| D(\alpha) |n\rangle \Big|_{m<n} = (-1)^{n-m}\alpha^{*\,n-m} e^{-\frac{1}{2}|\alpha|^2} \sqrt{\frac{m!}{n!}} L_m^{(n-m)}\left(|\alpha|^2\right). \tag{SS 290}$$

We may then write down the compact formula as stated in Eq. (5.1.73) for $\langle m| D(\alpha) |n\rangle$.

Sample solution to Problem 5.9

Using the completeness property of the Fock states,

$$\langle n| D(\alpha - \alpha')(-1)^{A^\dagger A} D(-\alpha + \alpha') |n\rangle$$

$$= \sum_{m=0}^{\infty} (-1)^m \langle n| D(\alpha - \alpha') |m\rangle \langle m| D(-\alpha + \alpha') |n\rangle$$

$$= e^{-|\alpha - \alpha'|^2} \sum_{m=0}^{\infty} (-1)^m \frac{n_<!}{n_>!} |\alpha - \alpha'|^{2(n_> - n_<)} \left[L_{n_<}^{(n_> - n_<)} \left(|\alpha - \alpha'|^2 \right) \right]^2,$$

$$\text{(SS 291)}$$

where we have noted that

$$\left. (-1)^{n_> - n_<} (-1)^{(n-m)\eta(n-m)} (-1)^{(m-n)\eta(m-n)} \right|_{m \neq n}$$

$$= (-1)^{n_> - n_<} \times \begin{cases} (-1)^{m-n} & \text{for } m \geq n \\ (-1)^{n-m} & \text{for } n < m \end{cases}$$

$$= 1. \qquad\qquad \text{(SS 292)}$$

Since

$$\langle n| D(\alpha - \alpha')(-1)^{A^\dagger A} D(-\alpha + \alpha') |n\rangle = (-1)^n e^{-2|\alpha - \alpha'|^2} L_n\left(4 |\alpha - \alpha'|^2\right)$$

$$\text{(SS 293)}$$

too, we have

$$\sum_{m=0}^{\infty} (-1)^{m-n} \frac{n_<!}{n_>!} k^{(n_> - n_<)} \left[L_{n_<}^{(n_> - n_<)}(k) \right]^2 = e^{-k} L_n(4k), \qquad \text{(SS 294)}$$

where $k = |\alpha - \alpha'|^2$.

Sample solution to Problem 5.10

Since

$$(-1)^{A^\dagger A} |n\rangle = \int \frac{(\mathrm{d}\gamma)}{\pi} |\gamma\rangle \underbrace{\langle\gamma^*|n\rangle}(-1)^n$$
$$= \langle-\gamma^*|n\rangle \, (-1)^n$$
$$= \left(\int \frac{(\mathrm{d}\gamma)}{\pi} |\gamma\rangle \langle-\gamma^*|\right) |n\rangle \qquad \text{(SS 295)}$$

is true for all Fock kets $|n\rangle$, Eq. (5.1.76) must be true. Therefore,

$$\rho_{\mathrm{w}}(\alpha) = 2\,\mathrm{tr}\Big\{\rho\, D(\alpha)(-1)^{A^\dagger A} D(-\alpha)\Big\}$$
$$= 2\int \frac{(\mathrm{d}\gamma)}{\pi} \mathrm{e}^{2\mathrm{i}\,\mathrm{Im}\{\alpha\gamma^*\}} \langle-\gamma^*+\alpha^*|\rho|\gamma+\alpha\rangle$$
$$= 2\sum_{m=0}^{\infty}\sum_{n=0}^{\infty} \frac{\rho_{mn}}{\sqrt{m!\,n!}}$$
$$\times \int \frac{(\mathrm{d}\gamma)}{\pi} \mathrm{e}^{2\mathrm{i}\,\mathrm{Im}\{\alpha\gamma^*\}} \mathrm{e}^{-\frac{1}{2}|\gamma-\alpha|^2-\frac{1}{2}|\gamma+\alpha|^2} (-\gamma^*+\alpha^*)^m (\gamma+\alpha)^n$$
$$= 2\sum_{m=0}^{\infty}\sum_{n=0}^{\infty} \frac{\rho_{mn}}{\sqrt{m!\,n!}}$$
$$\times \int \frac{(\mathrm{d}\gamma)}{\pi} \mathrm{e}^{-|\gamma|^2-|\alpha|^2+\alpha\gamma^*-\alpha^*\gamma} (-\gamma^*+\alpha^*)^m (\gamma+\alpha)^n, \quad \text{(SS 296)}$$

where

$$\int \frac{(\mathrm{d}\gamma)}{\pi} \mathrm{e}^{-|\gamma|^2-|\alpha|^2+\alpha\gamma^*-\alpha^*\gamma} (-\gamma^*+\alpha^*)^m (\gamma+\alpha)^n$$
$$= \mathrm{e}^{|\alpha|^2}\int \frac{(\mathrm{d}\gamma)}{\pi} \mathrm{e}^{-|\gamma|^2} \mathrm{e}^{-\beta(\alpha^*-\gamma^*)-\beta^*(\alpha+\gamma)} (-\gamma^*+\alpha^*)^m (\gamma+\alpha)^n \Big|_{\substack{\beta=\alpha \\ \beta^*=\alpha^*}}$$

$$= e^{|\alpha|^2} \left(-\frac{\partial}{\partial\beta}\right)^m \left(-\frac{\partial}{\partial\beta^*}\right)^n \int \frac{(\mathrm{d}\gamma)}{\pi} e^{-|\gamma|^2 - \beta(\alpha^* - \gamma^*) - \beta^*(\alpha+\gamma)} \Bigg|_{\substack{\beta=\alpha \\ \beta^*=\alpha^*}}$$

$$= e^{|\alpha|^2} \left(-\frac{\partial}{\partial\beta}\right)^m \left(-\frac{\partial}{\partial\beta^*}\right)^n e^{-|\beta|^2 - \alpha\beta^* - \alpha^*\beta} \Bigg|_{\substack{\beta=\alpha \\ \beta^*=\alpha^*}}. \tag{SS 297}$$

When $m \geq n$, we first do the differentiation with respect to β in Eq. (SS 297) and obtain

$$\left(-\frac{\partial}{\partial\beta^*}\right)^n \left[(\alpha^* + \beta^*)^m e^{-2\alpha\beta^*}\right]\Bigg|_{\beta^*=\alpha^*} \qquad (t = 2|\alpha|^2 + 2\alpha\beta^*)$$

$$= (-1)^n e^{2|\alpha|^2} (2\alpha)^{n-m} \left(\frac{\mathrm{d}}{\mathrm{d}t}\right)^n (t^m e^{-t})\Bigg|_{t=4|\alpha|^2}$$

$$= (-1)^n n! e^{-2|\alpha|^2} (2\alpha^*)^{m-n} L_n^{(m-n)}\left(4|\alpha|^2\right). \tag{SS 298}$$

When $m < n$, we just interchange α and α^* as well as m and n to get

$$e^{|\alpha|^2} \left(-\frac{\partial}{\partial\beta}\right)^m \left(-\frac{\partial}{\partial\beta^*}\right)^n e^{-|\beta|^2 - \alpha\beta^* - \alpha^*\beta}\Bigg|_{\substack{\beta=\alpha \\ \beta^*=\alpha^*}}\Bigg|_{m<n}$$

$$= (-1)^m m! e^{-2|\alpha|^2} (2\alpha^*)^{n-m} L_m^{(n-m)}\left(4|\alpha|^2\right). \tag{SS 299}$$

With $\alpha = |\alpha| e^{\mathrm{i}\phi}$, we have

$$e^{|\alpha|^2} \left(-\frac{\partial}{\partial\beta}\right)^m \left(-\frac{\partial}{\partial\beta^*}\right)^n e^{-|\beta|^2 - \alpha\beta^* - \alpha^*\beta}\Bigg|_{\substack{\beta=\alpha \\ \beta^*=\alpha^*}}$$

$$= (-1)^{n_<} 2^{n_> - n_<} n_<! e^{-2|\alpha|^2} |\alpha|^{n_> - n_<} e^{-\mathrm{i}(m-n)\phi} L_{n_<}^{(n_> - n_<)}\left(4|\alpha|^2\right),$$
$$\tag{SS 300}$$

which is the component of the double sum in Eq. (5.1.66).

Sample solution to Problem 5.11

Using Eq. (4.2.61) twice on Eq. (5.1.23),

$$\rho_{\mathrm{W}}(\alpha) = 2 \int \frac{(\mathrm{d}\alpha')(\mathrm{d}\alpha'')}{\pi^2} \rho\left(\alpha'^*, \alpha''\right) \langle \alpha''^* | D(\alpha)(-1)^{A^\dagger A} D(-\alpha) | \alpha' \rangle.$$
$$\tag{SS 301}$$

To evaluate the operator elements, we may first write the parity operator in its normal-order form using Eq. (4.2.93), namely

$$(-1)^{A^\dagger A} = \left[e^{-2A^\dagger A} \right]_{\mathrm{N}} .$$
(SS 302)

It then follows that

$$\langle \alpha''^* | \, D(\alpha)(-1)^{A^\dagger A} D(-\alpha) \, | \alpha' \rangle$$

$$= e^{i \left(\mathrm{Im}\{\alpha \alpha''^*\} - \mathrm{Im}\{\alpha \alpha'^*\} \right)} \underbrace{\langle \alpha''^* - \alpha^* | \, (-1)^{A^\dagger A} \, | \alpha' - \alpha \rangle}_{= e^{-2(\alpha''^* - \alpha^*)(\alpha' - \alpha)} \langle \alpha''^* - \alpha^* | \alpha' - \alpha \rangle}$$

$$= e^{\frac{1}{2} \left(\alpha \alpha''^* - \alpha^* \alpha'' - \alpha \alpha'^* + \alpha^* \alpha' \right)} e^{-\frac{1}{2}|\alpha'' - \alpha|^2 - \frac{1}{2}|\alpha' - \alpha|^2 - (\alpha''^* - \alpha^*)(\alpha' - \alpha)}$$

$$= e^{-2|\alpha|^2} e^{2(\alpha^* \alpha' + \alpha \alpha''^*)} e^{-\frac{1}{2}|\alpha''|^2 - \frac{1}{2}|\alpha'|^2 - \alpha''^* \alpha'}$$

$$= e^{-2|\alpha|^2} e^{2(\alpha^* \alpha' + \alpha \alpha''^*)} \langle -\alpha''^* | \alpha' \rangle .$$
(SS 303)

Sample solution to Problem 5.12

In terms of the real components, the operator

$$\alpha' \frac{\partial}{\partial \alpha} + \alpha''^* \frac{\partial}{\partial \alpha^*}$$

$$= \frac{1}{2} \left[(x' + ip') \left(\frac{\partial}{\partial x} - i \frac{\partial}{\partial p} \right) + (x'' - ip'') \left(\frac{\partial}{\partial x} + i \frac{\partial}{\partial p} \right) \right]$$

$$= \frac{1}{2} (x' + x'' + ip' - ip'') \frac{\partial}{\partial x} + \frac{i}{2} (x'' - x' - ip' - ip'') \frac{\partial}{\partial p} .$$
(SS 304)

The exponential operator, after acting on $f(\alpha, \alpha^*)$, clearly gives

$$e^{\alpha' \frac{\partial}{\partial \alpha} + \alpha''^* \frac{\partial}{\partial \alpha^*}} f \left(\frac{x + ip}{\sqrt{2}}, \frac{x - ip}{\sqrt{2}} \right)$$

$$= f \left(\frac{x + \frac{1}{2}(x' + x'' + ip' - ip'') + i(p + \frac{1}{2}(x'' - x' - ip' - ip''))}{\sqrt{2}}, \right.$$

$$\left. \frac{x + \frac{1}{2}(x' + x'' + ip' - ip'') - i(p + \frac{1}{2}(x'' - x' - ip' - ip''))}{\sqrt{2}} \right)$$

$$= f(\alpha + \alpha', \alpha^* + \alpha''^*) .$$
(SS 305)

Sample solution to Problem 5.13

For Gaussian functions,

$$\int dk \sqrt{\frac{a}{\pi}}\, e^{-ak^2+bk}\, \delta_{ix}(k) = \sqrt{\frac{a}{\pi}} \int \frac{dx'}{2\pi}\, e^{xx'} \underbrace{\int dk\, e^{-ak^2+(b+ix')k}}_{=\sqrt{\frac{\pi}{a}}\, e^{\frac{(b+ix')^2}{4a}}}$$

$$= e^{\frac{b^2}{4a}} \underbrace{\int \frac{dx'}{2\pi}\, e^{-\frac{1}{4a}x'^2+\frac{1}{2a}(2ax+ib)x'}}_{=\sqrt{\frac{a}{\pi}}\, e^{\frac{(2ax+ib)^2}{4a}}}$$

$$= \sqrt{\frac{a}{\pi}}\, e^{-a(ix)^2+b(ix)}. \tag{SS 306}$$

For a well-behaved function, it is sufficient to show that the above property is valid for powers of $k - c$, where c is the center of the function expansion:

$$\int dk\,(k-c)^n\, \delta_{ix}(k) = \left(\frac{d}{db}\right)^n \int dk\, e^{-ak^2+b(k-c)}\, \delta_{ix}(k)\Big|_{a=b=0}$$

$$= \left(\frac{d}{db}\right)^n e^{ax^2+b(ix-c)}\Big|_{a=b=0}$$

$$= (ix - c)^n. \tag{SS 307}$$

This reasoning is valid as long as the integral is understood as a limit.

Sample solution to Problem 5.14

Using Eq. (5.1.39), one quickly reaches the integral expression

$$\rho_P^{sq}(\alpha; \alpha'; r, \theta)$$

$$= \operatorname{sech} r\, e^{|\alpha|^2-|\alpha'|^2}\, e^{-\operatorname{Re}\{\alpha'^2 e^{-i\theta}\tanh r\}}$$

$$\times \int \frac{(du)}{\pi}\, e^{\frac{1}{2}e^{-i\theta}\tanh r\, u^2 + \frac{1}{2}e^{i\theta}\tanh r\, u^{*2}}\, e^{(\alpha-\alpha'\operatorname{sech} r)u^* - (\alpha^*-\alpha'^*\operatorname{sech} r)u}.$$

$$\tag{SS 308}$$

The integral above is strictly divergent except when $z = r = 0$, since the leading exponential function of x'^2 and p'^2 in this integral is $e^{\frac{1}{2}\text{Re}\{e^{i\theta}\tanh r\}(x'^2-p'^2)}$, where $u = (x' + ip')/\sqrt{2}$. Therefore, the squeezed coherent states are nonclassical. We can rewrite the P function as derivatives of a delta function by recognizing, for example, that

$$e^{\frac{1}{2}e^{-i\theta}\tanh r\, u^2}\, e^{-(\alpha^* - \alpha'^*\,\text{sech}\,r)u}$$

$$= \sum_{k=0}^{\infty} \frac{\left(\frac{1}{2}e^{-i\theta}\tanh r\right)^k}{k!} \left(\frac{\partial}{\partial \alpha^*}\right)^{2k} e^{-(\alpha^* - \alpha'^*\,\text{sech}\,r)u}$$

$$= e^{\frac{1}{2}e^{-i\theta}\tanh r\left(\frac{\partial}{\partial \alpha^*}\right)^2} e^{-(\alpha^* - \alpha'^*\,\text{sech}\,r)u}, \qquad \text{(SS 309)}$$

so that

$$\rho_{\text{P}}^{\text{sq}}(\alpha; \alpha'; r, \theta) = \text{sech}\,r\, e^{|\alpha|^2-|\alpha'|^2}\, e^{-\text{Re}\{\alpha'^2 e^{-i\theta}\tanh r\}}$$

$$\times e^{\text{Re}\{e^{i\theta}\tanh r\left(\frac{\partial}{\partial \alpha}\right)^2\}} \delta\left(\alpha - \alpha'\text{sech}\,r\right). \qquad \text{(SS 310)}$$

Sample solution to Problem 5.15

It is enough to derive one of the two interference terms in Eq. (5.1.100) using Eq. (5.1.108). We consider the interference term $\mathcal{N}\,|\alpha'\rangle\langle-\alpha'^*|$ of the state $\rho^{\text{o/e}} = |\alpha'_{\pm}\rangle\langle\alpha'_{\pm}|$. The operator elements for this term are given by

$$\mathcal{N}\,\langle m|\alpha'\rangle\langle-\alpha'^*|n\rangle = \mathcal{N}e^{-|\alpha'|^2} \frac{\alpha'^m\,(-\alpha'^*)^n}{\sqrt{m!\,n!}}. \qquad \text{(SS 311)}$$

With Eq. (5.1.108), the corresponding term for the P function of $\rho^{\text{o/e}}$ can then be calculated as

$$\mathcal{N}\,\pi\,e^{|\alpha|^2} \sum_{m=0}^{\infty}\sum_{n=0}^{\infty} \frac{\langle m|\alpha'\rangle\langle-\alpha'^*|n\rangle}{\sqrt{m!\,n!}} \left(-\frac{\partial}{\partial \alpha}\right)^m \left(-\frac{\partial}{\partial \alpha^*}\right)^n \delta(\alpha)$$

$$= \mathcal{N}\,\pi\,e^{|\alpha|^2-|\alpha'|^2} \sum_{m=0}^{\infty}\sum_{n=0}^{\infty} \frac{1}{m!\,n!} \left(-\alpha'\frac{\partial}{\partial \alpha}\right)^m \left(\alpha'^*\frac{\partial}{\partial \alpha^*}\right)^n \delta(\alpha)$$

$$= \mathcal{N}\,\pi\,e^{|\alpha|^2-|\alpha'|^2}\, e^{-\alpha'\frac{\partial}{\partial \alpha}+\alpha'^*\frac{\partial}{\partial \alpha^*}}\, \delta(\alpha), \qquad \text{(SS 312)}$$

which is one of the terms in Eq. (5.1.100).

Sample solution to Problem 5.16

With Eq. (4.2.20), we have

$$
\langle x_\vartheta | \alpha \rangle = \sum_{n=0}^{\infty} \mathrm{e}^{-in\vartheta} \langle x | n \rangle \langle n | \alpha \rangle \Big|_{x=x_\vartheta}
$$

$$
= \sum_{n=0}^{\infty} \langle x | n \rangle \langle n | \alpha\, \mathrm{e}^{-i\vartheta} \rangle \Big|_{x=x_\vartheta}
$$

$$
= \langle x | \alpha\, \mathrm{e}^{-i\vartheta} \rangle \Big|_{x=x_\vartheta}
$$

$$
= \frac{1}{\pi^{\frac14}} \mathrm{e}^{-\frac12 |\alpha|^2 - \frac12 (\alpha \mathrm{e}^{-i\vartheta})^2 - \frac12 x^2 + \sqrt{2}\alpha \mathrm{e}^{-i\vartheta} x}. \tag{SS 313}
$$

Taking a different route by using Eq. (4.2.61) on the other hand,

$$
\langle x_\vartheta | \alpha \rangle = \int \frac{(\mathrm{d}\alpha')}{\pi} \langle x | \alpha' \rangle \langle \alpha'^* | \mathrm{e}^{-i\vartheta A^\dagger A} | \alpha \rangle \Big|_{x=x_\vartheta}. \tag{SS 314}
$$

By writing the operator $\mathrm{e}^{-i\vartheta A^\dagger A}$ in normal-order form

$$
\mathrm{e}^{-i\vartheta A^\dagger A} = \left[\mathrm{e}^{(\mathrm{e}^{-i\vartheta}-1)A^\dagger A} \right]_N \tag{SS 315}
$$

with Eq. (4.2.93),

$$
\langle x_\vartheta | \alpha \rangle = \int \frac{(\mathrm{d}\alpha')}{\pi} \mathrm{e}^{(\mathrm{e}^{-i\vartheta}-1)\alpha'^*\alpha} \langle x | \alpha' \rangle \langle \alpha'^* | \alpha \rangle \Big|_{x=x_\vartheta}
$$

$$
= \int \frac{(\mathrm{d}\alpha')}{\pi} \mathrm{e}^{(\mathrm{e}^{-i\vartheta}-1)\alpha'^*\alpha} \frac{1}{\pi^{\frac14}} \mathrm{e}^{-\frac12|\alpha'|^2 - \frac12\alpha'^2 - \frac12 x^2 + \sqrt{2}\alpha' x}
$$

$$
\times \mathrm{e}^{-\frac12|\alpha'|^2 - \frac12|\alpha|^2 + \alpha'^*\alpha} \Big|_{x=x_\vartheta}
$$

$$
= \frac{1}{\pi^{\frac14}} \mathrm{e}^{-\frac12|\alpha|^2} \mathrm{e}^{-\frac12 x^2} \int \frac{(\mathrm{d}\alpha')}{\pi} \mathrm{e}^{-|\alpha'|^2 - \frac12\alpha'^2 + \sqrt{2}\alpha' x + \mathrm{e}^{-i\vartheta}\alpha'^*\alpha} \Big|_{x=x_\vartheta}. \tag{SS 316}
$$

The Gaussian integral can be easily evaluated with the help of Eq. (5.1.130), where $a = 1$, $b_1 = \sqrt{2}x$, $b_2 = e^{-i\vartheta}\alpha$, $c_1 = -1/2$ and $c_2 = 0$, giving us

$$I_{\text{gauss}}\left(1, \sqrt{2}x, e^{-i\vartheta}\alpha, -\frac{1}{2}, 0\right) = e^{-\frac{1}{2}\left(\alpha e^{-i\vartheta}\right)^2 - \sqrt{2}\alpha e^{-i\vartheta}x}. \qquad \text{(SS 317)}$$

Inserting this result back into Eq. (SS 316) also gives Eq. (5.1.132) as it should.

Sample solution to Problem 5.17

In terms of the real components of α, we express Eq. (5.1.113) as

$$\rho_{\text{P}}^{\text{Fock}}(\alpha; n) = \frac{\pi\, e^{|\alpha|^2}}{n!} \left|\frac{\partial}{\partial\alpha}\right|^{2n} \delta(\alpha)$$

$$= \frac{2\pi\, e^{|\alpha|^2}}{2^n n!} \sum_{k=0}^{n} \binom{n}{k} \left(\frac{\partial}{\partial x}\right)^{2k} \left(\frac{\partial}{\partial p}\right)^{2n-2k} \delta(x)\,\delta(p). \quad \text{(SS 318)}$$

Recalling that $\alpha\alpha'^* + \alpha^*\alpha' = xx' + pp'$, Eq. (5.1.112) becomes a product of single-variable integrals inasmuch as

$$\rho_{\text{W}}^{\text{Fock}}(x, p; n) = \frac{4\pi\, e^{-x^2-p^2}}{2^n\, n!} \sum_{k=0}^{n} \binom{n}{k} \int \frac{dx'\, dp'}{2\pi} e^{-\frac{1}{2}\left(x'^2+p'^2\right)+2\left(xx'+pp'\right)}$$

$$\times \left(\frac{\partial}{\partial x'}\right)^{2k} \left(\frac{\partial}{\partial p'}\right)^{2n-2k} \delta(x')\,\delta(p').$$

$$\text{(SS 319)}$$

Let us first evaluate the integral in x', which gives

$$\int \frac{dx'}{\sqrt{2\pi}} e^{-\frac{1}{2}x'^2+2xx'} \left(\frac{\partial}{\partial x'}\right)^{2k} \underbrace{\delta(x')}$$

$$= \int \frac{dl}{2\pi} e^{ilx'}$$

$$= \int \frac{dl}{2\pi} (i\,l)^{2k} \underbrace{\int \frac{dx'}{\sqrt{2\pi}} e^{-\frac{1}{2}x'^2+(2x+il)x'}}$$

$$= e^{\frac{1}{2}(2x+il)^2}$$

$$= \frac{1}{4^k} e^{2x^2} \left(\frac{d}{dx}\right)^{2k} \underbrace{\int \frac{dl}{2\pi} e^{-\frac{1}{2}l^2 + 2ixl}}$$

$$= \frac{1}{\sqrt{2\pi}} e^{-2x^2}$$

$$= \frac{1}{2^k \sqrt{2\pi}} H_{2k}\left(\sqrt{2}x\right). \tag{SS 320}$$

By inspection of Eq. (SS 320), the integral in p' is equal to $H_{2n-2k}\left(\sqrt{2}p\right)/2^{n-k}\sqrt{2\pi}$. Therefore,

$$\rho_W^{\text{Fock}}(\alpha; n) = 2 \frac{e^{-x^2-p^2}}{4^n n!} \sum_{k=0}^{n} \binom{n}{k} H_{2k}\left(\sqrt{2}x\right) H_{2n-2k}\left(\sqrt{2}p\right)$$

$$= 2(-1)^n e^{-x^2-p^2} L_n\left(2x^2 + 2p^2\right), \tag{SS 321}$$

which, after the replacements $\sqrt{2}x \longrightarrow x$ and $\sqrt{2}p \longrightarrow p$, results in the identity

$$\left(-\frac{1}{4}\right)^n \frac{1}{n!} \sum_{k=0}^{n} \binom{n}{k} H_{2k}(x) H_{2n-2k}(y) = L_n\left(x^2 + y^2\right). \tag{SS 322}$$

Sample solution to Problem 5.18

By the fundamental theorem of calculus,

$$\frac{\partial f(\alpha)}{\partial x_0} = -\frac{k}{2\pi} \int_{p_0-i\infty}^{p_0+i\infty} dt\, e^{\frac{k}{2}(t-p)^2} \left[\begin{array}{c} e^{\frac{k}{2}(x_0-x+iT)^2} f_{\text{GW}}(x_0 + iT, t) \\ \times e^{\frac{k}{2}(x_0-x-iT)^2} f_{\text{GW}}(x_0 - iT, t) \end{array} \right]\Bigg|_{0<T\longrightarrow\infty},$$

$$\tag{SS 323}$$

and since the function

$$e^{\frac{k}{2}(x_0-x\pm iT)^2} = e^{-\frac{k}{2}T^2 \pm ik(x_0-x)T} e^{\frac{k}{2}(x_0-x)^2} \tag{SS 324}$$

is rapidly decreasing in T, the above partial derivative is zero after taking the limit. The same is also true for $\partial f(\alpha)/\partial p_0$.

Sample solution to Problem 5.19

After introducing the new integration variables

$$x' = \frac{s - x_0}{i}, \quad p' = \frac{t - p_0}{i},$$ (SS 325)

the right-hand side of Eq. (5.2.10) turns into

$$k \int \frac{\mathrm{d}x'\,\mathrm{d}p'}{2\pi}\, e^{\frac{k}{2}\left[(x_0 - x + ix')^2 + (p_0 - p + ip')^2\right]} f_{\mathrm{GW}}(x_0 + ix',\, p_0 + ip')$$

$$= k\, e^{\frac{k}{2}\left[(x_0 - x)^2 + (p_0 - p)^2\right]} \int \frac{\mathrm{d}x'\,\mathrm{d}p'}{2\pi}\, e^{-\frac{k}{2}\left(x'^2 + p'^2\right) + ik\left[(x_0 - x)x' + (p_0 - p)p'\right]}$$

$$\times\, f_{\mathrm{GW}}(x_0 + ix',\, p_0 + ip')\,.$$ (SS 326)

By expressing the complex variables $\beta = (b_1 + ib_2)/\sqrt{2}$ and $\alpha = (x'' + ip'')/\sqrt{2}$ in terms of their real components, Eq. (5.2.1) becomes

$$f_{\mathrm{GW}}(b_1, b_2) = k \int \frac{\mathrm{d}x''\,\mathrm{d}p''}{2\pi}\, f(x'', p'')\, e^{-\frac{k}{2}\left[(b_1 - x'')^2 + (b_2 - p'')^2\right]},$$ (SS 327)

so that when $b_1 = x_0 + ix'$ and $b_2 = p_0 + ip'$,

$$f_{\mathrm{GW}}(x_0 + ix',\, p_0 + ip')$$

$$= k\, e^{\frac{k}{2}\left(x'^2 + p'^2\right)} \int \frac{\mathrm{d}x''\,\mathrm{d}p''}{2\pi}\, f(x'', p'')\, e^{-\frac{k}{2}\left[(x_0 - x'')^2 + (p_0 - p'')^2\right]}$$

$$\times\, e^{-ik\left[(x_0 - x'')x' + (p_0 - p'')p'\right]}.$$ (SS 328)

With this, Eq. (SS 326) simplifies into

$$k \int \frac{\mathrm{d}x'\,\mathrm{d}p'}{2\pi}\, e^{\frac{k}{2}\left[(x_0 - x + ix')^2 + (p_0 - p + ip')^2\right]} f_{\mathrm{GW}}(x_0 + ix',\, p_0 + ip')$$

$$= k^2\, e^{\frac{k}{2}\left[(x_0 - x)^2 + (p_0 - p)^2\right]} \int \frac{\mathrm{d}x''\,\mathrm{d}p''}{2\pi}\, f(x'', p'')\, e^{-\frac{k}{2}\left[(x_0 - x'')^2 + (p_0 - p'')^2\right]}$$

$$\times \underbrace{\int \frac{\mathrm{d}x'\,\mathrm{d}p'}{2\pi}\, e^{ik\left[(x'' - x)x' + (p'' - p)p'\right]}}_{= \delta(k(x'' - x))\,\delta(k(p'' - p))}$$

$$= f(x, p).$$ (SS 329)

Sample solution to Problem 5.20

The Q function for the Fock state $|n\rangle\langle n|$ is given by

$$
\begin{aligned}
\rho_Q^{\text{Fock}}(\alpha; n) &= \langle \alpha^* | n \rangle \langle n | \alpha \rangle \\
&= \frac{1}{n!} e^{-|\alpha|^2} |\alpha|^{2n} \\
&= \frac{1}{2^n n!} e^{-\frac{1}{2}(x^2+p^2)} \left(x^2+p^2\right)^n .
\end{aligned}
\tag{SS 330}
$$

Therefore

$$
\rho_Q^{\text{Fock}}(is, it; n) = \frac{(-1)^n}{2^n n!} e^{\frac{1}{2}(s^2+t^2)} \left(s^2+t^2\right)^n .
\tag{SS 331}
$$

To look for the Wigner function of this state, we set $k=2$ in Eq. (5.2.11) and find that

$$
\begin{aligned}
\rho_W^{\text{Fock}}(\alpha; n) &= \int \frac{ds\,dt}{\pi} e^{-\left[(s+ix)^2+(t+ip)^2\right]} \rho_Q^{\text{Fock}}(is, it; n) \\
&= \frac{(-1)^n}{2^n n!} e^{x^2+p^2} \int \frac{ds\,dt}{\pi} e^{-\frac{1}{2}(s^2+t^2)-2i(sx+tp)} \left(s^2+t^2\right)^n \\
&= \frac{e^{x^2+p^2}}{2^n n!} \left[\left(\frac{1}{2}\frac{\partial}{\partial x}\right)^2 + \left(\frac{1}{2}\frac{\partial}{\partial p}\right)^2\right]^n \underbrace{\int \frac{ds\,dt}{\pi} e^{-\frac{1}{2}(s^2+t^2)-2i(sx+tp)}}_{= 2\,e^{-2(x^2+p^2)}} \\
&= \frac{2\,e^{2|\alpha|^2}}{n!} \left|\frac{1}{2}\frac{\partial}{\partial \alpha}\right|^{2n} e^{-4|\alpha|^2},
\end{aligned}
\tag{SS 332}
$$

which leads to Eq. (5.1.71) after following the calculations that simplified Eq. (5.1.134) into Eq. (5.1.136).

To look for the P function, we set $k=1$, so that

$$
\begin{aligned}
\rho_P^{\text{Fock}}(\alpha; n) &= \int \frac{ds\,dt}{2\pi} e^{-\frac{1}{2}\left[(s+ix)^2+(t+ip)^2\right]} \rho_Q^{\text{Fock}}(is, it; n) \\
&= \frac{(-1)^n}{2^n n!} e^{\frac{1}{2}(x^2+p^2)} \int \frac{ds\,dt}{2\pi} e^{-i(sx+tp)} \left(s^2+t^2\right)^n
\end{aligned}
$$

$$= \frac{e^{\frac{1}{2}(x^2+p^2)}}{2^n\, n!} \left[\left(\frac{\partial}{\partial x}\right)^2 + \left(\frac{\partial}{\partial p}\right)^2\right]^n \underbrace{\int \frac{ds\, dt}{2\pi} \, e^{-i(sx+tp)}}_{=\, 2\pi\, \delta(x)\, \delta(p)}$$

$$= \frac{\pi\, e^{|\alpha|^2}}{n!} \left|\frac{\partial}{\partial \alpha}\right|^{2n} \boldsymbol{\delta}(\alpha),\qquad\qquad\qquad\text{(SS 333)}$$

consistent with Eq. (5.1.113).

Sample solution to Problem 5.21

Since

$$\rho_{\mathrm{W}}^{\mathrm{Fock}}(is, it; n) = 2(-1)^n\, e^{s^2+t^2}\, \mathrm{L}_n\!\left(-2s^2 - 2t^2\right),\qquad\text{(SS 334)}$$

we can set $k = 2$ in Eq. (5.2.11) to find that

$$\rho_{\mathrm{P}}^{\mathrm{Fock}}(is, it; n) = \int \frac{ds\, dt}{\pi}\, e^{-\left[(s+ix)^2+(t+ip)^2\right]}\, \rho_{\mathrm{W}}^{\mathrm{Fock}}(is, it; n)$$

$$= 4(-1)^n\, \pi\, e^{x^2+p^2} \int \frac{ds\, dt}{4\pi^2}\, e^{-2i(sx+tp)}\, \mathrm{L}_n\!\left(-2s^2 - 2t^2\right).$$

$$\text{(SS 335)}$$

The two-dimensional Fourier transform can be evaluated as follows:

$$\int \frac{ds\, dt}{4\pi^2}\, e^{-2i(sx+tp)}\, \mathrm{L}_n\!\left(-2s^2 - 2t^2\right)$$

$$= \sum_{k=0}^{n} \frac{1}{k!} \binom{n}{k} \underbrace{\int \frac{ds\, dt}{4\pi^2}\, e^{-2i(sx+tp)}\, \left(2s^2 + 2t^2\right)^k}_{}$$

$$= \left[-\frac{1}{2}\left(\frac{\partial}{\partial x}\right)^2 - \frac{1}{2}\left(\frac{\partial}{\partial p}\right)^2\right]\, \delta(2x)\, \delta(2p)$$

$$= \frac{1}{4}\mathrm{L}_n\!\left(-\frac{1}{2}\left(\frac{\partial}{\partial x}\right)^2 - \frac{1}{2}\left(\frac{\partial}{\partial p}\right)^2\right)\, \delta(x)\, \delta(p).\qquad\text{(SS 336)}$$

Hence, we may state the P function compactly as

$$\rho_{\mathrm{P}}^{\mathrm{Fock}}(\alpha; n)' = (-1)^n\, \pi\, e^{2|\alpha|^2}\, \mathrm{L}_n\!\left(\left|\frac{\partial}{\partial \alpha}\right|^2\right)\, \boldsymbol{\delta}(\alpha).\qquad\text{(SS 337)}$$

To show that both P functions in Eq. (SS 337) and in Eq. (5.1.113) represent the Fock state $|n\rangle\langle n|$, we compute the operator elements $\langle m|\rho|m'\rangle$ for integers m and m' using the original definition of the P function in Eq. (5.1.78). For the P function in Eq. (5.1.113), we first write

$$
\begin{aligned}
\left|\frac{\partial}{\partial\alpha}\right|^{2n}\delta(\alpha) &= \int\frac{(\mathrm{d}u)}{\pi^2}\left|\frac{\partial}{\partial\alpha}\right|^{2n}\mathrm{e}^{\alpha u^* - \alpha^* u} \\
&= \int\frac{(\mathrm{d}u)}{\pi^2}\left(-|u|^2\right)^n\mathrm{e}^{\alpha u^* - \alpha^* u} \\
&= \left(\frac{\partial}{\partial t}\right)^n\int\frac{(\mathrm{d}u)}{\pi^2}\mathrm{e}^{-t|u|^2 + \alpha u^* - \alpha^* u}\Bigg|_{t=0} \\
&= \frac{1}{\pi}\left(\frac{\partial}{\partial t}\right)^n\left(\frac{1}{t}\mathrm{e}^{-\frac{1}{t}|\alpha|^2}\right)\Bigg|_{t=0}.
\end{aligned}
\tag{SS 338}
$$

Therefore, along with the identity in Eq. (4.2.57),

$$
\begin{aligned}
&\langle m|\rho|m'\rangle \\
&= \int\frac{(\mathrm{d}\alpha)}{\pi}\langle m|\alpha\rangle\,\rho_{\mathrm{P}}^{\mathrm{Fock}}(\alpha;n)\langle\alpha^*|m'\rangle \\
&= \frac{1}{n!\sqrt{m!\,m'!}}\left(\frac{\partial}{\partial t}\right)^n\frac{1}{t}\int\frac{(\mathrm{d}\alpha)}{\pi}\alpha^m\alpha^{*m'}\mathrm{e}^{-\frac{1}{t}|\alpha|^2}\Bigg|_{t=0} \\
&= \frac{1}{n!\sqrt{m!\,m'!}}\left(\frac{\partial}{\partial t}\right)^n\frac{1}{t}\left(\frac{\partial}{\partial u}\right)^{m'}\left(-\frac{\partial}{\partial u^*}\right)^m\underbrace{\int\frac{(\mathrm{d}\alpha)}{\pi}\mathrm{e}^{-\frac{1}{t}|\alpha|^2 + u\alpha^* - u^*\alpha}}_{=\,t\,\mathrm{e}^{-t|u|^2}}\Bigg|_{\substack{t=0\\u=0\\u^*=0}} \\
&= \frac{1}{n!\sqrt{m!\,m'!}}\left(\frac{\partial}{\partial u}\right)^{m'}\left(-\frac{\partial}{\partial u^*}\right)^m\left(-|u|^2\right)^n\Bigg|_{\substack{u=0\\u^*=0}} \\
&= \frac{(-1)^{m+n}}{n!\sqrt{m!\,m'!}}\underbrace{\left[\frac{n!}{(n-m')!}u^{n-m'}\right]\Bigg|_{u=0}}_{=\,n!\,\delta_{m',n}}\underbrace{\left[\frac{n!}{(n-m)!}u^{*n-m}\right]\Bigg|_{u^*=0}}_{=\,n!\,\delta_{m,n}} \\
&= \langle m|n\rangle\langle n|m'\rangle.
\end{aligned}
\tag{SS 339}
$$

For the P function in Eq. (SS 337), making use of the identity

$$L_n(z) = \sum_{k=0}^{n} \binom{n}{k} \frac{(-z)^k}{k!} \qquad \text{(SS 340)}$$

and the result in Eq. (SS 338) we have just established,

$$\int \frac{(d\alpha)}{\pi} \langle m | \alpha \rangle \, \rho_P^{\text{Fock}}(\alpha; n)' \, \langle \alpha^* | m' \rangle$$

$$= \frac{(-1)^n}{\sqrt{m!\,m'!}} \sum_{k=0}^{n} \binom{n}{k} \frac{1}{k!} \left(-\frac{\partial}{\partial t} \right)^k \int \frac{(d\alpha)}{\pi} \, \alpha^m \alpha^{*m'} \, e^{(1-\frac{1}{t})|\alpha|^2} \Bigg|_{t=0}$$

$$= \frac{(-1)^n}{\sqrt{m!\,m'!}} \sum_{k=0}^{n} \binom{n}{k} \frac{1}{k!} \left(-\frac{\partial}{\partial t} \right)^k \frac{1}{t} \left(\frac{\partial}{\partial u} \right)^{m'} \left(-\frac{\partial}{\partial u^*} \right)^m$$

$$\times \underbrace{\int \frac{(d\alpha)}{\pi} \, e^{(1-\frac{1}{t})|\alpha|^2 + u\alpha^* - u^*\alpha}}_{= \frac{t}{1-t} \, e^{-\frac{t}{1-t}|u|^2}} \Bigg|_{\substack{t=0 \\ u=0 \\ u^*=0}}$$

$$= \frac{(-1)^{m+n}}{\sqrt{m!\,m'!}} \sum_{k=0}^{n} \binom{n}{k} \frac{1}{k!} \left(\frac{\partial}{\partial u} \right)^{m'} \left(\frac{\partial}{\partial u^*} \right)^m \left(-\frac{\partial}{\partial t} \right)^k \frac{e^{-\frac{t}{1-t}|u|^2}}{1-t} \Bigg|_{\substack{t=0 \\ u=0 \\ u^*=0}}.$$

$$\text{(SS 341)}$$

The remaining task is to evaluate the partial derivative in t. An introduction of the new variable $y = |u|^2/(1-t)$ yields

$$\left(\frac{\partial}{\partial t} \right)^k \left(\frac{1}{1-t} \right) e^{-\frac{t}{1-t}|u|^2} \Bigg|_{t=0}$$

$$= \frac{e^{|u|^2}}{|u|^{2(k+1)}} \left(y^2 \frac{d}{dy} \right)^k (y\,e^{-y}) \Bigg|_{y=|u|^2}$$

$$= \frac{e^{|u|^2}}{|u|^{2(k+1)}} \sum_{l=0}^{\infty} \frac{(-1)^l}{l!} \left(y^2 \frac{d}{dy} \right)^k y^{l+1} \Bigg|_{y=|u|^2}, \qquad \text{(SS 342)}$$

where

$$\left(y^2 \frac{\mathrm{d}}{\mathrm{d}y}\right)^k y^{l+1} = (l+1)\left(y^2 \frac{\mathrm{d}}{\mathrm{d}y}\right)^{k-1} y^{l+2}$$

$$\vdots$$

$$= \frac{(l+k)!}{l!} y^{l+k+1}$$

$$= y^{k+1}\left(\frac{\mathrm{d}}{\mathrm{d}y}\right)^k y^{k+l}, \qquad \text{(SS 343)}$$

which is again a result of Eq. (4.2.57). It follows that

$$\left(\frac{\partial}{\partial t}\right)^k \left(\frac{1}{1-t}\right) e^{-\frac{t}{1-t}|u|^2} = e^{|u|^2} \frac{y^{k+1}}{|u|^{2(k+1)}} \left(\frac{\mathrm{d}}{\mathrm{d}y}\right)^k \left(y^k e^{-y}\right)\Bigg|_{y=|u|^2}$$

$$= k!\, \mathrm{L}_k\!\left(|u|^2\right). \qquad \text{(SS 344)}$$

Finally, using the next identity for Laguerre polynomials,

$$\frac{z^n}{n!} = \sum_{k=0}^{n} \binom{n}{k}(-1)^k\, \mathrm{L}_k(z), \qquad \text{(SS 345)}$$

we get

$$\int \frac{(\mathrm{d}\alpha)}{\pi} \langle m|\alpha\rangle \, \rho_{\mathrm{P}}^{\mathrm{Fock}}(\alpha;n)' \langle \alpha^*|m'\rangle$$

$$= \frac{(-1)^{m+n}}{\sqrt{m!\,m'!}} \left(\frac{\partial}{\partial u}\right)^{m'} \left(\frac{\partial}{\partial u^*}\right)^{m} \sum_{k=0}^{n} \binom{n}{k}(-1)^k \mathrm{L}_k\!\left(|u|^2\right)\Bigg|_{\substack{u=0\\u^*=0}}$$

$$= \frac{(-1)^{m+n}}{n!\sqrt{m!\,m'!}} \left(\frac{\partial}{\partial u}\right)^{m'} \left(\frac{\partial}{\partial u^*}\right)^{m} |u|^{2n}\Bigg|_{\substack{u=0\\u^*=0}}$$

$$= \langle m|n\rangle \langle n|m'\rangle. \qquad \text{(SS 346)}$$

The first of the two identities from the hints for this problem can be obtained by simply applying the Leibniz rule of differentiation on the

Rodriguez formula for $L_n(z)$:

$$L_n(z) = \frac{1}{n!} e^z \left(\frac{d}{dz}\right)^n (z^n e^{-z})$$

$$= \frac{1}{n!} e^z \sum_{k=0}^{n} \binom{n}{k} \left[\left(\frac{d}{dz}\right)^{n-k} z^n\right] \left[\left(\frac{d}{dz}\right)^k e^{-z}\right]$$

$$= \sum_{k=0}^{n} \binom{n}{k} \frac{(-z)^k}{k!}. \tag{SS 347}$$

The second identity is then a consequence of evaluating its right-hand side with the right-hand side of the previous identity:

$$\sum_{k=0}^{n} \binom{n}{k} (-1)^k L_k(z) = \sum_{k=0}^{n} \binom{n}{k} (-1)^k \sum_{l=0}^{k} \binom{k}{l} \frac{(-z)^l}{l!}$$

$$= \sum_{l=0}^{n} \frac{(-z)^l}{l!} \sum_{k=l}^{n} \binom{n}{k}\binom{k}{l} (-1)^k, \tag{SS 348}$$

where the latter sum of products of binomial coefficients simplifies into

$$\sum_{k=l}^{n} \binom{n}{k}\binom{k}{l} (-1)^k = \frac{1}{l!} \sum_{k=l}^{n} \frac{n!}{(n-k)!(k-l)!} (-1)^k \qquad (m = k - l)$$

$$= \frac{(-1)^l}{l!} \sum_{m=0}^{n-l} \frac{n!}{(n-l-m)!m!} (-1)^m$$

$$= (-1)^l \binom{n}{l} \underbrace{\sum_{m=0}^{n-l} \binom{n-l}{m} (-1)^m}_{= \delta_{n,l}}$$

$$= (-1)^l \delta_{n,l} \tag{SS 349}$$

that would finally lead to the intended result.

Sample solution to Problem 5.22

The normally-ordered function of $F(A, A^\dagger)$ is given by

$$F_N\left(A, A^\dagger\right) = \int \frac{(\mathrm{d}\alpha'')}{\pi} \, \mathrm{e}^{-\frac{1}{2}|\alpha''|^2} \, \mathrm{e}^{\alpha'' A^\dagger} \, \mathrm{e}^{-\alpha''^* A}$$

$$= 2\left[\mathrm{e}^{-2A^\dagger A}\right]_N$$

$$= 2(-1)^{A^\dagger A}, \tag{SS 350}$$

which is two times the parity operator. On the other hand, the antinormally-ordered function of $F(A^\dagger, A)$ is

$$F_A\left(A, A^\dagger\right) = \int \frac{(\mathrm{d}\alpha'')}{\pi} \, \mathrm{e}^{\frac{1}{2}|\alpha''|^2} \, \mathrm{e}^{-\alpha''^* A} \, \mathrm{e}^{\alpha'' A^\dagger}$$

$$= \mathrm{e}^{-\frac{1}{2}\frac{\partial}{\partial A}\frac{\partial}{\partial A^\dagger}} \int \frac{(\mathrm{d}\alpha'')}{\pi} \, \mathrm{e}^{-\alpha''^* A} \, \mathrm{e}^{\alpha'' A^\dagger}$$

$$= \mathrm{e}^{-\frac{1}{2}\frac{\partial}{\partial A}\frac{\partial}{\partial A^\dagger}} \left[\delta(A)\right]_A . \tag{SS 351}$$

Sample solution to Problem 5.23

By writing the normal-order form $\left[A^{\dagger m} A^n\right]$ in terms of derivatives of a displacement operator and recalling Eq. (5.3.17),

$$\left\langle \left[A^{\dagger m} A^n\right]_N \right\rangle$$

$$= \left(\frac{\partial}{\partial\alpha}\right)^m \left(-\frac{\partial}{\partial\alpha^*}\right)^n \underbrace{\left\langle \mathrm{e}^{\alpha A^\dagger} \, \mathrm{e}^{-\alpha^* A} \right\rangle}_{= \mathrm{e}^{\frac{1}{2}|\alpha|^2} \, D(\alpha)} \Bigg|_{\substack{\alpha=0 \\ \alpha^*=0}}$$

$$= \int \frac{(\mathrm{d}\alpha')}{\pi} \rho_W(\alpha') \left(\frac{\partial}{\partial\alpha}\right)^m \left(-\frac{\partial}{\partial\alpha^*}\right)^n \mathrm{e}^{\frac{1}{2}|\alpha|^2+\alpha\alpha'^*-\alpha^*\alpha'} \Bigg|_{\substack{\alpha=0 \\ \alpha^*=0}} . \tag{SS 352}$$

We may now replace the derivatives with a Laguerre polynomial by noting that, for $m \geq n$,

$$
\left(\frac{\partial}{\partial\alpha}\right)^m \left(-\frac{\partial}{\partial\alpha^*}\right)^n e^{\frac{1}{2}|\alpha|^2 + \alpha\alpha'^* - \alpha^*\alpha'} \bigg|_{\substack{\alpha=0 \\ \alpha^*=0}}
$$

$$
= \left(-\frac{\partial}{\partial\alpha^*}\right)^n \left[\left(\frac{1}{2}\alpha^* + \alpha'^*\right)^m e^{-\alpha^*\alpha'}\right]\bigg|_{\alpha^*=0} \qquad \left(y = \alpha^*\alpha' + 2|\alpha'|^2\right)
$$

$$
= \frac{e^{2|\alpha'|^2}}{2^m}(-1)^n \alpha'^{n-m} \left(\frac{d}{dy}\right)^n \left(y^m e^{-y}\right)\bigg|_{y=2|\alpha'|^2}
$$

$$
= \left(-\frac{1}{2}\right)^n n!\, \alpha'^{*m-n} L_n^{(m-n)}\left(2|\alpha'|^2\right). \qquad \text{(SS 353)}
$$

For $m < n$, we interchange m and n and replace α'^* with $-\alpha'$ in the above result to obtain the corresponding expression, so that

$$
\left(\frac{\partial}{\partial\alpha}\right)^m \left(-\frac{\partial}{\partial\alpha^*}\right)^n e^{\frac{1}{2}|\alpha|^2 + \alpha\alpha'^* - \alpha^*\alpha'} \bigg|_{\substack{\alpha=0 \\ \alpha^*=0}}
$$

$$
= \left(-\frac{1}{2}\right)^{n_<!} n_<!\, |\alpha'|^{n_> - n_<}\, e^{-i(m-n)\phi} L_{n_<}^{(n_> - n_<)}\left(2|\alpha'|^2\right). \qquad \text{(SS 354)}
$$

Sample solution to Problem 5.24

Upon inserting the overcompleteness identity of the coherent states on both sides of the normal-order form, we have

$$
\left\langle \left[A^{\dagger m} e^{-A^\dagger A} A^n\right]_N \right\rangle
$$

$$
= \int \frac{d\alpha'\, d\alpha''}{\pi^2} \alpha'^{*m} e^{-\alpha'^*\alpha''} \alpha''^n \langle\alpha'^*|\alpha''\rangle \langle\alpha''^*|\rho|\alpha'\rangle
$$

$$
= \int \frac{d\alpha'\, d\alpha''}{\pi^2} \alpha'^{*m} \alpha''^n e^{-\frac{1}{2}|\alpha'|^2 - \frac{1}{2}|\alpha''|^2} \langle\alpha''^*|\rho|\alpha'\rangle
$$

$$= \left(\frac{\partial}{\partial \alpha}\right)^m \left(\frac{\partial}{\partial \alpha^*}\right)^n \int \frac{d\alpha'\,d\alpha''}{\pi^2} \underbrace{e^{-\frac{1}{2}|\alpha'|^2 - \frac{1}{2}|\alpha''|^2 + \alpha'^*\alpha + \alpha''\alpha^*} \langle \alpha''^* | \rho | \alpha' \rangle}_{= e^{|\alpha|^2} \langle \alpha'^* | \alpha \rangle \langle \alpha^* | \alpha'' \rangle} \bigg|_{\alpha=0}$$

$$= \left(\frac{\partial}{\partial \alpha}\right)^m \left(\frac{\partial}{\partial \alpha^*}\right)^n \left[e^{|\alpha|^2} \rho_Q(\alpha) \right] \bigg|_{\alpha=0}. \tag{SS 355}$$

With the Gauss–Weierstrass transform in Eq. (5.1.141), we can turn the right-hand side into a phase space average weighted with the P function $\rho_P(\alpha)$, that is

$$\left(\frac{\partial}{\partial \alpha}\right)^m \left(\frac{\partial}{\partial \alpha^*}\right)^n \left[e^{|\alpha|^2} \rho_Q(\alpha) \right] \bigg|_{\alpha=0}$$

$$= \int \frac{d\alpha'}{\pi} \rho_P(\alpha')\, e^{-|\alpha'|^2} \left(\frac{\partial}{\partial \alpha}\right)^m \left(\frac{\partial}{\partial \alpha^*}\right)^n \left(e^{\alpha\alpha'^* + \alpha^*\alpha'} \right) \bigg|_{\alpha=0}$$

$$= \int \frac{d\alpha'}{\pi} \rho_P(\alpha')\, \alpha'^{*m}\, e^{-|\alpha'|^2}\, \alpha'^n, \tag{SS 356}$$

which is consistent with Eq. (5.3.13) indeed.

Appendix: Squeezed Coherent States

A squeezed coherent state is the quantum state $|\alpha'; z\rangle \langle \alpha'^*; z^*|$ whose ket can be defined by the unitary transformation

$$|\alpha'; z\rangle = S(z) |\alpha'\rangle \qquad \text{(SCS 1)}$$

on the coherent ket $|\alpha'\rangle$, where the *squeeze operator* $S(z) = S(-z)^\dagger$ is of the form

$$S(z) = e^{\frac{1}{2}\left(z A^{\dagger 2} - z^* A^2\right)} \qquad \text{(SCS 2)}$$

that involves the complex squeeze parameter $z = r\, e^{i\theta}$, with $r \geq 0$ and $0 \leq \theta < 2\pi$.

If we define the three operators

$$L_0 = \frac{1}{2} A^\dagger A + \frac{1}{4}, \qquad L_+ = \frac{1}{2} A^{\dagger 2}, \quad \text{and} \quad L_- = \frac{1}{2} A^2, \qquad \text{(SCS 3)}$$

we find that these operators obey the commutation relations

$$[L_+, L_-] = -2L_0,$$
$$[L_0, L_\pm] = \pm L_\pm. \qquad \text{(SCS 4)}$$

For such operators, there exists, for any γ, the disentangling formula[*]

$$e^{\gamma L_+ - \gamma^* L_-} = e^{\frac{\gamma}{|\gamma|} \tan(|\gamma|) L_+} \left[\cos\left(|\gamma|\right)\right]^{-2 L_0} e^{-\frac{\gamma^*}{|\gamma|} \tan(|\gamma|) L_-}. \qquad \text{(SCS 5)}$$

---

[*]Refer to D. R. Truax, *Phys. Rev. D* **31**, 1988 (1985) for a proof.

It follows that the squeeze operator can be disentangled into

$$S(z) = e^{zL_+ - z^*L_-}$$

$$= e^{\frac{z}{|z|}(\tanh|z|)L_+} \left(\text{sech }|z|\right)^{2L_0} e^{-\frac{z^*}{|z|}(\tanh|z|)L_-}$$

$$= e^{\frac{1}{2}e^{i\theta}(\tanh r)A^{\dagger 2}} \left(\text{sech }r\right)^{A^\dagger A + \frac{1}{2}} e^{-\frac{1}{2}e^{-i\theta}(\tanh r)A^2}. \qquad \text{(SCS 6)}$$

Then,

$$|\alpha'; z\rangle = e^{\frac{1}{2}e^{i\theta}(\tanh r)A^{\dagger 2}} \left(\text{sech }r\right)^{A^\dagger A + \frac{1}{2}} |\alpha'\rangle e^{-\frac{1}{2}e^{-i\theta}(\tanh r)\alpha'^2}. \qquad \text{(SCS 7)}$$

From here, we can introduce the Fock basis, so that

$$|\alpha'; z\rangle = \sum_{n=0}^\infty e^{\frac{1}{2}e^{i\theta}(\tanh r)A^{\dagger 2}} |n\rangle \frac{\alpha'^n}{\sqrt{n!}} \left(\text{sech }r\right)^{n+\frac{1}{2}} e^{-\frac{1}{2}e^{-i\theta}(\tanh r)\alpha'^2 - \frac{1}{2}|\alpha'|^2}. \tag{SCS 8}$$

In this basis, the coherent-state wave function of $|\alpha'; z\rangle$, namely

$$\langle \alpha^* | \alpha'; z\rangle = \sqrt{\text{sech }r} \, e^{-\frac{1}{2}|\alpha|^2 - \frac{1}{2}|\alpha'|^2}$$

$$\times \exp\left(\frac{\alpha^*\alpha'}{\cosh r} + \frac{\tanh r}{2}\left(\alpha^{*2}e^{i\theta} - \alpha'^2 e^{-i\theta}\right)\right), \qquad \text{(SCS 9)}$$

can be derived very quickly.

We may also obtain the position wave function $\langle x | \alpha'; z\rangle$ from Eq. (SCS 9) by invoking the overcompleteness relation

$$\int \frac{(d\alpha)}{\pi} |\alpha\rangle \langle \alpha^*| = 1 \qquad \text{(SCS 10)}$$

for the coherent states, so that

$$\langle x | \alpha'; z\rangle = \int \frac{(d\alpha)}{\pi} \langle x | \alpha\rangle \langle \alpha^* | \alpha'; z\rangle$$

$$= \sqrt{\text{sech }r} \, e^{-\frac{1}{2}|\alpha'|^2 - \frac{1}{2}x^2 - \frac{1}{2}e^{-i\theta}(\tanh r)\alpha'^2}$$

$$\times \int \frac{(d\alpha)}{\pi} e^{-|\alpha|^2 - \frac{1}{2}\alpha^2 + \sqrt{2}\alpha x} e^{\alpha^*\alpha'\,\text{sech }r + \frac{1}{2}e^{i\theta}(\tanh r)\alpha^{*2}}. \qquad \text{(SCS 11)}$$

This two-dimensional Gaussian integral can be evaluated directly with the help of Eq. (5.1.130), for which we assign $a = 1$, $b_1 = \sqrt{2}x$, $b_2 = \alpha'\text{sech }r$,

$c_1 = -\frac{1}{2}$ and $c_2 = \frac{1}{2}e^{i\theta}\tanh r$ for the evaluation; Thereafter, we have

$$\langle x|\alpha';z\rangle = \frac{1}{\pi^{\frac{1}{4}}}\sqrt{\frac{\operatorname{sech} r}{1+w}}\, e^{-\frac{1}{2}|\alpha'|^2}\, e^{-\frac{1}{2}\alpha'^2\left[w^* + \frac{(\operatorname{sech} r)^2}{1+w}\right]}\, e^{\left(\frac{w}{1+w} - \frac{1}{2}\right)x^2 + \frac{\sqrt{2}\alpha'\operatorname{sech} r}{1+w}x},$$

$$\text{(SCS 12)}$$

where $w = e^{i\theta}\tanh r$. If we define the notations

$$\gamma(z) = \cosh r - e^{i\theta}\sinh r \qquad\qquad \text{(SCS 13)}$$

and

$$\zeta(z) = \frac{1 - e^{i\theta}\tanh r}{1 + e^{i\theta}\tanh r}, \qquad\qquad \text{(SCS 14)}$$

the term inside the square root in Eq. (SCS 12) can be expressed as

$$\frac{\operatorname{sech} r}{1 + w} = \frac{\zeta(z)}{\gamma(z)}. \qquad\qquad \text{(SCS 15)}$$

The exponent of the second exponential term in Eq. (SCS 12) is, in fact, a phase factor and to see this, we can also express the exponent in terms of $\gamma(z)$ and $\zeta(z)$:

$$-\frac{1}{2}\alpha'^2\left[w^* + \frac{(\operatorname{sech} r)^2}{1+w}\right] = -\frac{1}{2}\alpha'^2\left[e^{-i\theta}\tanh r + \frac{(\operatorname{sech} r)^2}{1 + e^{i\theta}\tanh r}\right]$$

$$= -\frac{1}{2}\alpha'^2\left(\frac{1 + e^{-i\theta}\tanh r}{1 + e^{i\theta}\tanh r}\right)$$

$$= -\frac{1}{2}\alpha'^2\left[\frac{\gamma(z^*)\zeta(z)}{\gamma(z)\zeta(z^*)}\right]. \qquad \text{(SCS 16)}$$

For the third exponential term, the coefficient of x^2 in its exponent is given by

$$\frac{w}{1+w} - \frac{1}{2} = \frac{1}{2}\frac{w-1}{1+w} = -\frac{1}{2}\zeta(z). \qquad\qquad \text{(SCS 17)}$$

With these new notations, the position wave function takes on the form

$$\langle x|\alpha';z\rangle = \frac{1}{\pi^{\frac{1}{4}}}\sqrt{\frac{\zeta(z)}{\gamma(z)}}\, e^{-\frac{1}{2}|\alpha'|^2 - \frac{1}{2}\alpha'^2\left[\frac{\gamma(z^*)\zeta(z)}{\gamma(z)\zeta(z^*)}\right]}\, e^{-\frac{\zeta(z)}{2}x^2 + \sqrt{2}\frac{\alpha'}{\gamma(z)}\zeta(z)x},$$

$$\text{(SCS 18)}$$

where its square-integrability is guaranteed by the fact that

$$\text{Re}\{\zeta(z)\} = \left|\frac{\zeta(z)}{\gamma(z)}\right|^2 \geq 0. \tag{SCS 19}$$

The reader is invited to check that this wave function is normalized and consistent with that of the usual coherent state $|\alpha'\rangle\langle\alpha'^*|$ when $z = 0$.

Let us examine the standard deviations of the position and momentum quadrature operators for these squeezed coherent states. This is done by calculating the four quantities $\langle X \rangle$, $\langle X^2 \rangle$, $\langle P \rangle$ and $\langle P^2 \rangle$ using the position wave function in Eq. (SCS 18). For the position quadrature operator X,

$$
\begin{aligned}
\langle X \rangle &= \int dx\, x\, |\langle x|\alpha';z\rangle|^2 \\[6pt]
&= \frac{e^{-|\alpha'|^2}}{\sqrt{\pi}} \left|\frac{\zeta(z)}{\gamma(z)}\right| \left|e^{-\frac{1}{2}\alpha'^2 \left[\frac{\gamma(z^*)\zeta(z)}{\gamma(z)\zeta(z^*)}\right]}\right|^2 \\[6pt]
&\quad \times \int dx\, x\, e^{-\text{Re}\{\zeta(z)\}x^2 + 2\sqrt{2}\text{Re}\left\{\frac{\alpha'}{\gamma(z)}\zeta(z)\right\}x} \\[6pt]
&= \frac{e^{-|\alpha'|^2}}{\sqrt{\pi}} \left|\frac{\zeta(z)}{\gamma(z)}\right| \left|e^{-\frac{1}{2}\alpha'^2 \left[\frac{\gamma(z^*)\zeta(z)}{\gamma(z)\zeta(z^*)}\right]}\right|^2 \\[6pt]
&\quad \times \frac{\partial}{\partial u} \int dx\, e^{-\text{Re}\{\zeta(z)\}x^2 + ux} \Bigg|_{u=2\sqrt{2}\text{Re}\left\{\frac{\alpha'}{\gamma(z)}\zeta(z)\right\}} \\[6pt]
&= e^{-|\alpha'|^2} \left|e^{-\frac{1}{2}\alpha'^2\left[\frac{\gamma(z^*)\zeta(z)}{\gamma(z)\zeta(z^*)}\right]}\right|^2 \frac{\partial}{\partial u}\, e^{\frac{u^2}{4\text{Re}\{\zeta(z)\}}} \Bigg|_{u=2\sqrt{2}\text{Re}\left\{\frac{\alpha'}{\gamma(z)}\zeta(z)\right\}} \\[6pt]
&= \sqrt{2}\,\text{Re}\left\{\frac{\alpha'\zeta(z)}{\gamma(z)}\right\} \frac{1}{\text{Re}\{\zeta(z)\}}, \tag{SCS 20}
\end{aligned}
$$

and

$$
\begin{aligned}
\langle X^2 \rangle &= e^{-|\alpha'|^2} \left|e^{-\frac{1}{2}\alpha'^2\left[\frac{\gamma(z^*)\zeta(z)}{\gamma(z)\zeta(z^*)}\right]}\right|^2 \left(\frac{\partial}{\partial u}\right)^2 e^{\frac{u^2}{4\text{Re}\{\zeta(z)\}}} \Bigg|_{u=2\sqrt{2}\text{Re}\left\{\frac{\alpha'}{\gamma(z)}\zeta(z)\right\}} \\[6pt]
&= \left\{2\left[\text{Re}\left\{\frac{\alpha'\zeta(z)}{\gamma(z)}\right\} \frac{1}{\text{Re}\{\zeta(z)\}}\right]^2 + \frac{1}{2\,\text{Re}\{\zeta(z)\}}\right\}, \tag{SCS 21}
\end{aligned}
$$

so that its uncertainty is

$$\delta X = \sqrt{\langle X^2 \rangle - \langle X \rangle^2} = \frac{1}{\sqrt{2 \operatorname{Re}\{\zeta(z)\}}}$$

$$= \sqrt{\frac{1}{2} \left[e^{2r} \left(\cos\frac{\theta}{2} \right)^2 + e^{-2r} \left(\sin\frac{\theta}{2} \right)^2 \right]}.$$

(SCS 22)

For the momentum quadrature operator P, the moments

$$\langle P \rangle = \frac{1}{i} \int dx \, \langle \alpha'^*; z^* | x \rangle \frac{\partial}{\partial x} \langle x | \alpha'; z \rangle = \frac{1}{i} \left[-\zeta(z) \langle X \rangle + \sqrt{2} \frac{\alpha'}{\gamma(z)} \zeta(z) \right],$$

(SCS 23)

where we note that $\langle P \rangle = \operatorname{Re}\{\langle P \rangle\}$, and

$$\langle P^2 \rangle = \int dx \left| \frac{\partial}{\partial x} \langle x | \alpha'; z \rangle \right|^2$$

$$= |\zeta(z)|^2 \langle X^2 \rangle - 2\sqrt{2} |\zeta(z)|^2 \operatorname{Re}\left\{ \frac{\alpha'}{\gamma(z)} \right\} \langle X \rangle + 2 \left| \frac{\alpha'}{\gamma(z)} \zeta(z) \right|^2,$$

(SCS 24)

are easily obtained from Eq. (SCS 20) and Eq. (SCS 21), with which we get

$$\delta P = \sqrt{\langle P^2 \rangle - \langle P \rangle^2} = \sqrt{\frac{|\zeta(z)|^2}{2 \operatorname{Re}\{\zeta(z)\}}}.$$

$$= \sqrt{\frac{1}{2} \left[e^{-2r} \left(\cos\frac{\theta}{2} \right)^2 + e^{2r} \left(\sin\frac{\theta}{2} \right)^2 \right]}.$$

(SCS 25)

The effects of quadrature squeezing can be visualized with the help of phase-space diagrams (refer to Fig. SCS 1). The uncertainties δX and δP shape the uncertainty ellipse for a Gaussian state. It is customary to associate the smaller of the two uncertainties with the "squeezed" quadrature, and the larger one with the "anti-squeezed" quadrature. The relative size of δX and δP depends strongly on the choice of θ.

According to the definition in Eq. (SCS 2) for the squeeze operator, it turns out that for the special case where z is real, $\theta = 0$ or $z = r > 0$ corresponds to a momentum-quadrature squeezing whence $\delta P = e^{-r}/\sqrt{2}$

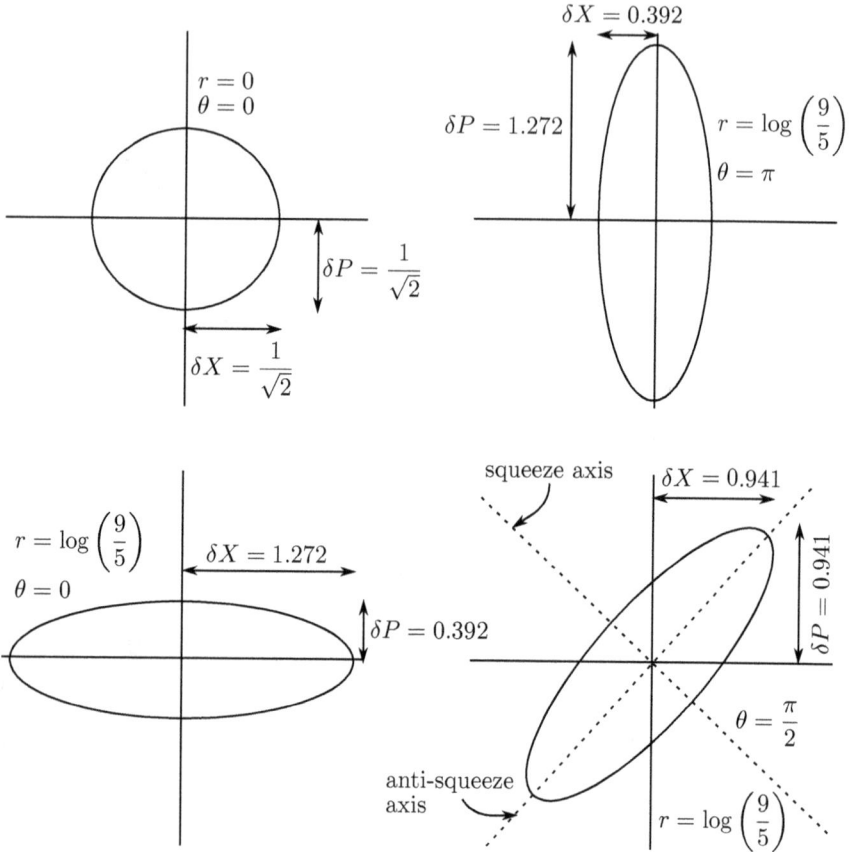

Fig. SCS 1 Uncertainty ellipses of squeezed coherent states for different squeeze magnitudes r and phases θ.

and $\delta X = e^r/\sqrt{2}$, whereas $\theta = \pi$ or $z = -r < 0$ corresponds to a position-quadrature squeezing by $\delta X = e^{-r}/\sqrt{2}$, implying that $\delta P = e^r/\sqrt{2}$. Indeed, a simple exercise of differentiation of $(\delta X)^2$ and $(\delta P)^2$ with respect to θ shows that these phase values give the extremal magnitudes of δX and δP for any r.

The product of δX and δP,

$$\delta X \delta P = \frac{1}{2}\frac{|\zeta(z)|}{\operatorname{Re}\{\zeta(z)\}} = \frac{1}{2}\sqrt{1 + (\sin\theta)^2[\sinh(2r)]^2}\,, \qquad \text{(SCS 26)}$$

is always greater than or equal to a half. This product equals a half when the squeeze parameter z is real, and only then. Such squeezed

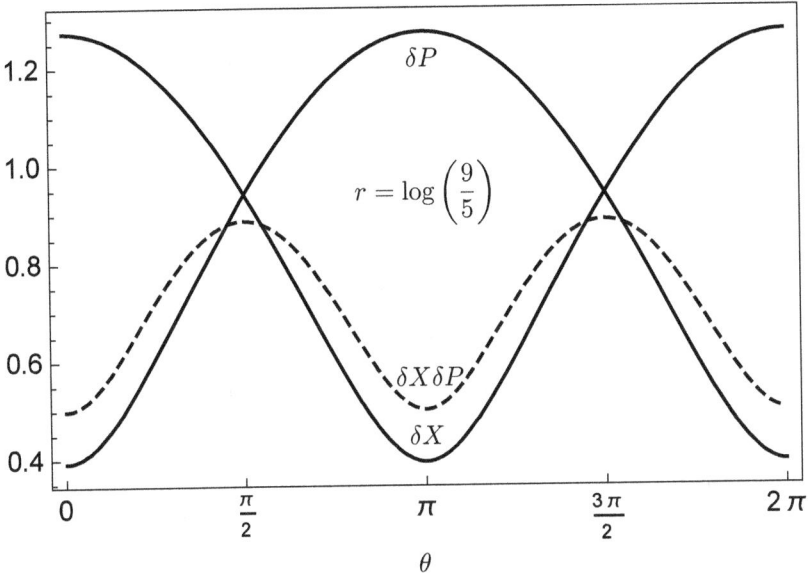

Fig. SCS 2   The behavior of the uncertainties at various values of θ.

coherent states are known as *minimum-uncertainty* states that saturate the Heisenberg uncertainty relation. For $\theta \neq 0$, it is the product of the respective uncertainties along the squeeze and anti-squeeze axes defined by θ that is minimum (refer to Fig. SCS 2).

Finally, we alert the reader that there exists an alternative parametrization for the squeezed coherent states. The definition stated as Eq. (SCS 1), is equivalent to a displacement on the vacuum state, followed by a squeeze operation — $S(z)D(\alpha')\,|0\rangle$. One may also perform these operations in reverse: a squeeze operation on the vacuum state, and then a displacement — $D(\alpha')S(z)\,|0\rangle$. These two prescriptions are not identical as $S(z)$ and $D(\alpha)$ do not commute. To quickly find out how they are related to each other, let us revisit Eq. (SCS 6) and recall the basic identities

$$F(A^\dagger A)\,A = A\,F(A^\dagger A - 1),$$

$$\text{and} \quad \left[A, F(A^\dagger)\right] = \frac{\partial}{\partial A^\dagger} F(A^\dagger)$$

so that $\left[A, e^{\frac{1}{2}e^{i\theta}(\tanh r)A^{\dagger 2}}\right] = A^\dagger e^{i\theta}\,(\tanh r)\,e^{\frac{1}{2}e^{i\theta}(\tanh r)A^{\dagger 2}},$ (SCS 27) which would lead to

$$S(z)A = e^{\frac{1}{2}e^{i\theta}(\tanh r)A^{\dagger 2}} \left(\operatorname{sech} r\right)^{A^{\dagger}A + \frac{1}{2}} e^{-\frac{1}{2}e^{-i\theta}(\tanh r)A^2} A$$

$$= (\cosh r) e^{\frac{1}{2}e^{i\theta}(\tanh r)A^{\dagger 2}} A \left(\operatorname{sech} r\right)^{A^{\dagger}A + \frac{1}{2}} e^{-\frac{1}{2}e^{-i\theta}(\tanh r)A^2}$$

$$= \left(A\cosh r - A^{\dagger}e^{i\theta}\sinh r\right) S(z) \qquad \text{(SCS 28)}$$

for $S(z)$ and A. With this, we have the relation

$$S(z)D(\alpha) = D(\alpha_{\mathrm{B}})S(z), \qquad \text{(SCS 29)}$$

where for $\alpha_{\mathrm{B}} = \alpha\cosh r + \alpha^* e^{i\theta}\sinh r$,

$$D(\alpha_{\mathrm{B}}) = \exp\!\big(\alpha\left(A^{\dagger}\cosh r - A\,e^{-i\theta}\sinh r\right) - \alpha^*\left(A\cosh r - A^{\dagger}e^{i\theta}\sinh r\right)\big)$$
$$\text{(SCS 30)}$$

is the *Bogolyubov*[*]*-transformed* displacement operator.

Therefore, one parametrization is related to the other simply by a Bogolyubov transformation ($A \to A\cosh r - A^{\dagger}e^{i\theta}\sinh r$, $A^{\dagger} \to A^{\dagger}\cosh r - A\,e^{-i\theta}\sinh r$) on the ladder operators in the displacement operator, or on the parameter $\alpha(\alpha \to \alpha_{\mathrm{B}})$. The choice of parametrization is irrelevant as far as the general properties of these Gaussian states are concerned.

---

[*]Nikolay Nikolayevich Bogolyubov (1909–1992).

Index

www.ingramcontent.com/pod-product-compliance
Lightning Source LLC
Chambersburg PA
CBHW061616220326

41598CB00026BA/3786